The IGS Geosynthetics Handbook

First Edition

Editor

George R. Koerner, PhD., PE

Published by

igs

With the Support Of

igs
Foundation

Table of Contents

Preface to 1st Edition

Geosynthetics have proven to be among the most versatile and cost-effective construction materials on the face of this earth. Their use has expanded rapidly into nearly all areas of civil engineering. This handbook overviews geosynthetics from the perspective of practice rather than design. It is intended to serve as a general reference in the field for those who are constructing facilities or structures that include geosynthetics.

The International Geosynthetics Society (IGS), which is the underwriter of this document, is dedicated to the establishment and development of resources that provide superior value to its membership and their customers. The IGS has long been committed to enhancing the availability of reference materials to the geosynthetic community. This inaugural edition of the *IGS Geosynthetics Handbook* published in 2025 is intended to fill a void by supplying generic and useful information about geosynthetics for consumption by the general public, rather than the geosynthetic technical/academic community.

This first edition of the *IGS Geosynthetics Handbook* is a living document which will see countless revisions going forward. It will always be a work in progress as a space in which IGS Members can inform and contribute on all matters related to geosynthetics. This document is a record that expands on the function and applicability of geosynthetics in an archival format for posterity. In this volume we demonstrate the use of geosynthetics as mandatory resources for modern-day Civil Engineering applications in the circular economy. Some of the chapters provide an overview of a geosynthetic application and address the important aspects of quality geosynthetic materials and installation. Other chapters present an extensive dissertation on the design aspects of critical applications. Together, they provide the reader with a fundamental understanding of geosynthetics in a comingled approach.

The IGS extends its sincere thanks to all the colleagues who contributed to this book, the Editor, and to the reviewers for their expertise and insights. One will note that this book does not read as if it was written by a single author—as it was not. After initial drafts were generated, IGS Technical and Education Committee members contributed many revisions.

Our sincere appreciation to the following originating authors and friends for sharing their time and talent. Their generous help has been indispensable in accomplishing this important task.

Handbook Editor: G. Koerner

Chapter 1:
Introduction to Geosynthetics
G. KOERNER

Chapter 2:
Geosynthetics in Roads and Pavements
E. CUELHO

Chapter 3:
Geosynthetics in Subsurface Drainage
B. CHRISTOPHER

Chapter 4:
Geosynthetics in Erosion and
Sediment Control
C.J. SPRAGUE and J.E. SPRAGUE

Chapter 5:
Geosynthetics in Reinforced Soil Systems
C. LAWSON

Chapter 6:
Geosynthetic Barriers in Seepage Control
Systems
K. Von MAUBEUGE

Chapter 7:
Geosynthetics in Environmental Protection
(Waste Containment)
K. ROWE

Chapter 8:
Geosynthetics Support Systems
G. KOERNER

While the use of geosynthetics is widespread, their use should be considered just beginning. It is envisioned that these materials will see staggering growth in the near future from an entire spectrum of conventional and new environmental, transportation and geotechnical engineering applications.

We acknowledge and thank the IGS Foundation and the IGS Premium Corporate Member companies for sustaining this effort. Our appreciation is extended to them for their continued and unwavering support.

Sincerely,

Samuel R. Allen
President, International Geosynthetics Society

George R. Koerner, Ph.D., P.E.
IGS Geosynthetics Handbook Editor, First Edition
Director, Geosynthetic Institute

Chapter 1

Introduction to Geosynthetics

1.0 Overview

The roots of geosynthetics can be traced to the 1920s when textiles were first utilized in road construction. The more modern concept of geosynthetics began to take shape in the late 1950s and early 1960s with the development of polymers. Polymers or "plastics" became widely used after World War II. Plastics were in high demand because they were inexpensive, versatile, sanitary, and easy to manufacture into many different forms such as geotextiles, geomembranes and related products.

Geosynthetics are synthetic products used in geotechnical applications. They are generally polymeric products used to solve engineering challenges. The polymeric nature of the products makes them suitable for use in or near the ground where high levels of durability are required. They can also be used in exposed applications.

Geosynthetics are available in a wide range of forms and materials. They are easy to manufacture, fabricate, transport and install. According to ASTM D4439 stated terminology for geosynthetics, a geosynthetic is defined as follows:

> **geosynthetic**, *n*—a planar product manufactured from polymeric material used with soil, rock, earth, or other geotechnical engineering related material as an integral part of a human-made project, structure, or system.

Since 1977, the time of the first geosynthetics conference in Paris, geosynthetics have emerged as exciting engineering materials in a wide array of civil engineering applications; e.g. transportation, geotechnical, geoenvironmental, hydraulics, and private development. The rapidity at which the related products have been and continue to be developed is nothing short of amazing. The reasons for this explosion of geosynthetic materials onto the civil engineering market are numerous and include the following:

> Quality-control manufactured in a factory environment.
> Can be installed rapidly and efficiently.
> Replace scarce natural raw material resources for sustainable solutions.
> Replace difficult designs using soil or other construction materials.
> Required by codes and regulations in many cases.
> Made heretofore impossible designs and applications possible.
> Actively marketed and are widely available.
> Technical database (both design and testing) is nicely established.
> Being integrated into the profession via generic specifications.
> Invariably cost competitive against soils or other construction materials.
> Their carbon footprint is very much lower than traditional solutions.
> Simple disposal or recycling after use.
> Geosynthetic projects can be deconstructed easily at the end of service life.

Industries most strongly influenced are transportation, geotechnical, environmental, hydraulics and private development engineering communities, although all soil, rock, and groundwater-related activities fall within the general scope of the various applications. This being the case, the term geosynthetics seems appropriate. *Geo*, of course, refers to the earth. The realization that the materials are almost exclusively from human-made products gives the second part to the name—*synthetics*. The materials used in the manufacture of geosynthetics are almost entirely from the plastics industry—that is, they are polymers made from hydrocarbons, although fiberglass, rubber, and natural materials are occasionally used. Interestingly, the case could easily be made that the entire technology could better be called *geopolymers*.

1.1 Basic Description of Geosynthetics

Lost to history are the initial attempts to reinforce soils; the adding of materials that would enhance the behavior of the foundation system was no doubt done long before our first historical records of this process. It seems reasonable to assume that first attempts were made to stabilize swamps and marshy soils using tree trunks, small bushes, and the like. These soft soils would accept the fibrous material until a somewhat stable mass was formed that had adequate properties for the intended purpose. It also seems reasonable to accept that either the continued use of such a facility was possible because of the properly stabilized nature of the now-reinforced soil (probably by trial-and-error), or was impossible due to a number of factors, among which were:

> insufficient reinforcement material for the loads to be carried;
> the pumping of the soft soil up through the reinforcement material; and
> the degradation of the fibrous material with time, leading back to the original unsuitable conditions.

Such stabilization attempts were undoubtedly continued with the development of a more systematic approach in which timbers of nearly uniform size and length were lashed together to make a mattressed surface. Such split-log "corduroy" roads over peat bogs date back to 3000 B.C. [1]. This art progressed to the point where the ridged surface was filled in smooth. Some of these systems were surfaced with a stabilized soil mixture or even paved with stone blocks. Here again, however, deterioration of the timber and its lashings over time was an obvious problem.

The concept of reinforcing soft soils has continued until the present day. The first use of fabrics in reinforcing roads was attempted by the South Carolina Highway Department in 1926 [2]. A heavy cotton fabric was placed on a primed soil subgrade, hot asphalt was applied to the fabric, and a thin layer of sand was placed on the asphalt. In 1935 results were published of this work, describing eight separate field experiments. Until the fabric deteriorated, the results showed that the roads were in good condition and that the fabric reduced cracking, raveling, and localized road failures. This project was certainly the forerunner of the separation and reinforcement functions of geosynthetic materials as we know them today. The separation and reinforcement of unsuitable soils is a major topic area of this book.

A second major topic area is that of providing an intermediate barrier between two dissimilar materials for the purpose of liquid (usually water) drainage and soil filtration. When requiring liquid flow across such a barrier, it must obviously be porous, yet the voids must not be open so much as to lose the retained soil; thus the necessity of using some sort of intermediate filter. Again the historical development of filtration provides an important background for understanding the work that followed. Naturally occurring sands and gravels which were found to be well graded had been used as filter material since ancient times. The idea of systematizing filtration criteria was originated by Karl Terzaghi and Arthur Casagrande in the 1930s and brought to use by Bertram [3] shortly thereafter. This idea of soil filters, even multiple-graded soil filters, is a target area for the geosynthetic materials described in this book—now for reasons of construction quality control and cost effectiveness.

A last major topic area is that of providing a waterproof barrier for preventing liquid or gas movement from a given containment area. Such liners have historically been made using low-permeability clay soils. The Roman aqueducts were lined in such a manner, and the technology undoubtedly preceded them by many years [4]. Liners made from bitumen and various cements have been used since the 1900s, but it was the synthetic rubber material "butyl" in the 1930s that ushered in polymeric liners [5]. Today, such liners (there are many different types) are regulatory-mandated for use in certain environmental related applications. Interestingly, the newest barrier materials are combinations of both geosynthetics and bentonite soil used as a composite material.

Thus, geosynthetic materials perform five major functions: (1) separation, (2) reinforcement, (3) filtration, (4) drainage, and (5) containment (of liquid and/or gas). The use of geosynthetics has basically two aims: (1) to perform better (e.g. with no deterioration of material or excessive leakage) and (2) to be more economical than using traditional materials and solutions (either through lower initial cost or through greater durability and longer life, thus reducing maintenance and replacement costs.

1.2 Types of Geosynthetics

There are twelve specific types of geosynthetics: (1) geotextiles, (2) geogrids, (3) geostraps, (4) geocells, (5) geonets, (6) geomembranes, (7) geosynthetic clay liners, (8) geopipes, (9) erosion control baskets and turf reinforcement mats, (10) geofoam, (11) geocomposites and (12) geo other. They are briefly described and discussed in the next sections.

1.2.1 Geotextiles

Geotextiles form one of the largest groups of geosynthetics. Although they are textiles in the traditional sense, they are engineered materials and are produced

Figure 1.1 Photograph of various Geotextiles (GTX). Credit: Geosynthetic Institute.

under strict quality control procedures to satisfy specified properties in conformance with engineered designs. They consist of synthetic fibers rather than natural ones such as cotton, wool, or silk. Thus biodegradation and subsequent short lifetime is not a problem. These synthetic fibers are made into flexible, porous fabrics by standard weaving machinery or are matted together in a random nonwoven manner. Some are also knit, woven and non-woven geotextiles which differ in the manufacturing process, appearance, properties and therefore uses. It should be acknowledged that standard or circular weaving equipment may be used to create geotextiles. Such different manufacturing processing will allow for such exotic applications as geotextile encased columns for deep foundation ground improvement and or geotextile tubes for formwork or coastal protection. The major point is that geotextiles are porous to liquid flow across their manufactured plane and also within their thickness, but to a widely varying degree. There are at least 100 specific application areas for geotextiles that have been developed; however, the fabric always performs at least one of four discrete functions: separation, reinforcement, filtration and/or drainage.

1.2.2 Geogrids

Geogrids represent a well-established product segment within geosynthetics. Rather than being a woven, nonwoven or knitted textile fabric, geogrids are plastics formed into a very open, gridlike configuration—i.e. they have large apertures between tensile elements in the multiple directions. Geogrids are formed in three ways: (1) stretched in one or two or more directions for improved physical properties, (2) made on weaving or knitting machinery by standard and well-established methods and then coated, or (3) made by bonding rods or straps together. There are many application areas, however, they function almost exclusively as reinforcement materials. Geogrids function as stabilization in the design of working platforms, rail, paved and unpaved roads and reinforcement in walls, slopes and embankments.

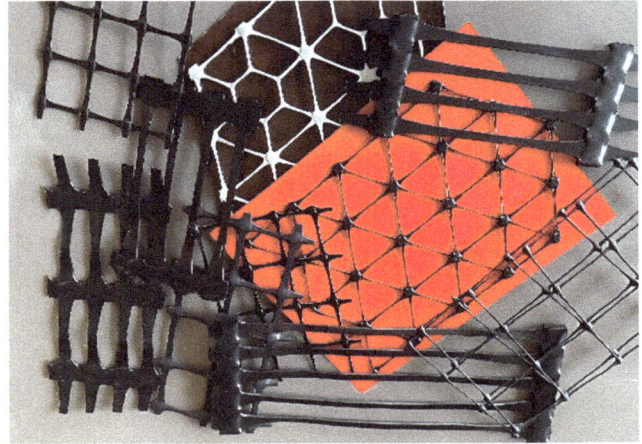

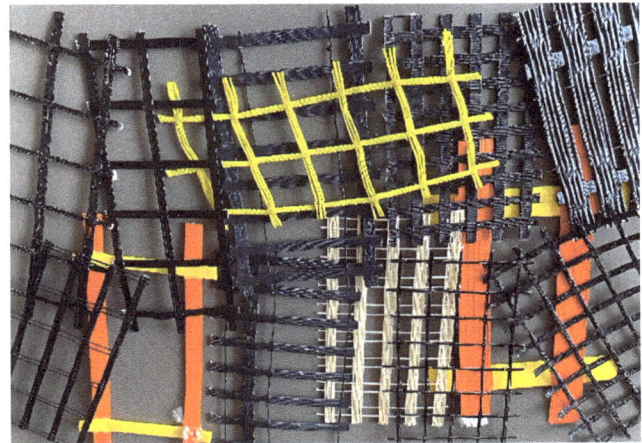

Figure 1.2 Photograph of various Geogrids (GGR). Credit: Geosynthetic Institute.

1.2.3 Geostrip

Geostrips, constitute another specialized segment within the geosynthetics area. They are formed by a continuous extrusion of parallel sets of polymeric strands of polyester, polyvinyl esters or nylon coated with a po yolefin. The high tenacity polyester is the load bearing element, while the sheath protects the yarns from installation damage and degradation. They can be made very strong by increasing the number of yarn bundles within the width constraint of any given strip. They are often used in mechanically stabilized earth walls

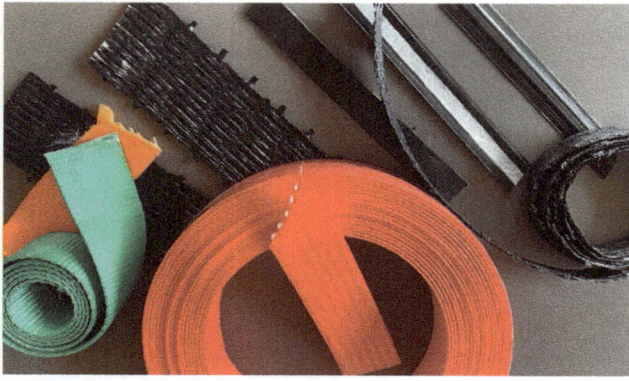

Figure 1.3 Photograph of various Geostrips (GST). Credit: Geosynthetic Institute.

Figure 1.4 Photograph of various Geocells (GCE). Credit: Geosynthetic Institute.

and slopes as reinforcement. They are wrapped around metallic bars imbedded within facing panels. They also can be used in tandem with synthetic connectors or sleeves embedded in the facing panels. Although they are considered extensible, the durability and strength characteristics of the geostrip exhibit inextensible behavior for easy installation. The high-strength geo-strip exhibit less extensibility than typical geogrids. Geostrips are suitable for highly aggressive environments and corrosive soil conditions and the connection to precast facing panels is simple, quick, and requires no tools. Consult EN ISO 10318-1 for further clarification and descriptions of these innovative and strong tensile inclusions.

1.2.4 Geocells

Geocells are expanding 3-D panels constructed of PE, PP, PET PVA and Aramids. They are made up of an inter-connected honeycomb-like networks made by linked strips of material together as a cellular confinement system. Geocells are stretched out on the subgrade and then filled with sand, aggregate, soil or concrete. The 3-d geocell confines the stabilized soil that would otherwise be unconfined and unstable under loading. Geocells are widely used in construction for erosion control, soil sta-bility, channel protection and reinforcement. Engineers found that sand-confinement systems performed bet-ter than conventional crushed stone sections and they could provide an expedient construction technique for access roads over soft ground, without being adversely affected by wet weather conditions. Geocells are folded and shipped to the job site in a collapsed configuration.

1.2.5 Geonets

A geonet is a geospacer consisting of parallel sets of ribs overlaying an integrally connected with a similar set of ribs at various angles. They constitute another specialized segment within the geosynthetics area. The relatively large apertures are formed into a netlike con-figuration. Their design function is completely within the drainage area where they are used to convey liquids of all types and gases.

Figure 1.5 Photograph of various Geonets or Geospacers (GN & GS). Credit: Geosynthetic Institute.

Figure 1.6 Photograph of various Geomembranes (GBR-P). Credit: Geosynthetic Institute.

Figure 1.8 Photograph of various Bituminous geomembrane (GBR-B or BGM). Credit: Geosynthetic Institute.

Figure 1.7 Photograph of various Geosynthetic Clay Liner (GBR-C or GCL). Credit: Geosynthetic Institute.

1.2.6 Geomembranes

Geomembranes represent the other largest group of geosynthetics described in this book. Their growth in the United States and Germany was stimulated by governmental regulations originally enacted in the early 1980s. The materials themselves are relatively thin, impervious sheets of polymeric material used primarily for linings and covers of liquid- or solid-storage facilities. This includes all types of landfills, reservoirs, canals and other containment facilities. Thus, the primary function is always containment as a liquid or vapor barrier or both. The range of applications, however, is very great, and in addition to the environmental area, applications are rapidly growing in geotechnical, transportation, hydraulic, mining and private development engineering.

1.2.7 Geosynthetic Clay Liners

Geosynthetic clay liners (GRB-C or GCLs) are an interesting juxtaposition of polymer and natural soil materials. They are rolls of factory-fabricated thin layers of bentonite clay sandwiched between two geotextiles or bonded to a geomembrane. Structural integrity of the composite is obtained by needle-punching, stitching or adhesive bonding. GCLs are used as a composite component beneath a geomembrane or by themselves in environmental and containment applications as well as in transportation, geotechnical, hydraulic and various private development applications.

1.2.8 Bituminous geomembrane (GBR-B or BGM)

Bituminous geomembranes (GBR -B of BGM) are factory produced materials in the form of rolls or sheets used to mitigate fluid loss. They are a specialty geomembrane made of strong watertight composite layers of petroleum not ccal based bitumen and synthetics (textile and films). This unique waterproofing system can be made in widths greater than 5 m. BGM liner consists of a stabilized (typically elastomeric) bitumen-based binder and an internal structure made from nonwoven, knit or woven textiles designed to have excellent physical, mechanical, chemical and endurance properties. BGM liners are available in different grades, depending on application conditions, geometry and environmental constraints.

This BGM liner caters to public works, mining, and civil infrastructure construction projects in a wide range of applications from environmental protection to hydraulic structures and transportation applications.

1.2.9 Geosynthetic Cementitious Composite Mats (GCCM)

Geosynthetic Cementitious Composite Mats (GCCM) are a factory-assembled geosynthetic composite consisting of a cementitious material contained within layer or layers of geosynthetic materials that becomes hardened when hydrated. The juxtaposition and containment of pozzolanic and synthetic materials is novel. They arrive on the job-site as rolls of factory-fabricated panels. They have been created specifically engineered for erosion control, containment and shelter applications. GCCM most often consists of a 3-dimensional matrix containing a specially formulated dry cementitious mix. Backing geomembranes or films ensures the GCCM can perform as fluid barriers. GCCMs can be hydrated by spraying or immersing in fresh or salt water. Once set, the fiber reinforcement prevents crack propagation within the cementitious mix. As a result, GCCMs provide a thin, durable, waterproof and lower-carbon alternative to traditional concrete.

Figure 1.9 Photograph of various Geosynthetic Cementitious Composite Mats (GCCM). Credit: Geosynthetic Institute.

Figure 1.11 Photograph of various GeoMats (EC-TRMs or GMA). Credit: Geosynthetic Institute.

Figure 1.10 Photograph of various Geopipe (plastic pipe) (GP). Credit: Geosynthetic Institute.

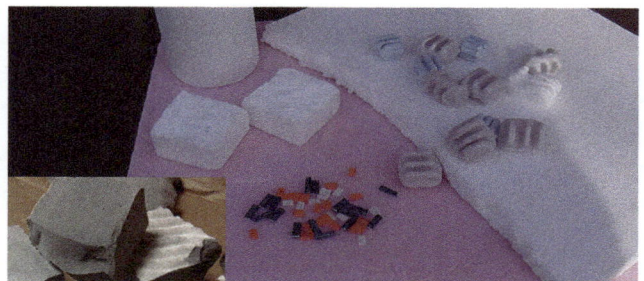

Figure 1.12 Photograph of various GeoFoams (GF). Credit: Geosynthetic Institute.

1.2.10 Geopipe

Geopipe or plastic pipe is a tubular section, or hollow cylinder, made of plastic. It is usually, but not necessarily, of circular cross-section, used mainly to convey substances which can flow such as liquids and gases (fluids), slurries, powders and masses of small solids. It can also be used for structural applications. Geopipes are used for the conveyance of drinking water, waste water, chemicals, heating fluid and cooling fluids, slurries, gases, compressed air, irrigation, and vacuum system applications. Plastic pipe systems fulfil a variety of service requirements such as pressurized or drainage. Product standards for plastics pipe systems are numerous. Quality controlled plastic pipes are capable of fulfilling the specific requirements for each application such as long lifetime with reliability and safety.

1.2.11 Geomat EC-TRM

Geomats fall into the erosion control application category. The erosion control has both temporary Erosion Control Blankets as well as Turf Reinforcement Mats which provide permanent solutions. Geomats are typically made from synthetic materials filaments that once tangled together form a three-dimensional layer which can be permeated and also provide drainage. Turf Reinforcement Mats (TRM) can be used for permanent applications to stabilize hillside soil and to promote vegetation growth. TRMs are placed in steep areas, or

where ground maintenance is challenging to access. Size, strength, and mat materials vary depending on the application. These materials can be of natural materials such as excelsior (wood fiber), coconut fiber or straw matrix or fully synthetic material made of polypropylene, polyethylene of polyamide (i.e. nylon).

1.2.12 Geofoam

Geofoam is a product created by polymeric expansion processes resulting in a "foam" that consists of many closed but gas-filled cells. The skeletal nature of the cell walls is the unexpanded polymeric material. The resulting product is generally in the form of large, but extremely light, blocks that are stacked side-by-side, providing lightweight fill in numerous applications. Although the primary function is dictated by the application, separation is always a consideration and geofoam will be included in this category rather than creating a separate one.

1.2.13 Geocomposites

Geocomposites consist of a combination of geotextiles, geogrids, geonets and/or geomembranes in a factory-fabricated unit. Also, any one of these four materials can be combined with another synthetic material (e.g. deformed plastic sheets or steel cables) or with soil. For

Figure 1.13 Photograph of various Geocomposites (GCO or GC). Credit: Geosynthetic Institute.

Figure 1.14 Photograph of various Geo Others and Appurtenances (GSY or AP). Credit: Geosynthetic Institute.

example, a geonet with geotextiles on both surfaces and a GCL consisting of a geotextile/bentonite/geotextile sandwich are both geocomposites. This exciting area brings out the best creative efforts of the engineer, manufacturer, and contractor. The application areas are numerous and growing steadily. They encompass the entire range of functions listed for geosynthetics: separation, reinforcement, filtration, drainage, and containment.

1.2.14 Geo Others and Appurtenances

The general area of geosynthetics has exhibited such innovation that many systems defy categorization. For want of a better phrase, *geo-others* describe items such as threaded soil masses, polymeric anchors, and encapsulated soil cells. As with geocomposites, the primary function of geo-others is product-dependent

and can vary. They can also be an appurtenance, fitting, accessory, part or other item associated with a particular geosynthetic system. Such elements lend themselves to all geosynthetic installations and typically involve a connection or detail.

1.3 Market Activity

To say that the geosynthetic market activity, as indicated by sales volume, is strong is decidedly an understatement. All existing application areas are seeing constant growth, albeit at different rates. The general motivators for such growth are increased benefit/cost ratio over conventional materials and solutions (for geotextiles and geogrids), and requirements by federal or state regulations (for geomembrane and geonets). Current geosynthetic sales are well over 10 billion US dollars annually with over one hundred active companies manufacturing the products (per ISO/TC 221 'Geosynthetics' Strategic Business plan).

While the total expenditure is impressive and indicates that geosynthetics are well-entrenched construction materials, the situation could, perhaps even should, be much larger than indicated. If one factors in the concept of sustainability, the carbon footprint of each type of geosynthetic in comparison to use of traditional materials is distinctively lower. The future will see this as an additional factor in furthering the use of geosynthetics.

It also should be mentioned that relatively few colleges and universities teach geosynthetics as a specialized course. It appears as though the only way that the subject is being introduced to students is as part of existing courses, which is a logical way to contrast geosynthetics with traditional materials. Geosynthetics can also, of course, be studied via professional courses taken after graduation. These are offered on intermittent schedules by numerous associations, institutes, and continuing education organizations. It appears as though the Internet will eventually be the educational outreach vehicle of the future for geosynthetics, particularly in the form of "webinars". Whatever vehicle form it takes, education in geosynthetics is still a major objective of IGS and should be a mission for all of us.

1.4 Polymeric Materials

The vast majority (well over 95%) of the geosynthetics discussed are made from synthetic polymers, broadly characterized as plastics. Thus a brief discussion on the topics of polymer composition, structure and identification is in order. This section is not meant to make a polymer engineer or a polymer scientist out of the reader, but only to afford an appreciation of (1) the wealth of information that is available, (2) the sophistication of the topic area, and (3) the need for at least a rudimentary understanding of geosynthetics at the molecular level, which will prove beneficial.

To begin with, recognize that the plastics industry is enormous. Worldwide sales are over $1.0 trillion per year and the distribution reflects both the strength and diversity of consumption. Fortunately, of the thousands of commercialized polymers in existence the area of geosynthetics utilizes very few. The following are the most commonly used in the manufacturing of geosynthetics:

> High density polyethylene (HDPE)—developed in 1941
> Linear low density polyethylene (LLDPE)—developed in 1956
> Polypropylene (PP)—developed in 1957
> Polyvinyl chloride (PVC)—developed in 1927
> Polyester (PET)—developed in 1950
> Expanded polystyrene (EPS)—developed in 1950
> Chlorosulphonated polyethylene (CSPE)—developed around 1965
> Thermoset polymers such as ethylene propylene diene terpolymer (EPDM)—developed in 1960
> Polyvinyl alcohol (PVA) – synthetized in 1924, It's also known as PVOH or PVA
> Nylon and Polyamides typified by amide groups (CONH) discovered 1931 with commercial production 1938

1.5 Brief Backgrounds of Polymer Manufacturing

The basic "feedstock" for almost all of the polymers used to make geosynthetics is ethylene gas. Almost all of the polymers mentioned previously are included in the various branches. Ethylene is reacted by a catalyst to form discrete particles, called "flake", in a huge refinery. To say that the chemistry involved in the reaction is complex is a vast understatement, as evidenced by Ziegler and Natta who shared the Noble Prize for their respective discoveries of the catalysts for polyethylene and for polypropylene in 1963. Subsequently, Flory received the Noble Prize in 1974 for understanding the physical chemistry of polymers. Polymer manufacturing and properties are a significant component of chemistry and materials engineering departments at every college.

The word *polymer* comes from the Greek *poly* meaning "many" and *meros* meaning "parts". Thus a polymeric material consists of many parts joined together to make the whole. Each part, or unit, is called a *monomer*, the molecular compound used to produce the polymer. It should be recognized that the monomers and the repeating molecular units are different. This is due to the polymerization process. The functionality (i.e. the number of sites at which a monomeric molecule can link with other monomer molecules) determines the type and length of the chain.

The molecular weight of a polymer is the degree of polymerization (i.e. the number of times a repeating unit occurs) multiplied by the molecular weight of the repeating unit. The average molecular weight and its statistical distribution are very important in the resulting behavior of the polymer, since increasing *average molecular weight* has several results: increased textile strength, increased elongation, increased impact strength, increased stress crack resistance, increased heat resistance, decreased flow behavior, and decreased processability. Narrowing the *molecular weight distribution* also has several results: increased impact strength, decreased stress crack resistance, decreased flow behavior, and decreased processability.

While most of the polymers used in the manufacture of geosynthetics are from one type of monomer, thus called *homopolymers*, there are other possibilities. A polymer made from two repeating units in its chain is called a *copolymer*. It is important to note here the manner of linking or joining the repeating units. This can be random, alternating, block, or branch (graft). Such copolymerization greatly expands the structural properties of the resulting polymer. Furthermore, it is possible to have three repeating units in the chain in what is called a *terpolymer*. It is easy to see that the options are essentially limitless, which explains why there are approximately 50,000 commercialized polymers in existence.

1.6 Polymer Identification

There are a number of possible ways to identify the specific polymer from which a material (in our case, a geosynthetic) has been made. Identification of the particular type of synthetic polymer, there are a number of *chemical analysis tests*. Such tests are finding a place in geosynthetic materials analysis for the following reasons:

> They are used in quality assurance and product certification.
> They are used to evaluate the estimated lifetime of field-retrieval samples.
> They are used in laboratory investigations into material degradation and subsequent lifetime prediction.
> They are valuable in the forensic analysis of field failures.
> They are used for research and development into new additive packages (stabilizers, antioxidants, plasticizers, and additives).
> They are used for new geosynthetic product development and application investigations.

No geosynthetic product is 100% of the polymer resin associated with its name. In all cases, the primary resin (from which the name is derived) is mixed, or formulated, with antioxidants, screening agents, fillers and/or other materials for a variety of purposes. The total amount of each additive in a given formulation varies widely—from a minimum of 1% to as much as 50%. The additives, either in particulate or liquid form, are used as ultraviolet (UV) light absorbers, antioxidants, thermal stabilizers, plasticizers, biocides, flame retardants, lubricants, colorants, foaming agents, or antistatic agents. The resulting mixture can be homogeneous or heterogeneous, depending upon the solubility parameters of the additives versus the primary resin polymer.

Common particulate additives include carbon blacks; various antioxidants; calcium carbonate; metallic powders and flakes; silicate minerals such as clay, talc, and mica; silica minerals such as quartz, diatomaceous earth, and novaculite; metallic oxides such as alumina, biocides, and other synthetic polymers. Common liquid additives include plasticizers, fillers, and colorants. Common fibrous additives (although rarely used in geosynthetic materials) include glass; carbon and graphite; cellulosic such as alpha cellulose; synthetic polymers such as nylon; metals such as steel fibers and strands; and boron.

Thus, an understanding of a polymeric material insofar as its formulation is concerned is a complex and formidable task, but a "doable" one. Unfortunately, it is rarely given a high priority in engineering curricula, the obvious exception being in a polymer (or materials) engineering program. Rarely (if at all) does a civil, mechanical or industrial engineer have any formal training in polymers, and even many chemical engineering programs are quite lean in this area. Future college curricula must be more attuned to the necessities of modern material systems, in which polymers play a key and ever-expanding role.

1.7 Functions

The different types of geosynthetics are uniquely crafted to deliver varying functional attributes. Some products may be designed to deliver a combination of functions but most have a primary function. The following section described the functions currently envisioned. Going forward the list will inevitably grow.

1.7.1 Filtration or Permeability

Filtration is a process to separate solids from liquids or gases using a filter medium (geotextile) that allows the fluid to pass while retaining the solid. Many researchers agree that the geotextile acts as a "catalyst" in stimulating the upstream soil to do the majority of the filtration by itself. Filtration is often considered a balance

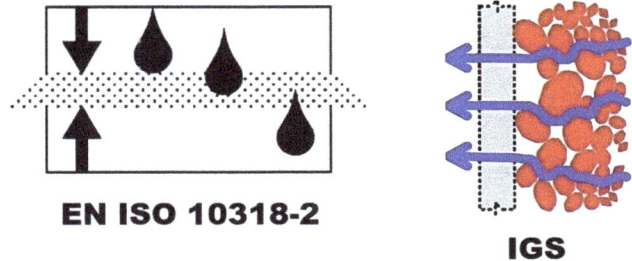

Figure 1.15 Symbol for Filtration or Permeability Diagrams

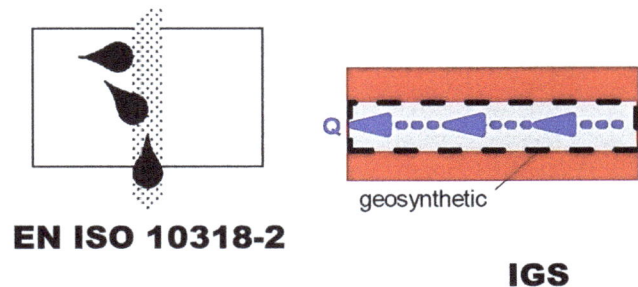

Figure 1.16. Symbol for Drainage Diagrams

between clogging and soil retention. The goal is to find an equilibrium between these two conditions. The geotextiles used in this application are generally checked for permeability, opening size, geometry, survivability and long-term clogging.

1.7.2 Drainage

Drainage is the act of carrying fluid through less permeable materials such as soil and rock. It is used to dissipate pore water pressure in and around structures. Geosynthetic drainage systems evolved from thick geotextiles to various drainage core types with thinner geotextiles bonded to the geonet cores. There is considerable development in all types of drainage cores. A drainage core of 10 mm (3/8 in.) thickness will drain the equivalent of 300 mm (12 in.) of medium sand having a hydraulic conductivity, i.e. permeability, of 0.1 cm/sec!!! The key to this interesting equivalency is that the velocity of flow in drainage cores is 20 to 30 times faster than in sand.

Some geosynthetic are being designed for capillary flow within their plane which is the spontaneous wicking of liquids in narrow spaces without the assistance of external forces.

1.7.3 Separation

To function as a separator, the geosynthetic must prevent soil with different particle size distributions from intermixing and causing the structural integrity to fail. Separation is a required function in many applications; however, it is vitally important to the layers of roads and pavements. Roadways have progressively stronger

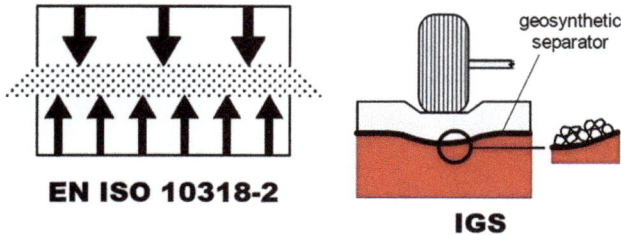

Figure 1.17 Symbol for Separation Diagrams

layers moving up through the cross section to support traffic over base and subgrade soil. The basic geosynthetic purpose is to preserve and enhance these layers. This is done by preventing adjacent dissimilar materials from intermixing. This in turn is accomplished from preventing the subgrade fines from migrating upward and the aggregate from being driven downward. It should be noted that the thinner the total roadway section, the greater the stress/contamination between the layers. This problem is made worse by the presence of moisture and the pumping action of traffic loading which all needs to be handled by the geosynthetic separator.

1.7.4 Barrier

Barriers are obstacles that prevents movement or access. Geosynthetic barriers are used to control fluid migration in manmade systems. Such barriers are usually Geomembranes, films, GCL's or Geocomposites (i.e. GCCM's). They are typically used as a barrier for liquids or gas, controlling the movement of fluids and providing containment on geotechnical engineering projects. Geomembranes can be especially useful where there is the potential for leakage of hazardous contaminants as this geosynthetic offers chemical-resistant properties and great durability over time.

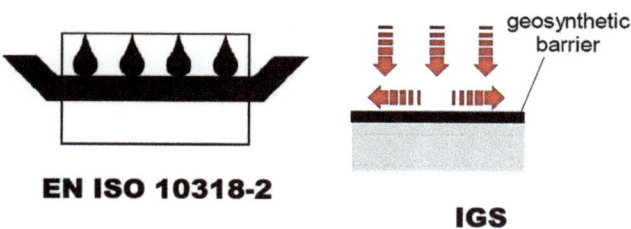

Figure 1.18 Symbol for Barrier Diagrams

1.7.5 Reinforcement

Reinforcement is the use of tensile stress-strain behavior of a geosynthetic to improve the mechanical performance of a system or structure made of soil or other construction material. The action or process of strengthening to provide additional stability in construction. Therefore, using geosynthetics to strengthen

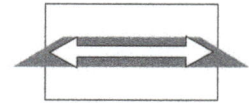

• **Reinforcement**

Figure 1.19 Symbol for Reinforcement Diagram

the soil/geosynthetic system, reinforce the ground to reduce ground deformation and or to increase the bearing capacity of the ground is common. The interaction of the soil with the geosynthetic elements is essential for good performance.Geosynthetic reinforcement consists of materials that can withstand tensile forces acting upon the soils from below, within and above their structures. Such reinforcement elements are usually Geogrids, Geotextiles or Geosynthetic Strips or Geocells.

1.7.6 Stress Relief

Roadways fatigue and fail due to traffic loading, environmental stresses and hydraulic challenges. Generally, the bitumen saturated geotextile is incorporated into the roadway to slow down the appearance of fissuring, limit infiltration of surface water and provide a stress absorbing interlayer with the roadway pavement. This can occur within new asphalt concrete pavements, beneath asphalt concrete overlays over asphalt concrete, Portland cement concrete or composite pavements and beneath chip seal surface treatments. This is the largest use of geotextile in geosynthetics with over *one billion* square yards (m²) employed to date in this 50-year young technology. The technology is quick and easy installation. In addition, it does not slow paving operations. Paving fabric interlayers effectively retard the development of both fatigue and reflective cracking through stress absorption/attenuation. In addition to geotextiles, other geosynthetics (geogrids and geocomposites) are also used to provide stress relief in pavements.

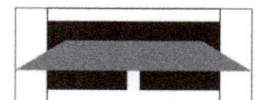

• **Stress Relief**

Figure 1.20 Symbol for Stress Relief Diagram

1.7.7 Protection

Protection or puncture resistance denotes the relative ability of a material to inhibit the intrusion of a foreign object. Puncture protection geosynthetic are generally geotextiles and geocomposites which are protecting the continuity of geomembrane barriers. The protection can be in the form of either static, dynamic or impact loading. The materials are generally rated on the basis of puncture strength and overall toughness. This major function of protecting the geomembrane from being pierced is accomplished by absorbing local stress and bridging voids in the adjacent soil.

• **Protection**

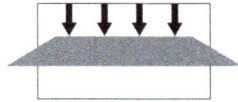

Figure 1.21 Symbol for Protection Diagram

1.7.8 Stabilization

Stabilization is the action or process of strengthening or enhancing subgrade and or unbound aggregate through confinement and restraint. This is achieved by good geosynthetic to soil interaction and through geo-synthetic to the soil size optimization which causes interlocking. Such stabilization elements are usually Geogrids or Geocells. Stabilization provides assures preservation of the full design section—creates a Permanent Road Foundation (PRF). This enables the use of a more open-graded, free-draining aggregate—higher drainage coefficient for added strength and replaces a more expensive and less effective foundation alternatives, like treated subgrade or bases. The solution also compensates for unforeseen subgrade weakness or wet areas and prevents total road reconstruction and all the associated costs and problems.

• **Stabilization**

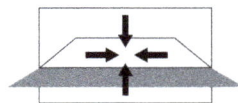

Figure 1.22 Symbol for Stabilization Diagram

1.7.9 Erosion Control

Erosion Control is the practice of preventing or controlling wind or water movement of soil in the environment. Such erosion control elements are usually Geomats, EC-TRM's, FFCRs/Geotextile matresses, Geotextiles, GCCMs and or Geocells which are placed along the surface of a slope or along channels. Erosion, and its control, is a massive function of geosynthetics. Applications are distinguished between slopes and channels with generic specifications based on both lab and field testing supporting best practices. There are many different temporary and permanent products available.

• **Erosion Control**

Figure 1.23 Symbol for Erosion Control Diagram

Figure 1.24 Symbol for Insulation Diagram

1.7.10 Insulation

Insulation is the action of separating a conductor from conducting bodies by means of nonconductors so as to prevent transfer of electricity, heat, energy or sound. Many geosynthetic materials provide insulation. Geo-foams are lightweight large blocks that are made by processing polystyrene into a foam. They are light-weight and primarily thermal insulators. They can also be foam glass aggregates used as ultra-light weight thermal barrier in road construction. As a results of the Geofoam's structure and very low unit weight, it has a very high "R" value.

1.7.11 Absorption and Adsorption

Absorption is a physical or chemical phenomenon or a process in which atoms, molecules or ions enter the liquid or solid bulk phase of a material. This is a different process from adsorption, since molecules undergoing absorption are taken up by the volume, not by the surface. There are a host of geosynthetics that exhibit such properties in regards to heat, water or energy. There are also a number of geosynthetics designed as absorption product for hydrocarbon and other priority pollutants such as Per- and Polyfluoroalkyl Substances (PFAS). It is interesting to contrast natural versus synthetic material absorption and adsorption characterization. Both are utilized heavily in geosynthetic technology.

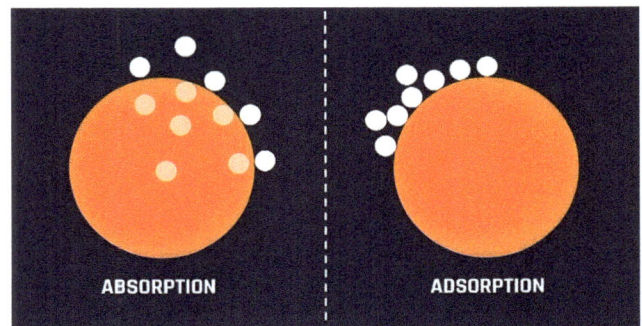

Figure 1.25 Symbol for Absorption and Adsorption Diagram

Table 1.1 Geosynthetics versus Primary Function

Function/Geosynthetic	GTX	GGR	GST	GCE	GN	GBR-P & GM	GBR-B & GM	GBR-C & GCL	GCCM	GP	GMA & EC-TRM	GF	GCO	GSY or AP
1. Filtration or Permeability	X										X		X	X
2. Drainage	X				X					X	X	X	X	X
3. Separation	X	X					X		X		X	X	X	X
4. Barrier	X					X	X	X	X				X	X
5. Reinforcement	X	X	X	X					X		X		X	X
6. Stress Relief	X	X					X				X	X	X	X
7. Protection	X						X		X		X		X	X
8. Stabilization	X	X	X	X							X		X	X
9. Erosion Control	X								X		X		X	X
10. Insulation	X										X		X	X
11. Absorption and Adsorption	X						X	X	X		X		X	X
12. Resistance versus Conductivity	X					X				X				
13. Appurtenances and Connection													X	X

GTX – geotextile
GST – geostrip
GN – geonet
GBR-C & GCL – geosynthetic clay liner
GP – geopipe
GF – geofoam
AP – appurtenances

GGR – geogrid
GCE – geocell
GBR-P &GM – geomembrane
GCCM – geosynthetic cementitious composite mat
GMA & EC-TRM – geomat
GCO – geocomposite

Figure 1.26 Symbol for Resistance and Conductivity

Figure 1.27 Symbol for Appurtenances and Connection Diagram

1.7.12 Resistance versus Conductivity

Resistance is a force that counteracts the flow of current or energy. In this way, it serves as an indicator of how difficult it is for current or energy to flow. Resistance is counterpointed with conductivity which is the measure of the ease at which an electric charge or heat can pass through a material. A conductor is a material which gives very little resistance to the flow of an electric current or thermal energy. There are many geosynthetics designed solely as conductors or resistors. As one can imagine, conductive geomembranes and geotextiles are critical to all electric leak location techniques Thermal conductivity is a key property that controls heat migration in a variety of applications including waste containment and subsurface insulation.

1.7.13 Appurtenances and Connections

Appurtenances, fittings, also known as connectors, attach one geosynthetic to another in order to extend or terminate continuity of the system. They also can be used to combine, divert or reduce supply. Appurtenances come in a variety of sizes to fit a wide range of products. They can be installed in the factory or the field as immovable or fixed objects which are critical to structure performance. Classic examples of these are seams, batten strip and pipe boots.

1.8 Contrasting Functions with Geosynthetic Type

The juxtaposition of the various types of geosynthetics just described with the primary function that the material is called upon to serve allows for the creation of a matrix which will be used throughout the book. In essence, this matrix is the "scorecard" for understanding the entire geosynthetic field and its design-related methodology. Table 1 illustrates the primary function that each of the geosynthetics can be called upon to serve. Note that these are primary functions and in many (if not most) cases there are secondary functions, and perhaps tertiary ones as well. For example, a geotextile placed on soft soil will usually be designed on the basis of its reinforcement capability, but separation and filtration might certainly be secondary and tertiary considerations. A geomembrane is obviously used for its containment capability, but separation will always be a secondary function.

The greatest variability from a manufacturing and materials viewpoint is the primary function will depend upon what is actually created, manufactured and installed. Table 1 clearly identifies each geosynthetic material vis-à-vis the primary function (usually by application) that is being served.

1.9 References

ASTM D4439 Standard Terminology for Geosynthetics, (2024) 100 Barr Harbor Drive, P.O. Box C700, West Conshohocken, PA 19428, USA, pp. 10, http://www.astm.org

Dewar, S., (1962) "The Oldest Roads in Britain," The Countryman, vol. 59, no. 3, pp. 547-555.

IGS, (2000) Recommended Descriptions of Geosynthetics Functions, Geosynthetics Terminology, Mathematical and Graphical Symbols, 4th edition 2000, pp. 17.

ISO 10318-2, (2015) Geosynthetics, Part 2 Symbols and pictograms

ISO Strategic Business Plan, (2023) – ISO/TC 221 'Geosynthetics' 2023 pp.10.

Kays, W. B., (1982) Construction of Linings for Reservoirs, Tanks and Pollution Control Facilities, 2nd ed., John Wiley and Sons.

Koerner, R. M., (2012), Designing with Geosynthetic, 6th Edition, Xlibris Corporation, Thorofare, NJ USA, pp.914.

Lauritzen, C. W., (1963) "Plastic Films for Water Storage," Journal of American Water Works Assoc., vol. 53, no. 2, February, pp. 135-140.

Mascia, L., (1982) Thermoplastics: Engineering. New York: Applied Science Publishers.

Rodriguez, F., (1996) Principles of Polymer Systems. 4th ed. New York: McGraw Hill.

Rosen, S. L., (1982), Fundamental Principles of Polymeric Materials. New York: Wiley.

Sperling, L. H., (1986), Introduction to Physical Polymer Science. New York: Wiley.

Chapter 2

Geosynthetics in Roads and Railways

2.0 Overview

Geosynthetics are known to provide valuable benefits in the design and construction of roads. The proper understanding and application of geosynthetics are crucial to achieve sustainability, cost-effectiveness, and durability goals. This chapter aims to provide the knowledge and tools necessary to incorporate geosynthetics successfully into roadway projects.

2.1 Geosynthetic Design Benefits

The primary goal of roadway design is to create a strong and durable structure that provides a smooth and safe driving surface over the entire design life. To do this, the structural layers within a roadway are designed to absorb and spread the applied traffic loads to delay and/or prevent damage to weaker underlying layers over time (Figure 2.1). The road surface can be made of unbound gravel layer, flexible pavement (asphalt, paving blocks) or rigid pavement (concrete). Choosing the appropriate road surface involves balancing initial costs, maintenance, durability, environmental impact, and specific project requirements. Each type of road surface has its strengths and weaknesses, and the decision should be tailored to the context of the road's intended use and local conditions.

In most applications, the natural subgrade is the weakest and most vulnerable part of the road. Tradi-

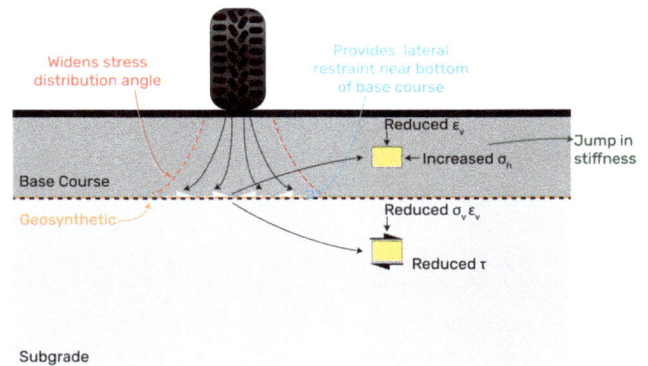

Figure 2.2 Improvement mechanisms in pavements (modified from Perkins et al 2010)

tional approaches either remove the weak subgrade and replace it with higher quality fill or simply bridge over weak areas with fill to create a working platform. These methods were easier to justify in times when resources were more plentiful and incentives for more efficient and eco-friendly designs were less important. Geosynthetics, however, provide a substantially more efficient, cost-effective, and durable solution to meet modern demands. Benefits are generally realized in terms of reduction of structural layer thicknesses and/or increased service life.

Specific to the subgrade, geosynthetic benefits include reduced rutting, decreased fines migration, greater uniformity in load transfer, limited need for removal and replacement, and reduced need for chemical stabilization. Unbound granular layers above the subgrade most often consist of a natural stone/sand mixture containing larger particles having looser gradation requirements (sub-base) or crushed angular gravels having a specific gradation of particles (base). Unbound aggregate layers have no tensile strength and rely on particle-to-particle friction and lateral confinement to restrain particle movement. The use of a geosynthetic with the granular layers restrains movement of unbound particles. Restraining particle movement reduces strains and increases the stiffness of the unbound materials within and around the geosynthetic (Figure 2.2).

In paved road applications, the flexible asphalt layer provides a continuous structural platform to further

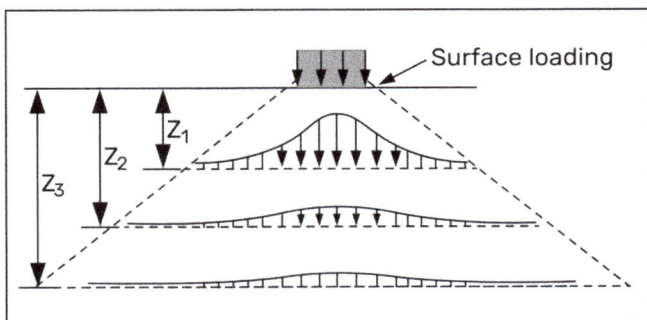

Figure 2.1 Distribution of stresses through the road cross section (Yoder and Witzak, 1975)

spread applied loads and protect underlying layers from water infiltration and erosion. The primary constituent in hot-mix asphalt is a well-graded aggregate mixture. Asphalt layers are susceptible to cracking from fatigue and temperature effects. The use of stiff geosynthetics within the asphalt layer can reduce and control cracking, provide stiffness improvements, and reduce particle movement over time, similar to unbound materials. This reduction in cracking and prevention of water infiltration helps protect and strengthen the asphalt layer and supporting structural layers from degradation over time.

Increased use of geosynthetics in pavement designs will result in the extended service life of the roadway (measured as traffic benefit ratio – TBR, as outlined in Section 2.4.1, below) and reduced impacts from maintenance operations and rehabilitation projects on road users and surrounding businesses. Simply stated, extending the pavement service life will lengthen the interval between major rehabilitation projects leading to fewer disruptions to traffic and thereby enhancing route safety. Alternatively, geosynthetics can be used to reduce the thickness of the subbase, base (measured as base course reduction factor – BCR, as outlined in Section 2.4.2 below) and asphalt layers, while still achieving the desired design life.

2.2 Geosynthetic Functions in Roads

Geosynthetics provide multiple functions within a roadway system, primarily: filtration, drainage, barrier, separation, stabilization, reinforcement, and stress relief (see Figure 2.3). When properly selected and constructed, individual geosynthetic products can provide one or more of these functions. Proper selection depends on the type of product and its material properties. This section provides an overview of the functions of geosynthetics in roadways to help users properly select products that can effectively provide these functions in the design.

It is well known that the presence of water within the road structure can be detrimental, from softening of fine-grained materials to frost heave in colder regions. Geosynthetics can be used to allow passage of water while simultaneously preventing the movement of soil particles (**filtration function**, Figure 1.14 – Chapter 1), direct the flow of water away from moisture susceptible soils (**drainage function**, Figure 1.15 – Chapter 1), or prevent water from infiltrating the road prism (**barrier function**, Figure 1.17 – Chapter 1). Subgrade soils that are sensitive to moisture or freeze-thaw conditions can be isolated or protected from water ingress using an impermeable geosynthetic (barrier). Geomembranes are most commonly used to provide this function by installing them beneath or around the sensitive soil to prevent fluctuations in moisture content over time. Special wicking geotextiles can also be used as

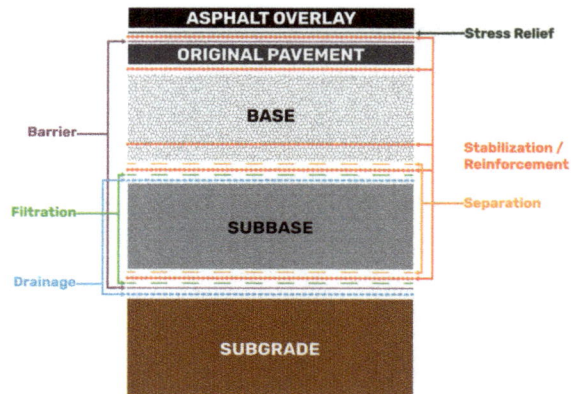

Figure 2.3 Location and functions of geosynthetics throughout the road cross section (modified from Zornberg, 2017)

a capillary break or to direct water away from sensitive soils, thus improving their performance in wet and cold conditions. Water control is most often addressed using geotextiles, geocomposites (oftentimes consisting of a geonet sandwiched between two geotextiles), geopipes, and geomembranes.

Another common use of geosynthetics in roads is geotextiles used as separators between the top of the subgrade and unbound granular layers above (**separation function**), as illustrated in Figure 1.16 – Chapter 1. The separation function prevents particle movement between adjacent layers (downward migration of granular particles from unbound layers into the subgrade and/or upward migration of fines promoted by water) especially in saturated conditions. This movement of materials across layers results in decreased structural support causing accelerated failure and shorter pavement life. Fines contents above 8 percent have been shown to significantly affect the load carrying capacity of stone layers (Jorenby and Hicks, 1986), as illustrated in Figure 2.4. The separation function is fundamentally advantageous over the life of the pavement as it helps preserve the original integrity of structural granular layers over time. The separation function is most effectively achieved using geotextiles as they prevent movement of the fines; however, a geogrid layer also prevents movement of larger granular materials downward into the subgrade.

The structural capacity of unbound layers can be improved using geosynthetics. This improvement is oftentimes characterized as stabilization, reinforcement, or stiffening. While these terms may seem synonymous, each is uniquely applied to geosynthetics of various types to describe their effect on structural capacity. The **stabilization function** (Figure 1.21 – Chapter 1) provided by geosynthetics in roadways is the ability of the geosynthetic to arrest movement of soils and aggregates within the structural layers from applied loads. In this way, a variety of products can provide this function with varying levels of success. A stabilized

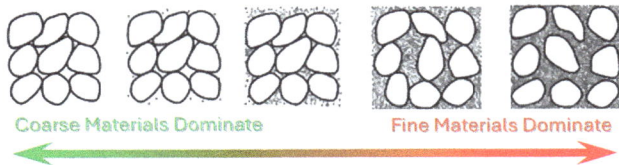

Figure 2.4 Effect of fines in aggregates (modified from Park and Santamarina, 2017)

road is less apt to allow particle rearrangement and movement from traffic loads. It can be achieved in conjunction with the separation function since it restricts movement of particles; however, the stabilization function focuses on preventing particle movement and degradation thus increasing shear strength of the aggregate while the separation function focuses on the prevention of intermixing between dissimilar materials over time. Stabilization includes mechanisms such as lateral confinement (the restriction of particles from moving horizontally beneath the applied load) and increases the aggregate shear capacity (due to particle restraint against translation and rotation). These mechanisms lead to a broadening of the applied load.

The **reinforcement function** (Figure 1.18 – Chapter 1) of the geosynthetic adds a tensile strength component to the unbound aggregate layer by absorbing some of the stresses of the applied load. The improvement in tensile strength contributes to the structural capacity of unbound aggregate layers by providing a tensile load carrying capacity, similar to the way rebar provides tensile strength to reinforced concrete after it has cracked. As illustrated in Figure 2.3, geosynthetics used as reinforcement can be introduced in various positions in the roadway cross-section.

Asphalt pavements experience fatigue from traffic loading and environmental stresses over their life, which commonly result in cracking. Geosynthetics are used to delay and mitigate cracking caused by these stresses. In this application, geosynthetics most commonly function as reinforcement, stress relief, and/or barriers in asphalt pavement applications. Geogrids and geocomposites placed between hot-mix layers can help strengthen and stiffen the asphalt (**asphalt reinforcement function**). When strategically placed in areas where cracks are expected, geosynthetics can absorb stresses to mitigate reflective cracking of the pavement surface. In this case, the geosynthetic increases the stiffness of the asphalt (i.e. increasing the tensile load carrying capacity with less strain). Cracking can also be mitigated using nonwoven geotextiles between asphalt layers. In this case, a relatively weak nonwoven geotextile layer helps dissipate stresses laterally (**stress-relief function**, Figure 1.19 – Chapter 1) rather than allowing cracking stresses to concentrate into a single location. Finally, geosynthetics impregnated with bitumen and tack also provide an impervious

layer restricting infiltration of water into and through the asphalt layer (**asphalt barrier function**).

2.3 Geosynthetic Road Applications

Selecting the most beneficial and cost-effective geosynthetic for a given application can be a daunting task given the vast number of geosynthetic styles, strengths, and manufacturing methods coupled with site complexities and specifications. There are an unlimited number of ways geosynthetics can be used to improve the performance of roads. Given the limited scope of this document, the broad scope of possibilities was narrowed down to a few common roadway applications. The five most popular applications for geosynthetics in roads are: 1) preventing intermixing between dissimilar layers (separation), 2) road construction on soft ground, 3) moisture control, 4) stabilization and reinforcement of unbound aggregate layers, and 5) asphalt interlayers. A brief description of each application, the mechanisms of improvement made by the geosynthetic, and selection of the most appropriate geosynthetic, is described below.

2.3.1 Preventing Intermixing between Dissimilar Layers (Separation)

Preventing dissimilar layers from mixing with one another (separation) is the most common and established use of geosynthetics in transportation applications. Using geosynthetics as separation is an inexpensive, proactive approach to preserve aggregate assets into perpetuity. While it is important to design the subbase or base

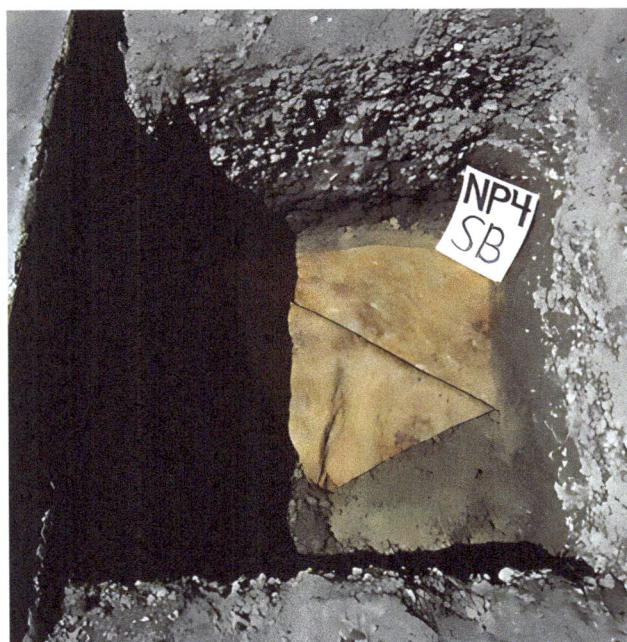

Figure 2.5 Effectiveness of separation after years in service (Collins and Holtz, 2005 – NP4 = 4.5 oz. nonwoven geotextile, SB = southbound test section)

course layers to have the proper gradation and additional sacrificial thickness to reduce or prevent the migration of fines, most of the time separation can be more efficiently and effectively achieved using a nonwoven or woven geotextile. A photo of an exhumed geotextile (Figure 2.5) reveals it is still doing its job after years in service. A relationship between aggregate loss and subgrade strength developed by Christopher and Holtz (1989) is shown in Figure 2.6. Nonwoven geotextiles are typically weaker so if structural support is also simultaneously desired, a stronger alternative such as an enhanced, high-modulus woven geotextile or composite geotextile and geogrid can be used to provide structural support. Effective separation can help maintain the structural integrity and permeability of unbound aggregates over time.

The primary and important benefit of incorporating separation layers, especially between fine-grained subgrade and the sub-base or base, is preserving the integrity of the structural layers over time, as illustrated by Figure 2.7. Preservation of higher quality structural fills provides substantial ecological benefits through reduction of rehabilitation, which reduces the need

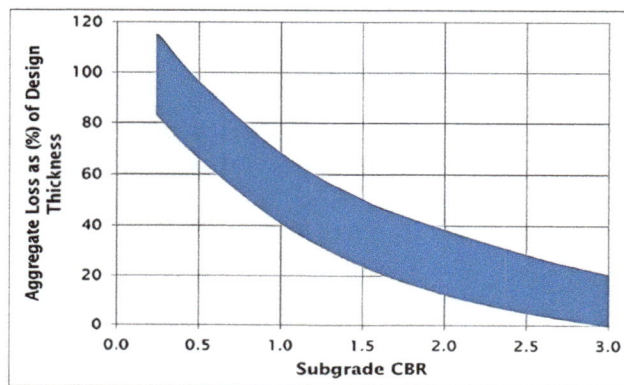

to mine and transport replacement materials, reducing disruptions to traffic, emissions, etc. The advantage geotextiles provide far outweighs their initial cost, which is relatively low. These materials are deep within the road structure and are unlikely to be disturbed or become disruptive to future construction activities. Filtration requirements should be evaluated to ensure the separation layer does not become clogged to the point of becoming a barrier, in which case water from upper layers may not be properly drained. Separation materials should allow for the free passage of water while restricting the movement of soil particles. For example, slit-film geotextiles, like those used for silt fences, should not be used as separators because they typically do not meet permittivity or filtration requirements.

Material requirements for this application include strength, puncture resistance, permittivity, and durability. Material properties and construction/installation guidelines are outlined and specified in the AASHTO M288 Specification (Geosynthetic Specifications for Highway Applications). A comparable international standard (ISO/TR 18228-2:2021(E)), which also draws from the AASHTO M288 standard, similarly outlines specifications and installation information for the separation application.

Nonwoven geotextiles can also be used in conjunction with other stabilization/reinforcement products (such as geogrids or geocells) to provide both separation

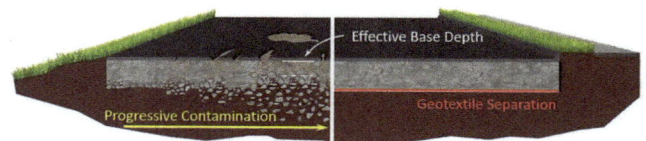

Figure 2.6 Anticipated loss of aggregate based on subgrade strength (from Christopher and Holtz, 1989)

Figure 2.7 Benefits of geosynthetics used as separation (modified from Solmax Technical Note, web access, 2024)

Figure 2.8 Combining separation and stabilization/reinforcement functions (left: separate materials (https://www.indiamart.com/proddetail/road-construction-geotextile-17009869333.html), right: a composite material (NAUE pavement reinforcement manual))

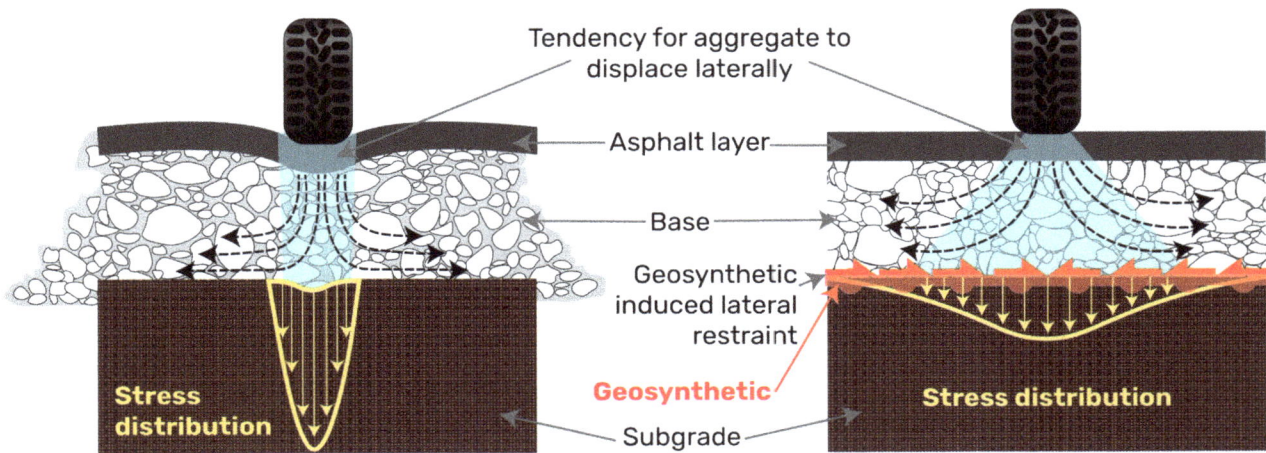

Figure 2.9 Stabilization of soft ground using geosynthetics (from Zornberg, 2017)

and strength improvements (Figure 2.8). High-strength geotextiles or geocomposites are also available to provide both benefits within a single product.

2.3.2 Road Construction on Soft Ground

The use of geosynthetics to improve the stability and strength of weak roadway foundations is one of the most cost-effective and beneficial uses of stabilization/reinforcement products in roadway construction. Planar geosynthetics used in this application provide a tensile element to the road structure to attract and absorb tensile stresses from construction traffic. Traditional methods of addressing this have been to remove and replace these areas, use chemical stabilization (e.g. lime, cement, polymers), or simply dump and spread in gravel until movement stops. The problem with these solutions is that they are costly, time-consuming, less eco-friendly, and not as effective when compared to geosynthetics. Geosynthetics placed directly atop the subgrade (beneath the granular base or subbase layer) help spread applied wheel loads more widely and reduce lateral spreading of granular materials. Doing so decreases applied stresses to the subgrade surface, increases the bearing capacity, and reduces losses of aggregate into the subgrade. This protects the subgrade from sustaining damage thereby helping advance construction efforts, while simultaneously providing long-term support over time. Geosynthetics in this application provide a more durable and sustainable solution when compared to traditional methods, mainly due to the reduced movement of materials to and from the construction site.

This application focuses on soft ground, typically CBR strengths from 1 to 3%, that is unable to support heavy construction equipment without sustaining significant damage. Geogrids, geogrid/nonwoven geotextile combinations, or high-modulus / high-strength

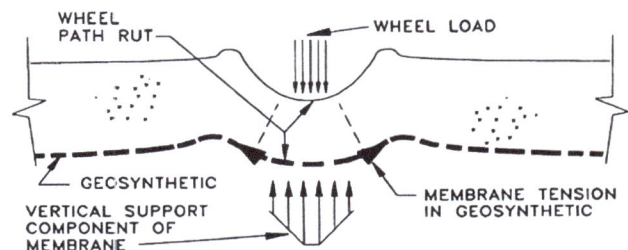

Figure 2.10 Tensioned-membrane effect (from Haliburton et al, 1981)

woven geotextiles are most often used for this application. Another option that is quite effective is a geocell, which can also be used in conjunction with nonwoven or high-strength geotextiles. Geosynthetic selection is based in part on the expected severity of the problem (how soft the subgrade is, truck load levels, amount of traffic, etc.). As traffic is applied, the aggregate within the unbound layers is forced downward and outward beneath the wheel load. If the subgrade support is low, localized failure of the subgrade may occur as the bearing capacity is exceeded. As illustrated in Figure 2.9, the stress distribution beneath the unstabilized base is narrow and concentrated beneath the applied load, whereas the stabilized base widens the stress distribution thereby reducing applied stresses on the subgrade and increasing its bearing capacity. Repeated loads can result in significant rutting. In this case, geosynthetics used at the bottom of the unbound layers are either 1.) stretched into a tight membrane as they are forced downward to retain base particles or 2.) retain particles through interlock thus increasing the shear capacity of the aggregate as shown in Figure 2.9. The first case is referred to as the tensioned-membrane effect (Figure 2.10). Geosynthetics should be selected to have adequate tensile strength for the first case and adequate stabilization properties (interlock) for the second case. In less severe cases, the most efficient geosynthetics can be selected based on design inputs.

Several design methods have been developed for this application over the past four decades. Early

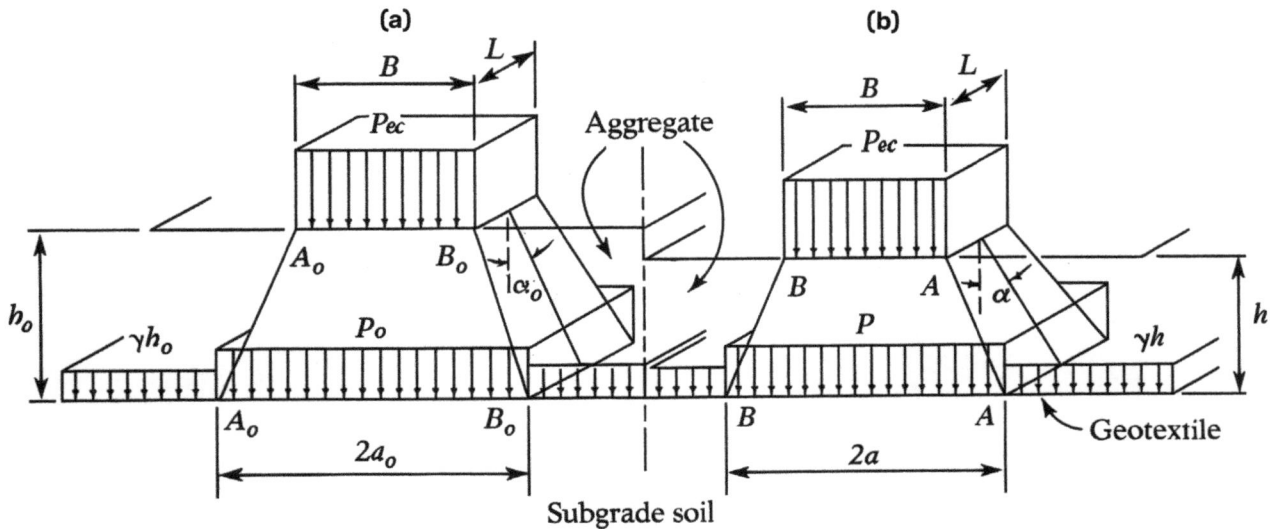

Figure 2.11 Load distribution by aggregate layer: a) case without geosynthetic, b) case with geosynthetic (Giroud and Noiray, 1981)

designs considered variables such as traffic level, axle load level, axle configuration, tire pressure, subgrade strength, and rut depth to determine the thickness of the base course with or without geosynthetics. Chapter 5 of the Geosynthetic Design and Construction Guidelines (Holtz et al, 2008) discusses these designs in detail including design examples. Subgrade support in these early designs was based on bearing capacity models, as illustrated in Figure 2.11. These designs were limited to geotextiles until the early 2000s when Tingle and Webster (2003) updated the design from Steward et al (1977) to include geogrids. Giroud and Han (2004a and 2004b) later developed a design method for geogrid reinforced unpaved roads based on limit-equilibrium bearing capacity theory. This method utilizes the wheel load, area of applied load on surface, number of axle passes, strength of the subgrade and base, geogrid properties, and rut depth to determine the thickness of the base (Eq. 2.1). The equation is used to determine the thickness of the base layer (h). This parameter is on both sides of the equation, so an iterative method must be used to solve it. This method was calibrated using a limited set of data based on laboratory test sections and field trials using biaxial geogrids. Further calibration work was done using data from Cuelho et al (2014) to determine material properties for a variety of geogrid types that related to their field performance. It was found that the stiffness of the junctions that join ribs or straps to one another to form the grid structure (i.e. junction stiffness, ASTM D7737) was most related to field performance for this application, rather than aperture stability modulus (ASTM D7864). The Giroud-Han design method was updated to reflect this as outlined in Cuelho and Perkins (2016). Geotextiles can also be used in this application and have been shown to work well since they combine both the separation and reinforcement functions into one product, as long as they have

sufficient surface roughness to properly interact frictionally with the aggregate base and remain anchored outside the wheel path.

$$h = \frac{0.868 + (0.661 - 1.006J^2)\left(\frac{r}{h}\right)^{1.5} \log N}{[1 + 0.204(R_E - 1)]} \left[\sqrt{\frac{\frac{P}{\pi r^2}}{\frac{s}{f_s}\left[1 - 0.9e^{-\left(\frac{r}{h}\right)^2}\right]N_c f_c CBR_{sg}}} - 1\right] r$$

Eq. 2.1

where,
h = compacted base course thickness {in}
J = aperture stability modulus {N-m/degree}
N = number of axle passes

$$R_E = \min\left(\frac{E_{bc}}{E_{sg}}, 5.0\right) = \min\left(\frac{3.48 CBR_{bc}^{0.3}}{CBR_{sg}}, 5.0\right)$$

P = tire load {lb}
r = radius of equivalent tire contact area {in}
s = maximum allowable rut depth {in}
f_s = reference rut depth {in}
CBR_{sg} = subgrade CBR strength {%}
f_c = CBR to undrained shear strength factor
N_c = bearing capacity factor

The primary mechanism at work in this application is restraining lateral movements of the aggregate beneath the load. Bonding or interlocking between the geosynthetic and surrounding aggregate particles is critical to effectively restrain lateral movement of the aggregate. Assuming good interlock and friction between the geosynthetic and the surrounding aggregate (Figure 2.12), the stiffness of the geosynthetic in the direction perpendicular to the direction of traffic (cross-machine direction) is important to resist the applied loads. Particle restraint (confinement) acts to stiffen the area

Figure 2.12 Geogrid-aggregate interlock (stabilization function). Images from (a) Industrial Fabrics; (b) Paramount Materials; (c) Naue; (d) Tensar

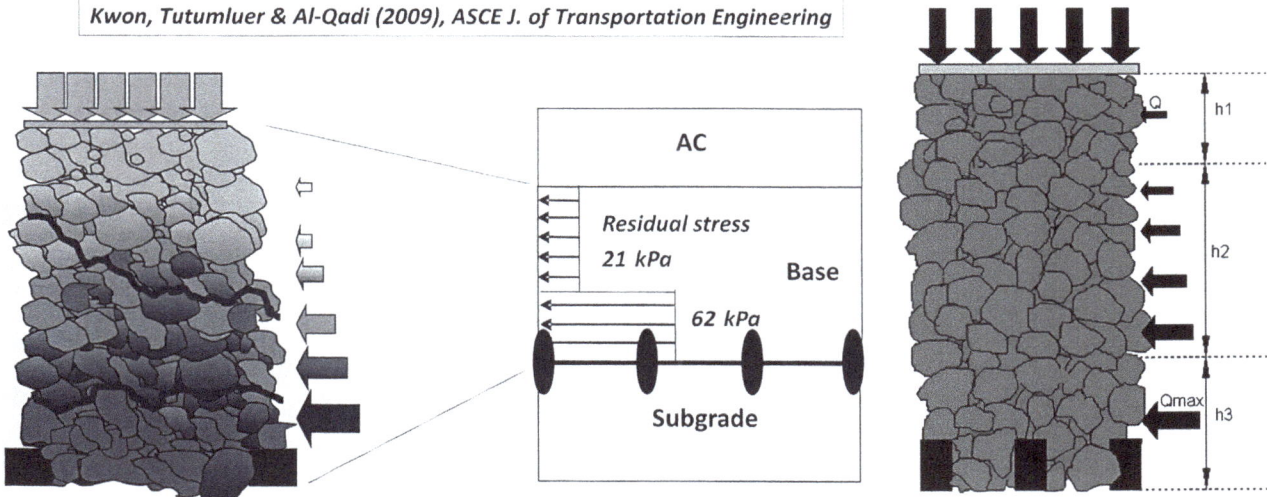

Kwon, Tutumluer & Al-Qadi (2009), ASCE J. of Transportation Engineering

AC

Residual stress
21 kPa

Base

62 kPa

Subgrade

Figure 2.13 Confinement of aggregate around geosynthetic (from Kwon et al, 2009 [let[and Kawalec, 2019 [right])

directly adjacent to the geosynthetic, and this restraint dissipates as you move upward away from the geosynthetic (Figure 2.13 – another alternative to these illustration options comes from the ISO T221 WG6 manual). To improve this effect, several geosynthetic layers can be used on a single project, sandwiched between layers of stone (illustration desired). Another alternative to multiple layers, that is also quite effective for this application, is geocells. The individual cell openings of a geocell restrict lateral movement of the infill material (granular fill) along the entire height of the geocell and extending some above the top of the cell structure (Figure 2.14). This creates a three-dimensional layer of stone pockets that are held together within the cellular structure. The cost of geocells may be higher but can be justified due to the efficient nature of these products

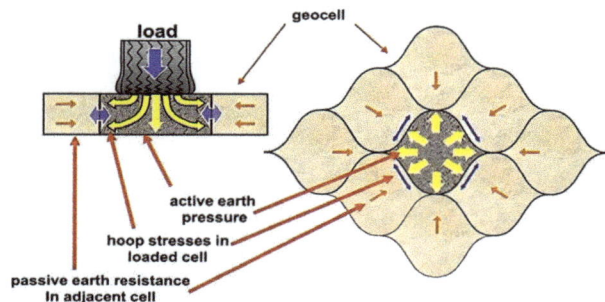

Figure 2.14 Confinement principle in geocell (modified from Geosynthetics, 2024)

$$Aperture\ Size \geq D_{50\ Fill}$$

AND

$$Aperture\ Size \leq 2 * D_{85}$$

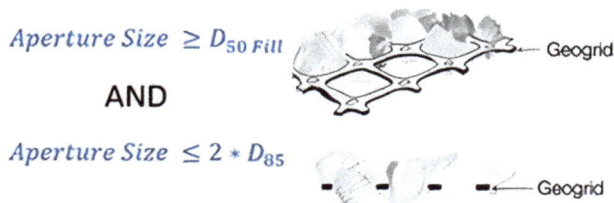

Figure 2.15 Geogrid aperture size requirements

to restrict movement of the aggregate layer under load. ASTM D8269 provides guidance on the use of geocells for roadway applications.

Restraining lateral movement of the unbound particles increases the lateral (confining) stresses making the composite geosynthetic/aggregate layer stiffer. Stiffer components attract load and distribute them over a wider area. This broadening of the area reduces stresses on lower layers. In this case, the subgrade, which is the weakest component in the cross-section is subjected to less stress, thereby decreasing damage which leads to rut. Interaction differs between different types of geosynthetics and the aggregate. Geotextiles interact through surface friction, geogrids interact through strikethrough (interlock) and surface friction, and geocells interact by confining the aggregate infill. Because this application relies on the ability of the geosynthetics to resist movement of the aggregate, the ability of the geosynthetic to do this is directly related to its performance. Geogrid apertures (holes formed by the structure of the geogrid) should be sized to adequately allow strikethrough of the stone but not large enough to allow the stones to easily pass through. Aperture size depends on the base aggregate gradation, as illustrated in Figure 12.5, as outlined in Holtz et al (2008).

2.3.3 Stabilization/Reinforcement of Unbound Aggregate Layers

Geosynthetics can be used to increase the life of the pavement, decrease the thickness of structural layers, or both. The mechanisms for geosynthetics used in this application are similar to those described above for soft ground. The principle difference between these two applications is the level of distress allowed to remain within the design goals. Soft ground applications (subgrade CBR less than about 3 percent) pertain mostly to working platforms on construction sites, temporary roads and detours, and unpaved and low-volume roads whereas firmer foundations are designed to support paved roads. This can involve an initial composite foundation with a geosynthetic to provide a proper foundation for the paved roads. Even for applications with firmer subgrade support, non-woven geotextiles should be used to ensure finer-grained subgrades do not migrate into the base aggregates during wetter seasons.

The primary benefit from geosynthetics is to stabilize the aggregate layer by restricting lateral movements of the aggregate over time and reducing accumulated deterioration of aggregate particles. This is achieved through the interlocking of aggregate particles with the geosynthetic that relies on the geosynthetic to absorb stresses and reduce strains thereby increasing stiffness. This stiffer layer not only reduces stress on underlying layers it resists small movement of aggregate particles over time. Lateral movement of gravel particles is directly related to vertical movements so restraining particles in the lateral direction directly affects rutting at the road surface. Performance of the geosynthetic in this application is directly related to its ability to restrict lateral movement of soil particles. Lateral movement of aggregate particles differs depending on the applied loads and the distance from the road surface. For thinner base course layers, it is best to place the geosynthetic at the subgrade-base course interface. For thicker base course layers, geosynthetics placed higher in the structural cross-section, but below about the top third of cross-section, have been shown to increase performance when compared to those placed lower in the cross-section (Al-Qadi et al, 2008; Saride and Baadiga, 2021). When further stability is needed, installing two layers of geogrid (one at the top third, and one at the subgrade-base course interface) will provide even greater benefit.

In an example design approach, benefits of the geosynthetic can be quantified within the structural number calculation in the AASHTO '93 pavement design equation (Eq. 2.2). The structural number is the sum of structural contributions from the asphalt, base, and subbase structural layers (Eq. 2.3). Because this application utilizes geosynthetics to improve the unbound aggregate layers, a special multiplier can be used to modify the base course term (a2D2m2) or subbase term (a3D3m3 to account for its increased strength (eq. 2.4). This term is referred to as the layer coefficient ratio (LCR) in GMA White Paper II (Berg et al, 2000). LCR can be determined using performance testing on full-scale models (see Section 2.4 below). Benefit calculated from

$$\log_{10}(W_{18}) = Z_R \times S_o + 9.36 \times \log_{10}(SN+1) - 0.20 + \frac{\log_{10}\left(\frac{\Delta PSI}{4.2-1.5}\right)}{0.40 + \frac{1094}{(SN+1)^{5.19}}} + 2.32 \times \log_{10}(M_R) - 8.07$$

Eq.2.2

where,

W_{18} = Anticipated traffic level
Z_R = standard normal deviate
S_o = standard error of traffic prediction
ΔPSI = difference in serviceability term
M_R = resilient modulus of subgrade {psi}
SN = structural number ($SN = a_1D_1 + a_2D_2m_2 + a_3D_3m_3$) Eq.2.3
                  (AC) (base) (subbase)

where,

a_i = structural coefficient
D_i = layer thickness
m_i = drainage coefficient

$$SN = a_1D_1 + (\mathbf{LCR}a_2)D_2m_2 + (\mathbf{LCR}_3a_3)D_3m_3 \quad Eq.2.4$$

this testing is oftentimes quantified in terms of the ability of the geosynthetic test section to accommodate a greater number of traffic loads to reach a specific surface distress when compared to an identical case without geosynthetics, referred to as the traffic benefit ratio (TBR). TBR can be used to directly evaluate the extension of life of pavement stabilized with geosynthetics; however, if the goal of the geosynthetic is to reduce thickness of the pavement design, two designs can be evaluated taking into consideration the benefits from the geosynthetic. A theoretical approach using the AASHTO '93 design equation can be used to adjust the thickness of structural layers to account for the improvements from geosynthetics. Base thickness reductions are typically less than 50%. The structural layer thickness of the unbound aggregate layers should not be less than about 6 in.

More advanced design methods such as the mechanistic-empirical pavement design guide (MEPDG) have been developed in more recent years. The M-E design methodology considers geosynthetic base stabilization through stiffness enhancements to the base course (Giroud et al, 2021). Mechanistic models are used within the M-E design process to augment the behavior of geosynthetics in the base and asphalt layers. Other methods can be used to adjust the calculated life of the pavement. Geosynthetics have a positive contribution to damage models, drainage, thickness calculations, structural stability, fatigue, rut models, resilient modulus and Poisson's ratio of unbound aggregates and fatigue cracking models of asphalt. M-E design models should be calibrated using laboratory testing, large-scale accelerated pavement tests, and performance trends of monitored field test sections (see Section 2.4 below).

Adequate aperture opening size and/or surface roughness is necessary to effectively engage the geosynthetic and base aggregate, as described in the previous section.

2.3.4 Moisture Control

Water within the road structure has a detrimental effect on performance. Primary water sources include precipitation, nearby canals or waterways, and ground water (Figure 2.16). Damage from excess water typically comes in the form of reduced strength of wetted layers, formation of ice lenses causing frost heave, which, when melted, leave large weak voids, and expansion and contraction of sensitive clays which cause large movements and cracking of the surface. Water from precipitation is typically controlled through an impermeable pavement surface that is sloped toward the road shoulder; however, pavement surfaces that are cracked will allow water to infiltrate into underlying layers. Runoff should be directed away from the road cross section where it can be stored, allowed to infiltrate, or channeled to a nearby waterway. Structural aggregate layers should be designed with adequate permeability to allow water passage through them and away from the roadway.

Geosynthetics can be an effective way to prevent water from infiltrating the surface or improving the drainage and flow of water through and away from the road cross section (Figure 2.17). Specific drainage products (e.g. geopipes, geocomposites made of geonets with nonwoven geotextiles laminated on either side) can be used to improve or replace free draining

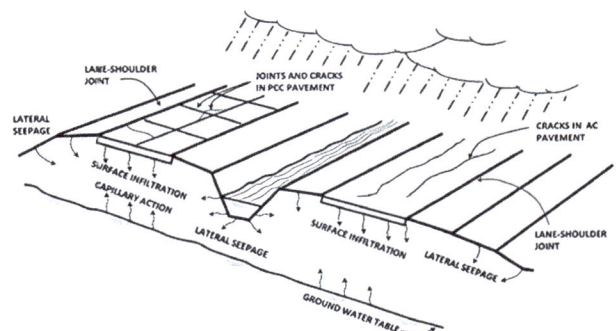

Figure 2.16 Illustration of water sources in pavements (from Hall and Crovetti, 2022 after Cedergren, 1972).

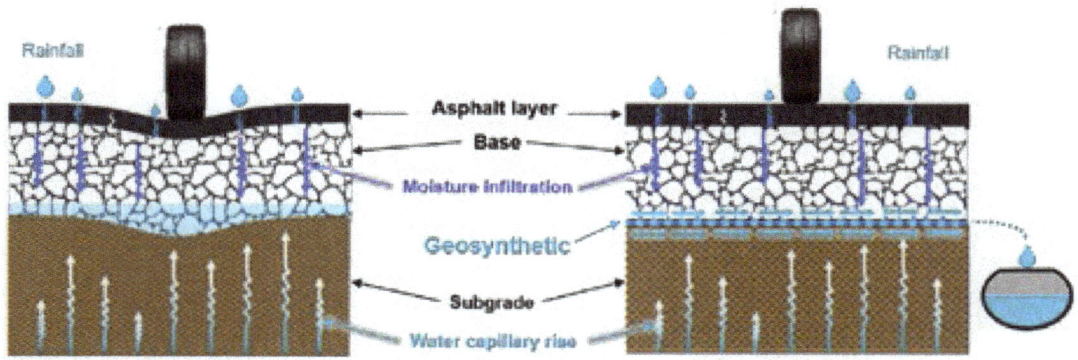

Figure 2.17 Geosynthetics used to improve drainage (from Zornberg et al, 2018)

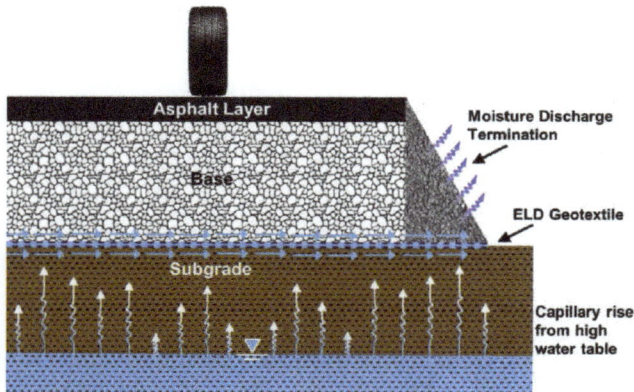

Figure 2.18 Enhanced lateral drainage (ELD) geotextiles (from Zornberg et al, 2018)

Figure 2.19 Highway edge drain (photo from National Highway Institute, GeoTechTools, Geo Institute).

base materials. Special water-wicking geotextiles (also known as enhanced lateral drainage (ELD) geotextiles) are also available to improve lateral movement of water away from the road cross section (Figure 2.18). These products have the added benefit of improving the strength properties of subgrade through suction of excess pore-water. Bitumen impregnated geotextiles in the asphalt layer can be used to prevent water infiltration through the pavement surface. Alternatively, a geomembrane or GCL can be used to prevent ground water from seeping in from below. This can be particularly helpful in areas with expansive soils, where water fluctuations can cause the roadway to expand and contract causing upheaval, swales, and an overall increase roughness.

Water from capillary action can be prevented using a capillary break. This can be made of an open-graded aggregate mix or a geosynthetic (e.g. drainage geocomposite) where the large open structure interrupts the upward movement of water. Alternatively, a wicking geotextile can be used for this purpose by intercepting the capillary water and directing it through its cross-section.

Longitudinal edge drains can be constructed adjacent to the roadway to help direct and carry water away from the road. Any drainage feature used in the cross-section can be tied into the edge drain to ensure a good outlet. A nonwoven geotextile is often used to line the trench that will be filled with an open-graded drainage aggregate and drainage pipe (Figure 2.19).

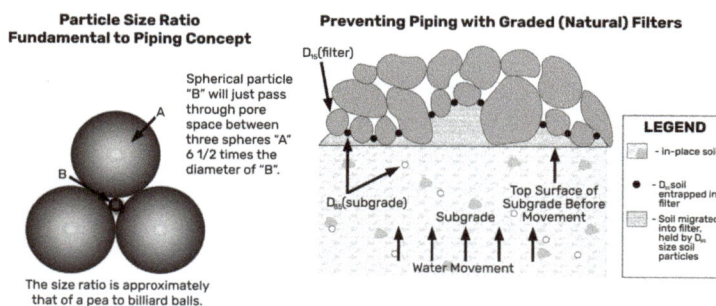

Figure 2.20 Natural filter criteria (from Holtz et al, 2008 after Cedergren, 1989)

Drainage needs are most often predicated on the condition of the pavement surface, the permeability characteristics of the base aggregate layer, and the sensitivity of the subgrade. The AASHTO '93 pavement design method (AASHTO, 1993) outlines the requirements for adequate drainage. Geosynthetic drainage products should have adequate hydraulic capacity to ensure good performance over time. This can be achieved by assessing opening size, permittivity, durability, strength, and transmissivity (for geonets and geocomposites). Filtration requirements should also be assessed to ensure the geotextile allows passage of water without clogging (Figure 2.20). Selection of geosynthetics used for drainage purposes can be accomplished following the specifications and guidance set out by AASHTO M288 (AASHTO, 2017) or design information in Chapter 2 of the Geosynthetic Design and Construction Guidelines (Holtz et al, 2008).

Geosynthetics are an efficient means to improve drainage because they are less expensive than the equivalent fill needed to achieve the same result, reduce the need for high-quality stone to be quarried, washed, graded, and transported to the site, thereby reducing time, money, resources, and emissions.

2.3.5 Asphalt Interlayers

Asphalt pavements are designed to provide a strong, durable, smooth, and impermeable driving surface. Environmental distresses and fatigue loading cause deterioration and cracking over time. Cracks allow passage of water which is the primary cause of most pavement deterioration, reduced structural capacity, and premature failure (Marienfeld and Baker, 1999). There are a variety of geosynthetic products designed to slow or mitigate these distresses. Geosynthetic interlayers are designed to reduce fatigue cracking (reinforcement function), decrease reflective cracking (reinforcement and stress-relief functions), and prevent water infiltration (barrier function). Asphalt interlayers can be used on new pavements, beneath asphalt overlays, or beneath chip seal surface treatments. They are often used to prevent cracking within existing pavement layers from migrating upward into a new asphalt overlay.

Asphalt reinforcement geosynthetics are strong and stiff – specifically designed to absorb tensile stresses thereby increasing the stiffness of the asphalt. Increased stiffness improves the load carrying capacity by providing a tensile element within the asphalt. The tensile strength of asphalt is relatively low, which oftentimes leads to cracking distresses. Additionally, the individual stone particles within the asphalt matrix tend to move over time due to traffic loading. This movement is similar to unbound aggregate although much slower due to the binding effects of the bitumen. Geosynthetics used for reinforcement are made of high strength

Figure 2.21 Composite geogrid-geotextile interlayer being installed beneath asphalt overlay (Zornberg, 2017b)

and stiff geogrids, or geogrids coupled with a nonwoven geotextile (geocomposites), as shown in Figure 2.21. The geosynthetic is placed between asphalt layers. Stone particles within the hot-mix interlock with the grd structure and an adhesive shear bond is formed between the geosynthetic and the cementitious properties of the asphalt. Applied loads are then transferred through the bonded matrix to provide additional strength and stiffness, which will delay or prevent cracking of the asphalt layer over time. This increased stiffness of the asphalt layer will absorb more of the applied load thereby reducing loads to lower layers.

Existing cracks in asphalt pavement will oftentimes "reflect" into a new asphalt overlay (Figure 22). Transferring these stresses into a strong material is what is done in the reinforcement function (described above). An alternatve strategy is to dissipate applied stresses over a greater area to slow or prevent reflective cracking. Special nonwoven geotextiles (and composites) can be used to relieve tensile stress by spreading them over a larger area rather than concentrating stresses at a single point (Figure 2.22). An asphalt core showing how the existing crack was arrested by the interlayer is shown in Figure 2.23. Because the geosynthetic materials are impregnated with bitumen when installed, they provide an impermeable barrier to prevent infiltration of water through the paved surface. Geosynthetic interlayers have also been used on composite pavements to control reflective cracking.

Geosyrthetic interlayers must be placed properly to be effective. Their structure allows for the absorption of asphalt cement tack coat. It is critical to apply sufficient tack coat to saturate the geotextile and/or composites and bond the interlayer system together with the pavement overlay to form a continuous moisture barrier. Lack of sufficient tack may result in debonding of the overlay. Too much tack coat can cause excess liquid asphalt to bleed into the overlay and compromise the bonding between the asphalt and the asphalt interlayer. The optimum bonding coat rate for different types of asphalt interlayer to achieve the optimum shear bond strength has been investigated and an optimum tack coat rate was proposed for different types of asphalt interlayer by Correia et al (2024). When properly constructed, interlayers will not affect rehabilitation and recycling efforts (e.g. milling). Good installations include applying a uniform asphalt cement tack coat onto a dry pavement surface (without streaking, skipping, or dripping), immediately installing the interlayer onto the warm tack coat (material should be taught and have no wrinkles or folds), and brooming or rolling to ensure good contact between the geosynthetic interlayer and pavement surface (Figure 2.24). Special attention should be paid when installing glass grids on milled surfaces to ensure the rough surface does not damage the geosynthetic.

Another beneficial use of geosynthetic interlayers is with chip seals. In this case the interlayer is applied with tack before the chip sealing process (Figure 2.25). The flexibility of the geosynthetic interlayer beneath the

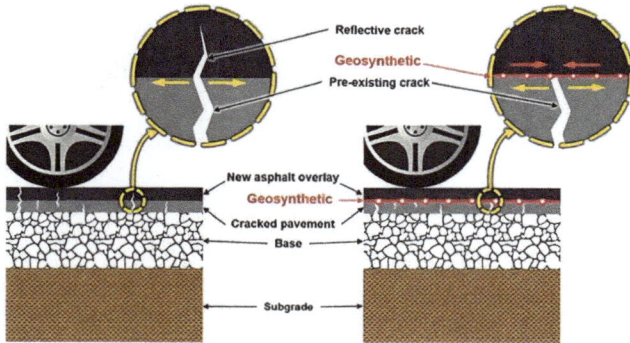

Figure 2.22 Benefit of geosynthetic interlayer (from Zornberg, 2017)

Figure 2.23 Interlayer used to mitigate reflective cracking (Beak et al, 2008)

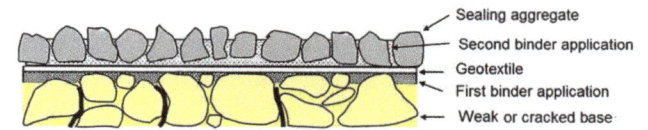

Figure 2.25 Anatomy of a chip seal with interlayer geotextile (Austroads, 2019)

Figure 2.26 Interlayer chip seal bridging existing cracks in pavement surface

Figure 2.24 Proper application of tack coat and installation of pavement interlayer (courtesy Petromat)

stone chip layer provides a bridge over surface cracks thereby slowing water infiltration (Figure 2.26).

Product selection, specifications, and construction/installation guidelines for this application can be aided through the AASHTO M288 (AASHTO, 2017) and Chapter 6 of the Geosynthetic Design and Construction Guidelines (Holtz et al, 2008). This is an example only. Local specifications should be referred to and followed, as appropriate.

2.4 Performance Measurement

Geosynthetics are known to improve the structural capacity and performance of roads. Many research and testing programs have been conducted over the past several decades to document these benefits; however, the availability of <u>reliable</u> and <u>non-proprietary</u> performance data leaves engineers to rely more heavily on proprietary software and/or their own judgment rather than definitive results from research and testing efforts. Engineers should look for test data that, at a minimum, describes in detail the way in which the tests were done; properties of subgrade, geosynthetic, subbase, base, and asphalt layers; QA/QC measures taken during construction to ensure uniformity and consistency from test to test; a description of the loading apparatus; distress measurements; and a summary of performance attributes. AASHTO R50-09 (AASHTO, 2018) generally outlines the requirements to determine performance properties that can be appropriately applied to pavement design. While performance can be measured in the field on actual roadways, the timeline is typically too long and variability in the site conditions make direct comparisons between design iterations unreliable. Performance properties are best determined from controlled full-scale tests on realistic road cross sections constructed in the laboratory. Information from these tests is used to help engineers more easily incorporate beneficial products into designs and to more accurately quantify life extension or material savings to substantiate cost-benefit analyses.

A common method used to quantify performance in the laboratory is a cyclic plate load (CPL) test. CPL tests were recently standardized in ASTM D8462. This test method is based on GMA White Paper II (Berg et al, 2000). CPL tests are constructed to full-scale and cyclically loaded using a stationary plate to apply cyclic load pulses to simulate traffic (Figure 2.27). Each load pulse is equivalent to 1 ESAL (equivalent single-axle load) as defined by AASHTO (AASHTO, 1993). A similar, and typically larger, alternative is an accelerated pavement test using a rolling wheel load (Figure 2.28). Also built under controlled conditions, these tests tend to be more realistic because they are trafficked using a rolling wheel rather than a stationary plate. The choice between test types lies generally in a balance between

how realistic construction and trafficking is, the time to get the results, and cost. The primary features of these two test types are outlined in Table 1. Rapid, reliable, and objective methods should be used to quantify the pavement performance improvement provided by geosynthetics.

Test sections can be constructed to determine

Figure 2.27 Example cyclic plate load test apparatus (courtesy of TRI Environmental, Inc.)

Figure 2.28 Example accelerated pavement testing apparatuses (a): courtesy of TRI Environmental, Inc.; (b): Army Corps of Engineers, ERDC)

Table 2.1 General attributes of cyclic plate and rolling wheel load tests

Attribute	Cyclic Plate Load Test	Rolling Wheel Load Test
Size/Boundaries	> Large rigid box at least 2 m wide	> Channelized test area > With or without concrete lining > Wide enough and deep enough to accommodate wheel loading without adversely affecting the test results
Load application	> Stationary cyclic plate > 40 kN load (1 ESAL/cyclic pulse) > 305 mm circular rigid plate	> Rolling wheel > Single or dual wheel assembly > 600-800 kPa tire pressure > 40 kN load (1 ESAL/wheel pass), can be increased up to 80 kN for multiple ESALs/wheel pass
Construction	> Smaller equipment that fits within the boundary walls	> Full size construction equipment
Test Setup	> 1 test per iteration > Compare to separate unreinforced control to determine performance	> Multiple test sections tested simultaneously, including control
Cost	$$,$$$	$$$,$$$
Timeline	Weeks	Months
Pros & Cons	> Standardized test (ASTM D8462) > Footprint allows for rapid construction at realistic size > Loading @ 1 Hz, ~86k pulses/day > Loading from stationary, rigid plate tends to be more aggressive than rolling wheel > More difficult to construct HMA > Two separate simulations necessary to determine performance	> Realistic because constructed at full-scale > Longer construction timeline > Trafficking 10-20k passes/day > Rolling wheel load is realistic > More expensive than CPL tests

benefit from geosynthetics in terms of life extension or structural improvement. Life extension is measured through increased accommodation of load cycles from traffic. Structural improvements are measured primarily through stiffness measures. While these two measures are related, they can be used differently by designers depending on the required inputs. Accelerated tests should be constructed carefully and as close to as-built pavement cross sections as possible, keeping in mind that test sections that differ significantly from the design of interest may not provide an accurate performance measure for that particular situation. Materials should be prepared identically across test sections that will be compared to one another. Variables should be controlled to allow for the direct comparison of improvements associated with the geosynthetic. Small changes in moisture content, thicknesses of layers, strengths, elapsed time, and temperature are known to affect the behavior and performance of roads. The size and energy of construction equipment should be optimized as much as possible to achieve the most realistic result.

In general, the more stress that is transmitted to the geosynthetic, the greater benefit they provide. Performance depends on how much load is absorbed by the geosynthetic. A single geosynthetic can have significantly different performance characteristics dependent on how it is tested. The following four sub-sections outline the most common performance metrics used by engineers for geosynthetic roadway design.

2.4.1 Traffic Benefit Ratio

Traffic benefit ratio (TBR) is a measure of the life extension provided by geosynthetics when compared to an identical design without geosynthetics. In this case, life is expressed by the number of traffic load cycles that can be accommodated by a pavement to reach a pre-determined damage level. It is calculated as the ratio of the number of load cycles accommodated by the stabilized test section to the number of load cycles accommodated by the unstabilized (control) test section (Eq. 2.5). An illustration showing examples of TBR calculations is shown in Figure 2.29. The output parameter calculated from ASTM D8462 (CPL test) is S-TBR$_i$, the TBR based on a stationary cyclic plate. TBR can be expressed as a single value associated with a certain level of pavement distress or across a range of expected distresses. TBR can be used directly to show benefit in life extension of a roadway or to back-calculate the structural contribution by applying the benefit to improvements to the base aggregate layer. Reasonable TBR values are up to about 10 but can be as high as 70 in some cases.

$$TBR_i = \frac{N_{stabilized}}{N_{unstabilized}} \qquad Eq.\ 2.5$$

Where,

TBR_i = traffic benefit ratio at rut level i

$N_{stabilized}$ = number of load cycles accommodated by the stabilized test section to reach rut level i

$N_{unstabilized}$ = number of load cycles accommodated by the unstabilized test section to reach rut level i

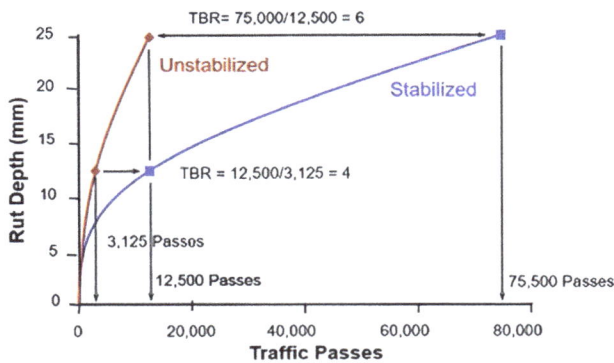

Figure 2.29 Example TBR calculations

2.4.2 Base Course Reduction Factor

The base course reduction (BCR) factor is the percentage of base layer thickness that can be reduced within a geosynthetic reinforced test section to match an equally performing section made without geosyn-

thetics (Figure 2.30). BCR is calculated directly using the equation in Figure 2.30 for pavement systems that have identical life. Calculating a precise value for BCR is difficult using test sections because it is difficult to determine the exact thicknesses of base to test. BCRs of up to 30% are typical but can be as much as 50% in some cases.

2.4.3 Layer Coefficient Ratio

The layer coefficient ratio (LCR) is used to modify the structural number (SN) within the AASHTO 1993 flexible pavement design equation (Eq 2.2 in Section 2.3.3), much like ECR. The structural number is used to quantify the contribution from each of the structural layers within the pavement. Customarily, LCR is a multiplicative term that increases the structural contribution of the base layer. LCR can be directly determined from BCR using Eq. 2.6 or back-calculated using the AASHTO '93 design equation.

$$LCR = \frac{1}{\left(1 - \frac{BCR}{100}\right)} \qquad Eq.\ 2.6$$

2.4.4 Modulus Improvement Factor

The modulus improvement factor (MIF) is a measure of the stiffness improvements from geosynthetics. It is simply the ratio of the stiffness of the stabilized layer over the stiffness of the unstabilized layer (typically the base aggregate layer), as shown in Eq. 2.7. Stiffness is measured by applying a load and measuring the displacement. This is most often done using static loads but can also be done using quasi-static or dynamic loading apparatuses. MIF is used in mechanistic-empirical design (similar to LCR) by modifying the structural contribution of the layer being improved by the geosynthetic – typically the base course.

$$BCR = (D_U - D_R)/D_U * 100$$

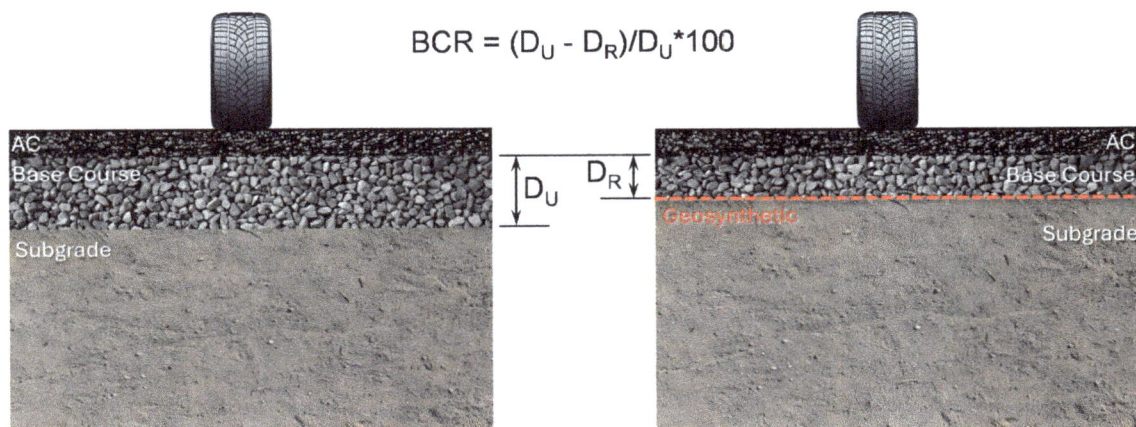

Figure 2.30 Comparison of pavement systems with identical life used to calculate BCR

$$MIF = \frac{E_{reinforced}}{E_{unreinforced}} \qquad \text{Eq. 2.7}$$

where,

$E_{reinforced}$ is the stiffness of the test section with geosynthetic reinforcement

$E_{unreinforced}$ is the stiffness of the test section without geosynthetic reinforcement

2.5 Railway Ballast Improvements Using Geosynthetics

Geosynthetics can be used in rail applications to address challenges from soft saturated subgrades such as intermixing between dissimilar layers, track construction over soft ground, moisture control, and stabilization/reinforcement of subballast and ballast aggregates - similar to roads. The primary difference between these applications is primarily the upper ballast layer that supports the rail structure. The most common issues associated with ballast are:

1. particle displacement – the movement and rearrangement of particles due to applied dynamic train loads and vibration, resulting in excessive settlement and loss of lateral stability in the track structure;
2. particle degradation – (also known as ballast breakage) the breakdown and reduction

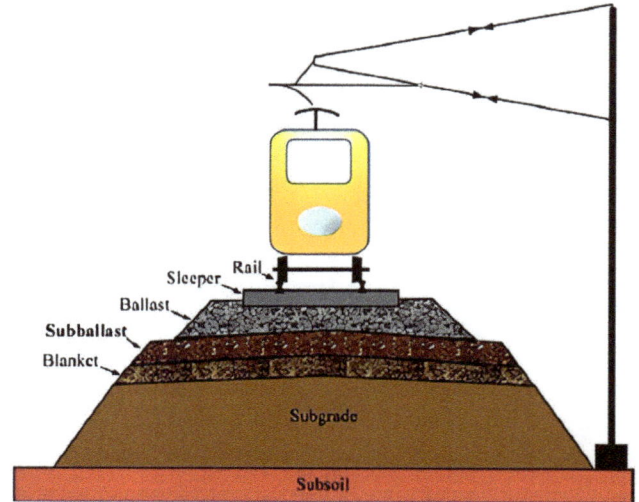

Figure 2.31 Typical railway structural cross-section (modified from Signes et al, 2016)

of particles, resulting in reduced strength, increased fines, and decreased drainage; and

3. ballast fouling - accumulation of fines within the ballast layer due to aggregate particle degradation and breakdown, and from deposits such as coal dust, or from the pumping of fines from the subgrade, resulting in reduced strength and decreased drainage.

A typical cross-section of the rail support structure consists of a subballast layer which lies directly atop a cushion layer or the subgrade, and an upper ballast layer

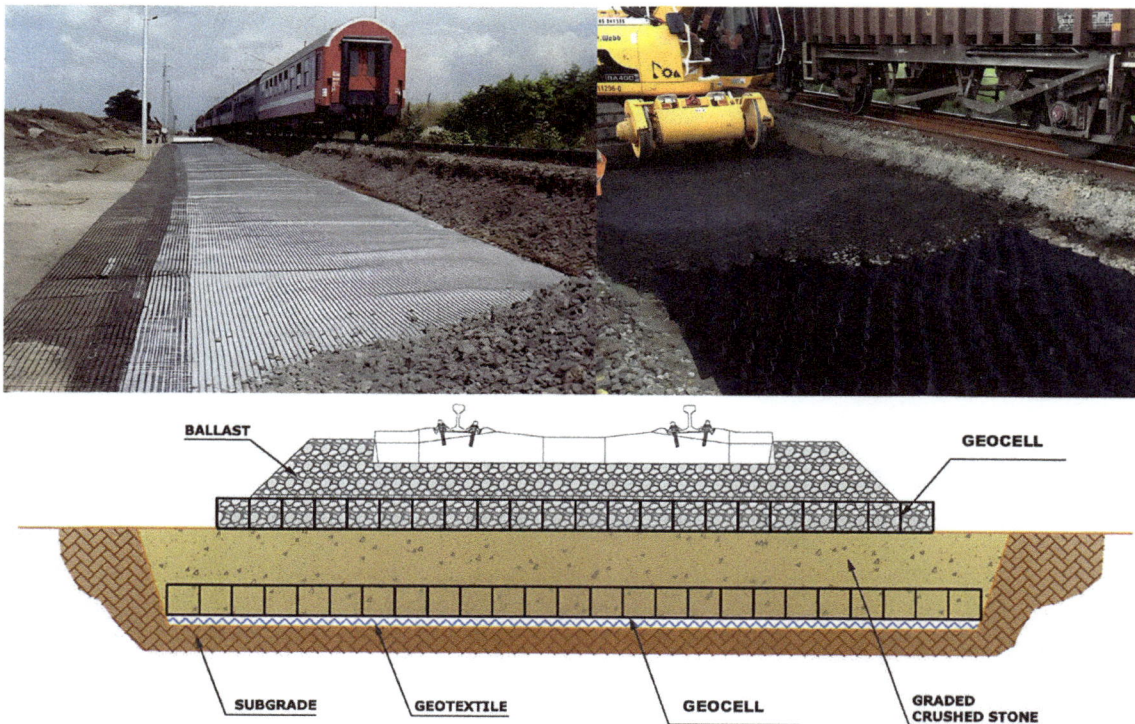

Figure 2.32 Geosynthetics used in rail construction (top left: IndiaMart; top right: Network Rail; bottom: PRS)

directly beneath the track structure, as shown in Figure 2.31. The subballast layer provides structural support, frost protection, separation between the subgrade and ballast layers, and drainage. The thickness and strength of the subballast and ballast layers are designed to prevent failure of the subgrade by absorbing and distributing stresses from rail traffic. The ballast layer is made from open-graded, angular stone with high voids to allow free passage of water. This open unbound composition allows for frequent maintenance to bring the track structure into operational compliance.

Geosynthetics such as geogrids, geotextiles, geocomposites (geogrid/geotextiles) and geocells have been used in rail applications to improve performance and extend the time between maintenance efforts, similar to roads (Figure 2.32). Geosynthetic reinforcement products such as woven geotextiles, geogrids, and geocells have been used within the subballast layer to increase its strength and spread the load over a greater area thereby increasing the bearing capacity of the subgrade. Drainage products have also been used to facilitate and improve drainage.

The use of geogrids and geocells within the ballast layer improves load capacity, reduces the frequency of ballast redistribution, and extends the life of the ballast from reduced degradation. These products interlock with the ballast to prevent movement of individual stones relative to one another. This particle restraint helps prevent particle breakdown and provides greater structural support. The three-dimensional structure of geocells provides a cellular network that works together to provide a more uniform load distribution (commonly known as the "mattress effect"), thereby increasing the stiffness of the ballast. Care should be taken not to damage installed geosynthetics during ballast maintenance activities.

2.6 Geosynthetic Storage, Handling, and Installation Tips

Geosynthetics are engineered products that depend on remaining intact to provide the expected benefit. While some damage is inevitable during construction, minimizing this will help ensure a durable and effective design over time. General principles are outlined below for common applications. Manufacturer recommendations and local requirements should be followed. Consensus standards are available that provide guidelines for the identification and packaging of rolled geosynthetics by the manufacturer and for the handling and storage of geosynthetics by the end user (example: ASTM D4873 outlines the guidelines for "the identification and packaging of rolled geosynthetics by the manufacturer and for the handling and storage of geosynthetics by the end user.").

Handling, transportation, and storage of geosynthetics should be done carefully to avoid damage. Many geosynthetic rolls have an inner core that help maintain the shape of the roll during transfer. Consult manufacturer recommendations to avoid breaking, distorting, or creating a safety hazard when loading/unloading, lifting, or transporting rolls. Stacks of rolls may need to be limited to avoid crushing of lower rolls. Storage of geosynthetics should be done in a way to avoid sunlight, high heat (>60 °C), moisture (especially with geosynthetics that can hold water, which would make them extremely heavy), chemicals, or damage from wildlife. Refer to and follow local requirements and manufacturer recommendations to achieve the best results. Cutting can be done using scissors, razor blades, or hot knives. Some geosynthetics (especially geogrids) can be sharp when cut, so gloves should be worn to avoid injury.

Proper installation begins with site preparation. The construction site should be cleared of large obstacles and sharp debris, and larger voids should be filled to prevent puncture, tearing, or distortion of the geosynthetic during installation and throughout the life of the project. Planar geosynthetics are typically rolled out on the prepared surface in the direction of traffic and tensioned by hand to remove slack and wrinkles. Small deposits of fill or sandbags can be used to temporarily hold the geosynthetic in place. On soft ground (CBR < 1%), staples can be used to temporarily anchor the material to prevent movement during construction. If the width of the roll does not cover the area to be improved, adjacent deployment of geosynthetic should be overlapped to provide continuous coverage (consult local and/or manufacturer requirements). Depending on the application and the anticipated loads, it may be required to sew the adjacent edges of geotextiles together to ensure continuity. Folding or cutting the material can be done to conform to curves in the road.

Placement and construction of the aggregate layer should be done in a way to avoid distortion and/or damage to the geosynthetic. Driving directly on the geosynthetic, especially in areas where the subgrade is soft, should be avoided. Lift thicknesses should be thick enough to prevent puncture or disturbance from construction equipment, and low-pressure equipment should be used for thinner layers. End-dumping is commonly used to deposit the material directly atop the geosynthetic, where low ground pressure equipment can be used to carefully spread the aggregate. Turning or sudden stops should be avoided on the first lift of aggregate. If the geosynthetic is damaged during construction, it can typically be repaired by removing the damaged piece and replacing with a new piece that is sufficiently overlapped in all directions (typically >0.5 m).

2.7 References

AASHTO (1993) *AASHTO Guide for Design of Pavement Structures*, American Association of State Highway and Transportation Officials, Washington, D.C.

AASHTO (2018) *Standard Practice for Geosynthetic Reinforcement of the Aggregate Base Course of Flexible Pavement Structures*, AASHTO R50-09, American Association of State Highway and Transportation Officials, Washington, D.C.

AASHTO (2017) *Standard Specification for Geosynthetic Specification for Highway Applications*, AASHTO M288-17, American Association of State Highway and Transportation Officials, Washington, D.C.

Al-Qadi, I.L., Dessouky, S.H., Kwon, J., and Tutumluer, E. (2008) Geogrid in Flexible Pavements: Validated Mechanism, Proceedings: Transportation Research Board Annual Meeting, Washington, D.C.

ASTM D7737, Standard Test Method for Individual Geogrid Junction Strength, ASTM International, West Conshohocken, PA.

ASTM D7864, Standard Test Method for Determining the Aperture Stability Modulus of Geogrids, ASTM International, West Conshohocken, PA.

ASTM D8269, Standard Guide for the Use of Geocells in Geotechnical and Roadway Projects, ASTM International, West Conshohocken, PA.

ASTM D8462, Standard Test Method for Cyclic Plate Load Tests to Evaluate the Structural Performance of Roadway Test Sections with Geosynthetics, ASTM International, West Conshohocken, PA.

Austroads (2019) Guide to Pavement Technology Part 4K: Selection and Design of Sprayed Seals, AGPT04K-18.

Collins, B.M., and Holtz, R.D. (2005) *Long-term Performance of Geotextile Separators, Bucoda Test Site – Phase III*, Washington State Transportation Center (TRAC), Washington State Department of Transportation, WA-RD 595.1.

Correia, N.S., Silva, M.P.S., Shahkolahi, A., 2024. Optimum tack coat rate for different asphalt geosynthetic interlayers to achieve optimum shear bond strength. Geotextiles and Geomembranes 52, 778–789.

Cuelho, E.V., Perkins, S.W. (2016) Geosynthetics Subgrade Stabilization – Field Testing and Design Method Calibration, *Transportation Geotechnics*, Volume 10, 22-34.

Cuelho, E., Perkins, S., Morris, Z. (2014) *Relative Operational Performance of Geosynthetics Used As Subgrade Stabilization*, Final Report to the Montana Department of Transportation, FHWA/MT-14-002/7712-251, p. 328.

Giroud, J.P., Han, J. (2004a) Design Method for Geogrid-Reinforced Unpaved Roads, Part I – Development of Design Method. *Journal of Geotechnical and Geoenvironmental Engineering*,130(8):775–86.

Giroud, J.P., Han, J. (2004b) Design Method for Geogrid-Reinforced Unpaved Roads, Part II – Calibration and Applications. *Journal of Geotechnical and Geoenvironmental Engineering*, 130(8):787–97.

Giroud, J.P., Han, J., Tutumluer, E., Dobie, M.J.D. (2021) The Use of Geosynthetics in Roads, *Geosynthetics International*.

Giroud, J.P. and Noiray, L. (1981) "Design of Geotextile Reinforced Unpaved Roads," *Journal of the Geotechnical Engineering Division*, ASCE, Vol. 107, No. GT9, pp. 1233-1254.

Haliburton, T.A., Lawmaster, J.D., McGuffey, V.C. (1981) Use of Engineering Fabrics in Transportation Related Applications, Federal Highway Administration, FHWADTFH61-80-C-00094.

Hall, K.T., Crovetti, J.A. (2022) Subsurface Drainage Practices in Pavement Design, Construction, and Maintenance, NCHRP Synthesis 579, Transportation Research Board, Washington, D.C.

Holtz, R.D., Christopher, B.R., and Berg, R.R. (2008) Geosynthetic Design and Construction Guidelines, National Highway Institute, Federal Highway Administration, FHWA-NHI-07-092, Washington, DC.

Jorenby, B.N., and Hicks, R.G. (1986) Base Course Contamination Limits, *Transportation Research Record*, No. 1095, Washington D.C., p. 86-101.

Kawalec, J. (2019) Stabilisation with Geogrids for Transport Applications – Selected Issues, MATEC Web of Conferences 265, 01001.

Kwon, J., and Tutumluer, E. (2009) Geogrid Base Reinforcement with Aggregate Interlock and Modeling of Associated Stiffness Enhancement in Mechanistic Pavement Analysis. *Transportation Research Record*, No. 2116, pp. 85–95.

Kwon, J., Tutumluer, E., Al-Qadi, I.L. (2009) Validated Mechanistic Model for Geogrid Base Reinforced Flexible Pavements, *Journal of Transportation Engineering*, Vol. 135, Issue 12.

Marienfeld, M.L. and Baker, T.L. (1999) Paving Fabric Interlayer as a Pavement Moisture Barrier, E-Circular E-C006, Transportation Research Board, Washington, D.C.

Perkins, S.W., Christopher, B.R., Cuelho, E.V., Eiksund, G.R., Hoff, I., Schwartz, C.W., Svanø, G., and Watn, A. (2004) *Development of Design Methods for Geosynthetic Reinforced Flexible Pavements*, U.S. Department of Transportation, Federal Highway Administration, Washington, DC, FHWA Report Reference Number DTFH61-01-X-00068, 263p.

NCHRP (2003) *Development of NCHRP 1-37A Design Guide, Using Mechanistic Principles to Improve Pavement Design*, National Cooperative Highway Research Program, Project 1-37A Report, National Research Council, Washington, DC.

Saride, S. and Baadiga, R. (2021) New Layer Coefficients for Geogrid-Reinforced Pavement Bases, *Indian Geotechnical Journal*, 51(1), p. 182-196.

Steward, J., Williamson, R., Mohney, J. (1977) *Guidelines for Use Of Fabrics In Construction And Maintenance of Low-Volume Roads*, USDA, Forest Service, Portland, OR. Also Reprinted as Report No. FHWA-TS-78-205, Federal Highway Administration, Washington, D.C.

Tamrakar, P., Kwon, J., Wayne, M.H. and Lee, H.S. (2021) Calibration of Pavement ME Rutting Model for Geogrid Stabilized Roadways, *Transportation Geotechnics*, Vol. 31.

Tingle, J.S., Webster, S.L. (2003) Corps of Engineers Design of Geosynthetic Reinforced Unpaved Roads, *Transportation Research Record*, No. 1849, National Academy of Sciences, p. 193–201.

Zornberg, J.G. (2017a) Functions and Applications of Geosynthetics in Roadways, *Transportation Geotechnics and Geoecology*, TGG 2017, St. Petersburg, Russia: 298-306.

Zornberg, J G. (2017b) Part 1: Functions and Applications of Geosynthetics in Roadways, *Geosynthetics Magazine*, Advanced Textiles Association, February, 2017.

Zornberg, J.G., Roodi, G.H., Sankaranarayanan, S., Hernandez-Uribe, L.A. (2018) *Geosynthetics in Roadways: Impact in Sustainable Development*, Proceedings of the 11th International Conference on Geosynthetics, Seoul, Korea.

Chapter 3

Geosynthetics in Subsurface Drainage

3.1 Background

Control of subsurface liquids and gases is critical to the performance of landfills, buildings, pavements, embankments, retaining walls, and many other structures. Drains are used to relieve hydrostatic and gas pressure in waste or soil, or against retaining walls, slabs, and underground tanks. This prevents loss of internal shear strength of soil or waste, and maintains stability in slopes, embankments, and road structures. A properly functioning drain must readily accept liquids or gases seeping from the surrounding material while simultaneously preventing clogging. These dual, sometimes opposing, requirements result in subsurface drains that must be carefully designed and constructed to ensure optimal performance.

Hydrostatic pressure plays a critical role in the design of subsurface structures and soil-structure interaction. For example, below-grade structures such as foundations, retaining walls, basements, elevator pits, and subgrade parking facilities are subjected to hydrostatic loads from groundwater. The stability of a landfill cap may be compromised by unvented gas pressures generated from the degradation of stored waste. These pressures can be substantial and can lead to structural distress or failure if not adequately accounted for in design.

Hydraulic hydrostatic pressure is the isotropic pressure exerted by a fluid at equilibrium due to the force of gravity. This pressure increases linearly as a function of its depth according to the following relationship:

$$P = \gamma_w \cdot h$$

where: P is the hydrostatic pressure; γ_w is the unit weight of water (9.81 kN/m³); and h is the depth below the water surface.

When water accumulates in a poorly drained soil mass, either due to precipitation, flooding, or inadequate surface and subsurface drainage, porewater pressure increases. Porewater pressure is of particular concern in cohesive soils with low permeability, where it dissipates slowly. Increased hydrostatic pressure leads to

reductions in shear strength, bearing capacity, and slope stability. In extreme cases, this can result in hydraulic uplift, basal heave, or slope failure.

Poor grading, clogged subsurface drains, and impermeable surface layers, as well as soil with low hydraulic conductivity, exacerbate the accumulation of water and, consequently, hydrostatic pressure. When porewater drains from the soil, the contact between soil particles is not interfered with by water and thus increases, leading to a corresponding increase in effective stress and, consequently, an improvement in the soil's strength.

Proper accounting of hydrostatic pressure involves subsurface investigations and soil characterization, including piezometric monitoring, soil permeability testing (e.g. constant-head or falling-head tests), and groundwater modeling. The design must incorporate pressure estimates for both steady-state and temporary conditions. Hydrostatic forces are critical in the:

> design of earth-retaining structures and basement walls (active, passive, and at-rest earth pressures including hydraulic pressures),
> evaluation of seepage and piping potential,
> determination of safety factors in slope stability models,
> design of uplift resistance for covers, slabs, and other horizontal elements, and
> engineering of subsurface drainage systems to relieve or intercept groundwater pressures.

3.2 Geosynthetic Drains

When used for the drainage/transmission function, a geosynthetic acts as a drain, allowing liquids or gases to flow in the plane of the geosynthetic. Geosynthetic drains have sufficient flow capacities to allow liquids or gases to freely flow from or through the drained material, for example, in pavement edge drains, slope interceptor drains, retaining wall drains, and landfill drains.

There is a wide variety of geosynthetic drains available to the design engineer, providing preferential flow paths for liquids and gases to drain. Geosynthetic drains function similarly to traditional gravel drains but offer significant advantages in constructability, consistency (as they are fabricated before delivery to the construction site, and are considerably easier to transport to and across the construction site), and hydraulic performance. Geosynthetic drains typically consist of a drainage core (such as a biplanar or triplanar geonet, cuspated geomat, polymeric spacer, or perforated pipe) encapsulated by a suitable geotextile on one or both sides, providing soil filtration and preventing clogging. Figure 3.1 shows several types of prefabricated geosynthetic drains.

Geosynthetic drains can be equivalent to, and often more effective than, gravel drains in managing fluid flow. A relatively thin geosynthetic drainage layer can handle the same or more flow as a significantly thicker layer of gravel. This is due to the open void structure of geosynthetics, which allows for faster water flow through the material. Due to their high flow capacity, less material is required compared to gravel.

Aggregate drains rely on the spaces between gravel stones and blocks to allow fluid to flow. For example, a significant portion of a gravel drain is occupied by the gravel itself, leaving less space for water storage compared to geosynthetics. The gravel is heavy and bulky, making installation more challenging, especially in large areas or where access is limited. While gravel relies on a large volume of material to achieve sufficient flow, geosynthetics can achieve the same flow with a much smaller volume. This is demonstrated in Figure 3.2.

Geosynthetic drains may also offer a more efficient and potentially more cost-effective solution for managing fluid flow compared to traditional granular drains, especially in situations where space is limited or high flow rates are needed. In addition, on-site changes

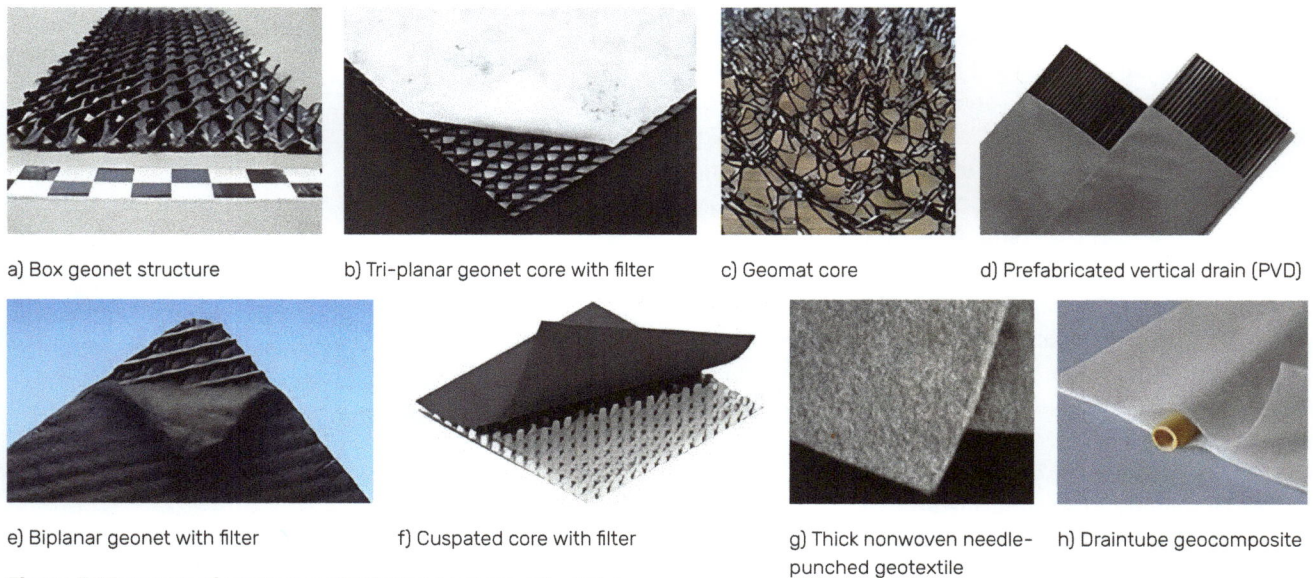

a) Box geonet structure

b) Tri-planar geonet core with filter

c) Geomat core

d) Prefabricated vertical drain (PVD)

e) Biplanar geonet with filter

f) Cuspated core with filter

g) Thick nonwoven needle-punched geotextile

h) Draintube geocomposite

Figure 3.1 Examples of common prefabricated geocomposite drains

300 of these... = 1 of these

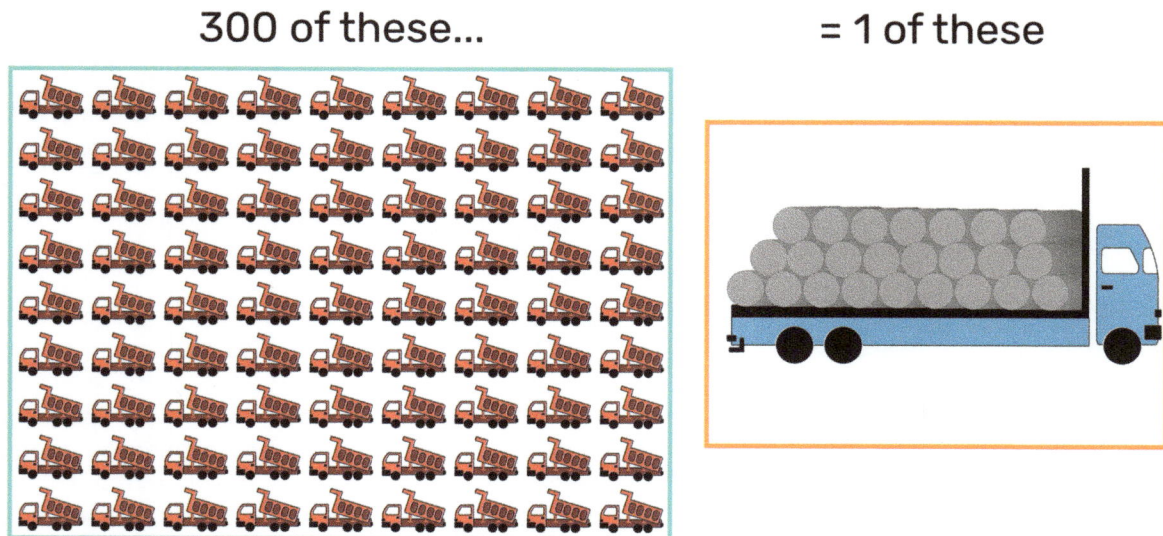

Figure 3.2 Geosynthetic drain to drainage block comparison (from R. Heritage, D. Shercliff, 2022)

or material re-distribution needs are accommodated much more easily with geosynthetic drains, as moving a produced product roll of geosynthetic drain is much more efficient than relocating a substantial thickness of drainage aggregate.

The in-plane flow rate per unit width of these products is measured using the following consensus-based standards:

> ISO 12958-1, Geotextiles and geotextile-related products – Determination of water flow capacity in their plane, Part 1: Index test
> ISO 12958-2, Geotextiles and geotextile-related products – Determination of water flow capacity in their plane, Part 2: Performance test
> ASTM D4716, Standard Test Method for Determining the (In-plane) Flow Rate per Unit Width and Hydraulic Transmissivity of a Geosynthetic Using a Constant Head

The measured flow rate per unit width must meet project-specific flow requirements under anticipated normal stress and hydraulic gradients.

While geotextile drains have much lower flow capacities than geocomposite drains (on the order of 2 to 3 orders of magnitude less), the in-plane drainage characteristics of some geotextiles (e.g. thick nonwoven needle-punched geotextile) can provide benefit by allowing dissipation of porewater pressures in fine grain clay and silt type soils, allowing seepage to exit (e.g. during thaw events and consolidation), gas venting, and irrigation applications. For example, at the base of roadway embankments or in wet soil during compaction. Some geotextiles are manufactured to perform a drainage mechanism driven by capillary action rather

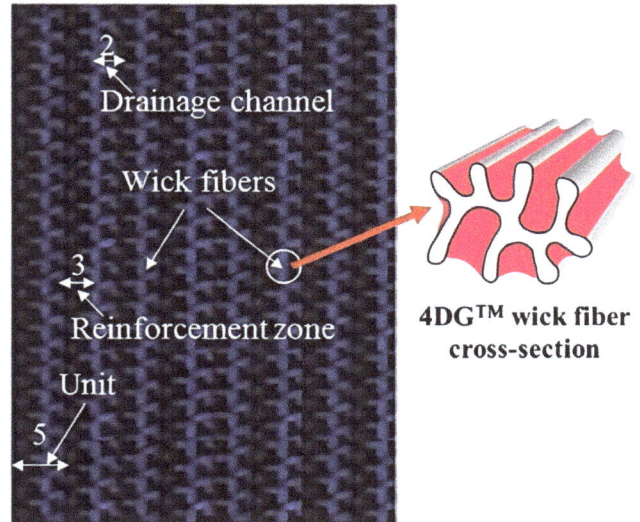

Figure 3.3 Wicking Geotextile (Guo et al, 2016)

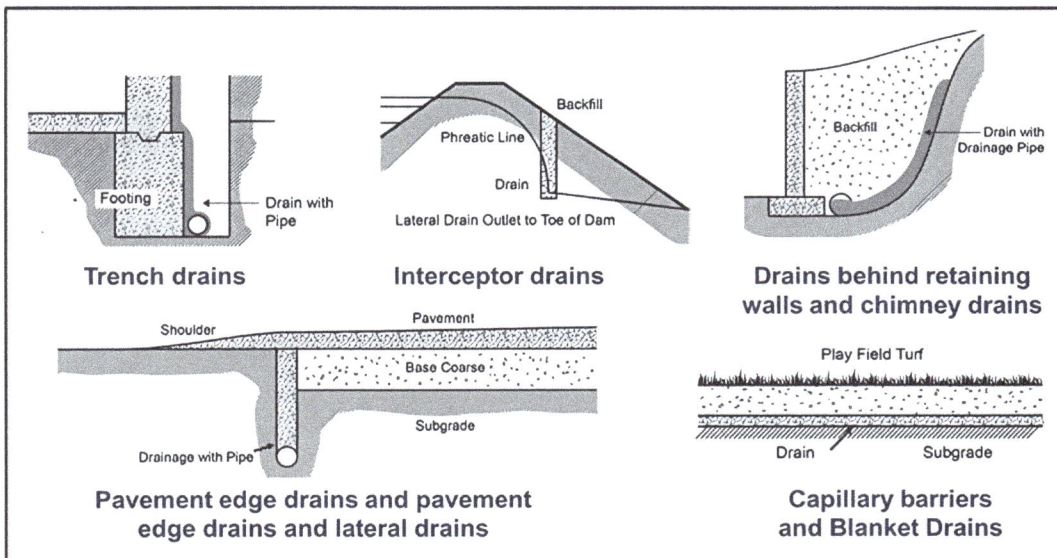

Figure 3.4a Uses of geosynthetic drains (FHWA-NHI-132013)

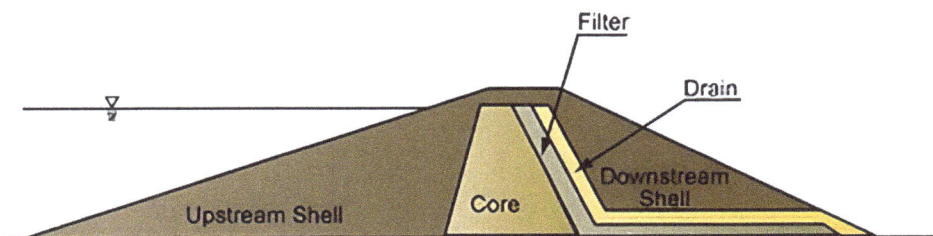

Figure 3.4b Uses of geosynthetic drain in Chimney Drain in Dam (Montana DOT, 2010)

than gravity flow. These wicking geotextiles are typically characterized by specific engineered fiber geometry and can drain both free water and capillary water under saturated and unsaturated conditions.

The in-plane flow capacity of geotextile drains is significantly influenced by compressive forces, incomplete saturation, and hydraulic gradients, all of which must be considered to determine their suitability for a given application.

Some geocomposites available on the market include discrete draining elements, such as small diameter pipes, instead of a continuous draining core (see Figure 3.1h). In this type of geocomposite, the geotextile acts as a filter and also contributes to carrying the water flow. Generally, the water flow capacity provided by geotextiles is significantly smaller than that offered by discrete drainage elements.

As geosynthetic drains are fabricated under controlled conditions before delivery to a site, they are generally very consistent in their properties and performance. They typically range in thickness from 5 to 25 mm or greater and have in-plane flow capacities between 0.0001 and 0.01 m³/s per linear width of drain, equivalent to 25 to 100 mm of gravel. Due to their high flow capacity, they are used to replace or supplement conventional aggregate drainage systems. Geosynthetic drains provide a readily available material with well-defined filtration and hydraulic flow properties, as well as easy installation, resulting in construction efficiencies. Most granular drainage layers wrapped in a geotextile, serving only the drainage function, could be replaced by a geosynthetic drainage geocomposite. (see Figure 3.4).

Additionally, drainage geocomposites are routinely used for leachate collection at the base of waste management facilities, in leak detection layers (also referred to as secondary drainage layers), as well as in gas venting and rainwater drainage systems for capping/closure projects (see Figure 3.6).

3.2.1 Advantages, Cost Considerations, and Sustainability

Geosynthetic drains offer significant advantages over conventional graded granular drainage systems. They significantly reduce the usage of more costly, less sustainable drainage aggregate due to a reduced volume of required material and reduced or eliminated excavation of aggregate. They also enable expedient construction, making labor resources very efficient. For example, geocomposite drains used for pavement edge drains typically represent a tenth to a third of the cost of conventional geotextile-wrapped gravel drains with a pipe, on a cost-per-linear-meter-installed basis (Berg et al, 2009). Considering that they are premanufactured

systems that are readily available with known filtration and hydraulic properties, they also provide greater reliability.

Improved sustainability is also a significant factor when using geosynthetic drains. Sustainable features include reducing aggregate quantities, resulting in substantial reductions in carbon units due to lower mining, processing, and hauling requirements. In the case of trench drains, there is also a reduction in excavation and disposal of excavated material.

The IGS environmental sustainability calculator can be used to compare the carbon units for conventional aggregate drains versus geosynthetic drains. One may find out how to access this calculator at the following address: https://www.geosyntheticssociety.org/sustainability/calculator/.

This calculator provides an example template comparing a landfill cap cover drainage system consisting of a 50 cm thick aggregate drain vs a geosynthetic drain alternative. The resulting carbon calculation demonstrates a 67% reduction in kg CO_2e when using the geosynthetic drain design.

Gutierrez and Conesa (2020) demonstrated the effectiveness of high-quality drainage geocomposite in reducing carbon emissions compared to conventional solutions. They showed that even with very long transportation distances for geosynthetics, CO_2 emissions are typically lower than those associated with using aggregates to deliver a drainage function. This case history may be found on the International Geosynthetics Society website.

Additionally, geosynthetics producers may provide Environmental Product Declarations (EPDs), which quantify the environmental sustainability of products. This information then provides a basis for evaluating the sustainability and reduced environmental impact of buildings and other construction projects that incorporate geosynthetics. EPDs also help manufacturers improve processes and material sourcing, as well as identify the potential re-use and recycling of geosynthetic materials at the end of their service life.

3.2.2 Basic Drainage Design Considerations (all applications)

There are four basic design considerations for the design and selection of geosynthetic drains:

1. Adequate filtration design of the geotextile covering the core in the drainage geocomposite or used for the geotextile drain without clogging or piping.
2. Adequate inflow/outflow capacity under design loads to provide maximum anticipated seepage during design life. (Note: for design

of the geocomposite drain, this is a function of both the core and the geotextile working together and not just the geotextile alone.)

3. Construction considerations that impact design and drainage performance, especially during installation, including damage, contamination, or both, that may occur during construction.

4. System performance considerations for the specific project requirements.

Resources for geosynthetic filtration and drainage design considerations include the following:

ISO/TR 18228-3: 2021, Design using geosynthetics - Part 3: Filtration

ISO/TR 18228-4: 2022, Design using geosynthetics - Part 4: Drainage

3.1.2.1 Filtration Design and Installation Considerations

Successful geotextile filter applications rely on the filter's ability to retain the soil structure while allowing water to pass through its plane perpendicularly. Specifically, the geotextile must retain the soil particles (retention criterion), while allowing water to pass (permeability criterion), throughout the service life of the structure (clogging resistance criterion and durability requirements). To perform effectively, the geotextile must also survive the installation process (survivability or constructability criterion).

Specific criteria to meet these design requirements, design details, and construction requirements are provided in Designing with Geosynthetics (Koerner). A number of other international criteria exist, and designers, of course, should also refer to local codes and references.

In general, geotextile filters are selected by a designer based on the grain size(s) of the material to be drained, permeability requirements, clogging resistance, and physical property requirements. Designers and installers are careful to ensure the filter's optimal performance by being aware of conditions that may affect its performance. These include:

> A geotextile filter and the soil to be filtered should be in intimate contact for proper performance. Without intimate contact, fine soil particles eroded from soil surfaces may be carried in suspension to the geotextile surface, forming a "cake" that reduces the filter's performance, potentially even blocking the designed flow.

> Extended exposure of the geotextile filter to UV radiation may degrade the filter, thereby compromising its performance before it is even installed. It's important to install the geotextile

filter upon opening its package or provide adequate cover to protect it from UV exposure. The installed geotextile filter should be completely intact, without any physical damage, and free from mechanical stress.

> Finally, the filter designer must be aware of certain environmental factors that are challenging for geotextiles, such as biologically rich applications, including leachate collection layers, environments with specific soil chemistries that promote the formation of excessive precipitate, and water slurries with high concentrations of solids.

3.1.2.2 Drainage Design and Installation Considerations

To determine the needed flow capacity of the geosynthetic drain, the designer first estimates the maximum seepage flow into the system using standard geotechnical and hydraulic engineering methods for the given application. Then, the geosynthetic drain core and filter geotextile, or geotextile drain, are selected based on these seepage requirements. Consideration is given to system performance factors, such as the distance between drain outlets, the hydraulic gradient of the drains, and the potential for blockage due to small animals burrowing, freezing, and other factors. When using geosynthetics to drain earth-retaining structures and abutments, the drain location and pressures on the wall or abutment are taken into account.

The specific flow capacity or flow rate per unit width of a geosynthetic drain per unit of hydraulic gradient:

$$Q = q / B$$

where

Q is the specific flow rate per unit width of the geosynthetic drain (l/s/m or m^3/s/m);

q is measured flow rate for a geosynthetic drain specimen of width (l/s or m^3/s);

B is the width of the geosynthetic drain in the flow capacity test (m);

Some drainage specifications refer to the hydraulic transmissivity of the geosynthetic. The hydraulic transmissivity is the flow capacity under a specific hydraulic gradient.

$$\theta = (q/B)/i$$

where

θ is the transmissivity of the geosynthetic drain (l/s/m or m3/s/m);

The performance flow capacity of the geosynthetic drain is determined by applying a set of design reduction factors to a flow capacity measurement collected during a laboratory test. These factors account for in-application conditions that reduce the flow rate of a geosynthetic drain over its design life.

$$q_a = {q_{ultl}}/{(RF_{IN} \times RF_{CR-Q} \times RF_{CC} \times RF_{BC} \times RF_L)}$$

Where:

Q_a is available long-term flow rate for the geosynthetic drain.

$_{Qultl}$ is the short-term flow rate obtained from laboratory tests with the appropriate boundary conditions.

RF_{IN} is the Reduction Factor for the intrusion of filter geotextiles into the draining core due to tensile creep of the gtx, occurring after the short-term test.

RF_{CR-Q} is the Reduction Factor for the compressive creep of the geosynthetic drain; RF_{cc} is the Reduction Factor for chemical clogging of the draining core.

RF_{BC} is the Reduction Factor for biological clogging of the draining core.

RF_L is the Reduction Factor for overall uncertainties on laboratory data and field conditions.

The available long-term flow rate of the geosynthetic (q_a) can then be compared with the long-term flow rate required of the drainage system ($q_{req'd}$). Therefore, the composite material should be evaluated by an appropriate laboratory model (performance) test, under the anticipated design loading conditions, including the adjacent material (e.g. soil) type, with a safety factor (FS) applied for the design life of the project such that the flow rate value is an allowable result. That is:

$$FS = {q_a}/{q_{reqd}}$$

Another consideration for drainage geocomposites is that a portion of the filter geotextile may be covered by or bonded to the core material. The reduced area must be accounted for in evaluating inflow capacity by either testing the permeability of the geotextile on the geocomposite drain or by using the following equation:

$$q_{reqd} = q_{GTX}(A_g/A_t)$$

where:

q_{reqd} = flow rate required for design

q_{GTX} = flow rate through the geotextile

A_g = geotextile area available for flow

A_t = total geotextile area.

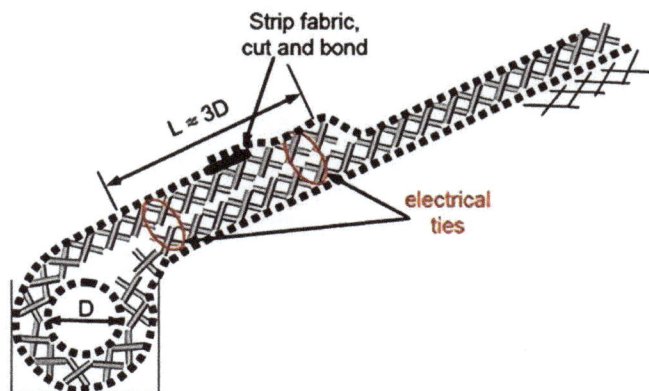

Figure 3.5 GRI recommended details for installing a drainage geocomposite connection to a perforated pipe in a trench drain

As with all geosynthetic applications, proper installation of the geosynthetic drain is key to achieving optimal product performance. It is common for geosynthetics manufacturers to provide installation instructions tailored to a specific drainage application. By following these installation guidelines and best practices, the effectiveness and lifespan of the geosynthetic drain system may be maximized, contributing to successful and sustainable engineering projects.

Additional resources for installation include GRI GN2 and GRI GC13, *Standard Guide for "Joining and Attaching Geonets and Drainage Composites*, see figure 3.5. This guide promotes several best practices, including the removal of any filter geotextile between drainage pipes and cores, to allow for the efficient transmission of fluids or gases from one drain to another. It also provides general guidance for connections on slopes and other field design configurations.

Installation best practices will depend on the specific application. In general, planar geocomposite drains can overlap with the core geonet portion, which is tied, and the geotextile portion is thermally bonded, stitched, or seamed as required by the project specifications. Geotextiles or geotextile filters in a geocomposite drain and,should be covered within a maximum time determined by the design engineer, based on site-specific conditions, to minimize UV exposure. When using sheet drains, inspect the use of low ground pressure vehicle tires and/or other points of contact with the geocomposite to ensure they do not exhibit signs of damage, loose and/or angular equipment, exposed wires, or other deleterious items that could damage the geocomposite. No sharp turns or sudden braking should occur when operating low-ground-pressure vehicles on top of the geocomposite.

3.3. Common Applications for Geosynthetic Drains

Geosynthetic drains are utilized across various industries, primarily in civil engineering, environmental engineering, and construction. Here are some of their most common applications:

1. **Landfills and waste containment:** Geosynthetic drains play a crucial role in landfill construction, particularly in the collection and containment of leachate. A drainage core-geotextile filter composite is used to facilitate drainage while ensuring the separation of waste layers. The geotextile component filters fine particles, while the core provides a channel for leachate flow, preventing contamination of surrounding soil and groundwater.

2. **Road construction and highways:** Geosynthetic drains are widely used in road construction for drainage and reinforcement purposes. They prevent the mixing of subsoil with the aggregate base, maintaining the structural integrity of the road. Additionally, geosynthetics enhance drainage in highway shoulders, thereby reducing the risk of pavement damage caused by water accumulation.

3. **Slope stabilization:** For embankments and slopes, geosynthetic drains are used to improve stability and prevent erosion. By draining excess water away, they reduce porewater pressure and enhance soil shear strength, thereby reducing the risk of landslides or slope failures.

4. **Tunnels and underground structures:** Geosynthetic drains provide effective waterproofing and drainage solutions in tunnel construction. A drainage system directs groundwater away from the tunnel structure, preventing seepage, corrosion, and damage to concrete linings.

5. **Retaining walls:** Vertical geosynthetic drains are installed behind retaining walls to manage hydrostatic pressure. The drain efficiently channels water away from the structure, improving its stability and reducing maintenance costs.

6. **Mining operations:** In the mining industry, geosynthetics are utilized in tailings storage facilities and drainage systems. They help collect and channel seepage, ensuring safe and environmentally compliant operations.

7. **Green and blue roofs:** Geosynthetic drains are also used in green roof systems to manage water drainage and provide separation between soil and waterproof membranes. This ensures proper plant growth while protecting the roof structure.

3.3.1 Drainage for Containment Applications

Properly designed containment projects, such as waste landfills, reservoirs, and ponds, will include elements of drainage. A range of related applications for geosynthetic drains in containment facilities exists. For example, geosynthetic drains may vent or drain subsurface water and gases, prevent rainwater intrusion over a capping barrier, drain away collected leachate at the bottom of a landfill, or serve as a leak detection layer to indicate potential compromises in containment. Table 1 provides a summary of applications, along with the critical functions they perform and the essential parameters needed for design. Figure 3.6 provides examples of geosynthetics in a landfill application, including the use of geosynthetic drains.

Table 3.1 From Landfill Applications for Geocomposites (Thiel, Richardson, Narejo, modified)

Containment facility applications for geosynthetic drains	Critical functions performed
Cover system drainage layer	Preserve veneer stability
	Reduce surface erosion potential
Gas removal and seep collection below covers	Preserve veneer stability
	Enhance environmental containment by controlling random seeps
Primary leachate collection and removal	Maintain a low hydraulic head on the primary liner system by providing efficient leachate collection and removal.
Secondary leachate collection and removal	Relatively quick and efficient conveyance of fluids that leak past the primary liner to the sump
Leachate recirculation	Distribution of recirculated fluids into waste mass to accelerate biodegradation
Facility internal drainage function	Preserve slope stability by providing a drainage sink for porewater fluids.

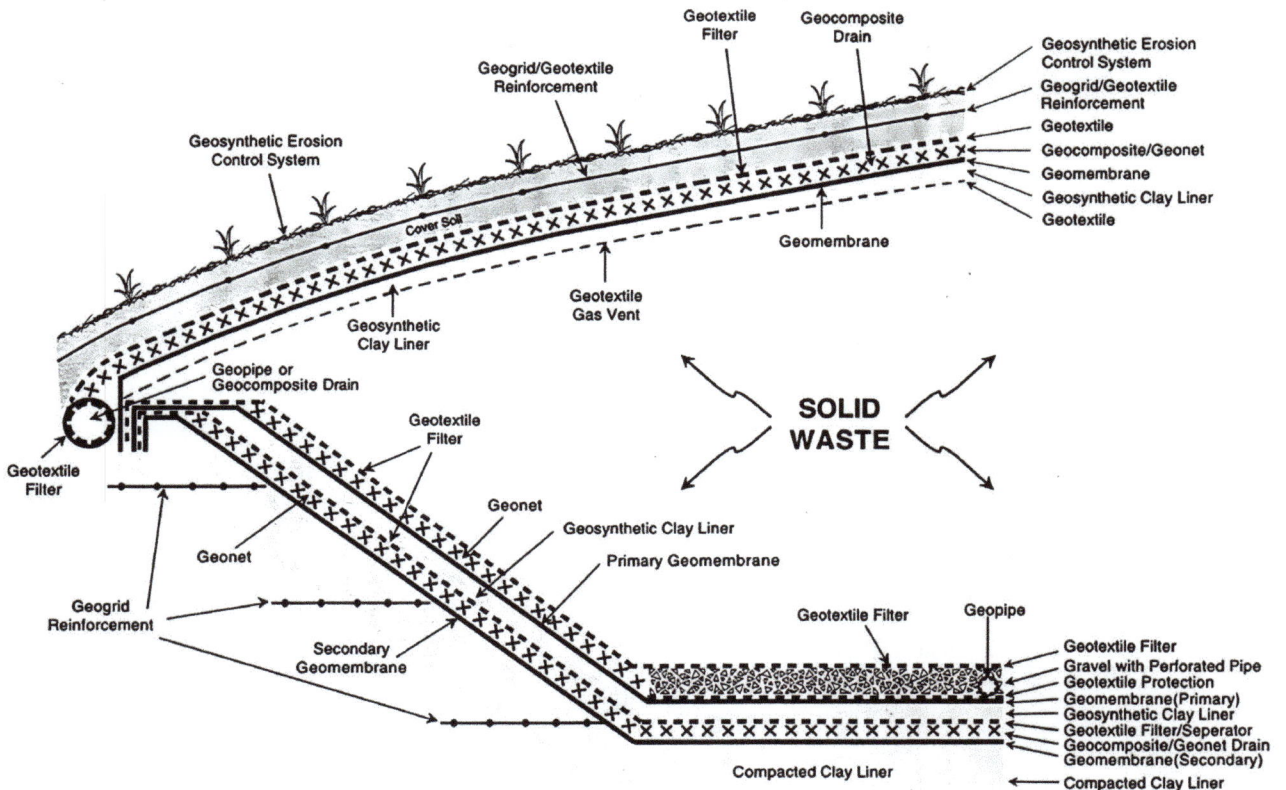

Figure 3.6 Geosynthetic materials in landfills, including geosynthetic drains and filters (after Koerner 2016)

3.3.2 Structural Drainage

Applications of geosynthetic drains in structural drains include:

> chimney drains and back drains for concrete cantilever walls,
> chimney drains and base drains for reinforced soil walls,
> reinforced soil slopes and unreinforced soil slopes;
> chimney and toe drain for earth dams and levees, to provide seepage control, and
> pavement edge drains and horizontal lateral drains for drainage of the base course and subgrade (e.g. see Fig. 3.4a and 3.4b).

Geosynthetic drains are much easier to construct than graded granular drains, which often require multiple layers (i.e. drainage and filter-sized aggregates). When using geosynthetic drains to drain retaining structures and abutments, the drain location and pressures on the wall or abutment must be considered in the design. Care should be taken to evaluate the slip plane that the drain may create, as the geosynthetic drain-to-soil interface friction angle (ϕ_{GCO-D}) will likely be less than that of the soil (ϕ_{soil}). ASTM D5321, Standard Test Method for Determining the Shear Strength of Soil-Geosynthetic and Geosynthetic-Geosynthetic Interfaces by Direct Shear, and ISO 12957-1, Geosynthetics — Determination of friction characteristics, Part 1: Direct shear test, provide interaction test methods using a large-scale direct shear test.

Driven by concern over hydrostatic pressures that may compromise the structural integrity of reinforced wall structures, geosynthetic drains are often placed vertically behind the face of retaining walls to prevent water from seeping through the face. Often, geosynthetic drains with only one-sided flow are used for this application to prevent water from seeping through the concrete face. However, placing a drain in that position does not necessarily completely relieve the hydrostatic stress at the back of the wall face, as the drain may not lower the water head down to the base of the wall. Also, depending on the permeability of the backfill (i.e. less than 0.002 cm/sec based on Terzaghi et al 1996 and Cedergren 1989), a vertical drain may increase seepage forces on the back of the wall from infiltration water during rain events due to seepage forces that are now directed toward the back of the wall face. Unless free-draining backfill is used, it is essential that a geosynthetic drain also be located away from the wall face and be appropriately inclined so that it can intercept seepage before it impinges on the back of the wall (e.g. a chimney drain, as shown in Figure 3.4a). Infiltration water will also preferentially flow toward the chimney drain away from the wall face. The inclusion of a chimney drain is especially recommended for hillside construction applications due to the potential for seepage to occur through retained soil and rock seams, faults, and joints during rain events (and snow melts) that may not be apparent during subsurface exploration and construction. A geosynthetic drain can also be placed at the base of a wall to promote further downward flow (i.e. away from the face), collect and drain outlet water from the chimney drain, and collect and drain groundwater out of the wall to reduce drawdown effects. For chimney drain and base drain applications, only geocomposite drains that allow two-sided flow (i.e. flow into the drains from both sides) should be used.

Designing the outlets is a crucial drainage detail. It ensures the system removes all collected water without developing excess hydrostatic pressures. Details must provide sufficient collector pipe diameters and outlet spacing. Geocomposite drains can be wrapped around collector slotted pipes or placed into a collector trench at the base of the wall. Free flow from the outlets must be assured. The outlet should also be covered with a wire mesh to prevent rodents from entering the pipe, and erosion protection systems, such as headwalls, should be constructed to avoid erosion below the wall.

Another approach for improving the performance of fill in steep slopes or retaining walls is to modify the seepage regime by using geosynthetic reinforcements with in-plane drainage (i.e. geosynthetic drains with sufficient strength to provide reinforcement). This approach addresses water seepage from existing slopes. It enables the dissipation of porewater pressure that may develop in high-moisture-content marginal soils (e.g. soils with a high fines content, especially wet native soils) (see Figure 3.7). This approach provides both improved stability and allows acceleration of settlement of marginal soils (just like wick drains) and is especially effective when the geosynthetic drain also provides reinforcement.

Porewater pressures generated during construction within reinforced, poorly draining fill can be dissipated if the geosynthetic provides both reinforcement and lateral drainage. The use of these types of geosynthetic

drains may allow the use of locally available fill, such as for land reclamation projects (e.g. hydraulically dredged materials).

A problem frequently reported for embankments using unreinforced, compacted cohesive soils is the development of tension cracks at the surface. When soils are reinforced with geosynthetic drain layers, it has been observed that the crack only propagates down to the first layer at the top of the fill, and water, which usually accumulates in the crack, drains out of the reinforcement at the face of the embankment slope (e.g. see Christopher et al, 1998). Geosynthetic drain reinforcements (i.e. those having sufficient long-term reinforcement design strength) and geotextile drains bonded to reinforcement geosynthetics) can also prevent destabilizing seepage forces from developing from flow configurations within the fill (Christopher et al, 1998). This has been verified through several instrumented full-scale structures and supported by analytical analysis (Soong & Koerner, 1999).

In addition to addressing stability problems, geosynthetic drain layers can also improve construction in wet soils, which typically require drying to achieve the desired compaction levels and associated design strength. By allowing the porewater pressure induced during compaction to dissipate in the geosynthetic layers, compaction improvements can be achieved in low-plasticity silts and clays placed at a wetter than optimum moisture content. Test pads should be used to determine the level of compaction improvement and the time required for the dissipation of pore pressure generated during compaction (see Christopher et al, 1998). Test pads can be constructed to determine the effectiveness of the geosynthetic in dewatering the soil. Nonwoven geotextile drains are effective in nearly saturated to saturated soils, as the porewater pressure during compaction is generally sufficient to move the water through the geotextile. Several instrumented case histories have confirmed significant improvements

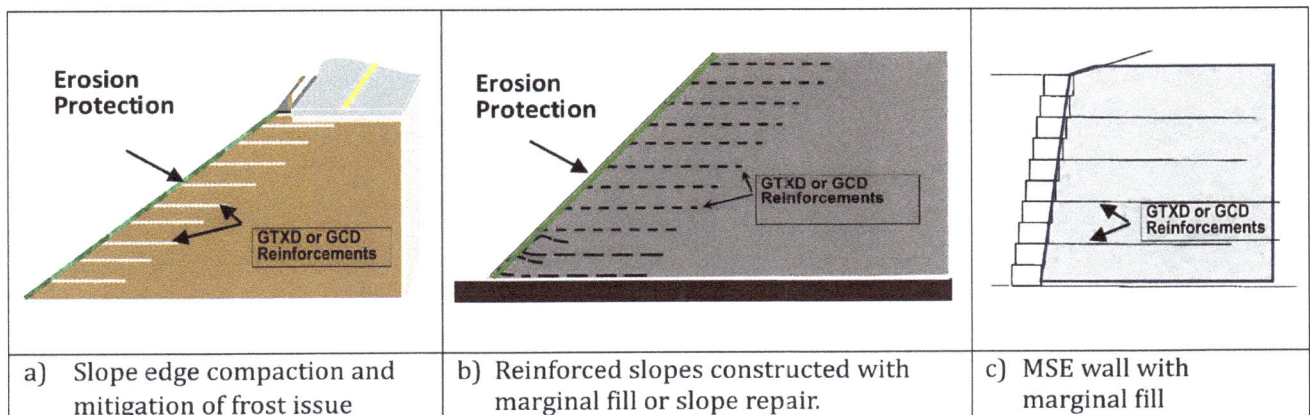

| a) Slope edge compaction and mitigation of frost issue | b) Reinforced slopes constructed with marginal fill or slope repair. | c) MSE wall with marginal fill |

Figure 3.7 Geosynthetic drains to mitigate seepage issues in slopes and walls. (In this figure GTXD is a geotextile drain and GCD is a geocomposite drain)

in terms of both the rate of settlement and strength gain (e.g. Berg et al, 2009, following Wayne et al, 1996). The more recently developed wicking geotextiles remove water through capillary action, removing moisture even before compaction and following frost events.

A widely used structural application is the use of geocomposite drains to replace conventional trench drains. For trench drain design, only geocomposite drains that allow two-sided flow should be used. The geocomposite drain is placed in a sand-filled trench as shown in Figure 3.6 with equally spaced prefabricated pipe outlets. For roadway edge drains, the geocomposite drain should be placed to the outsde of the trench and backfilled with sand on the inside between the base material and the drain (see Figure 3.8). This detail provides the ability to meet filtration requirements under dynamic, pulsating water flow by reducing the risk of a void between the drain and the side of the trench under the pavement section, which often occurs due to rough cutting during excavation. Therefore, sand should be infilled between the geocomposite edge drain and base layers to provide intimate contact for the filter geotextile of the edge drain. The sand backfill gradation should be designed to be compatible with the base and subbase layer, following standard filter criteria. The geotextile should be designed to be compatible with the inflow from the sand and the subgrade material on the backside of the edge drain.

Another pavement application is the use of a geocomposite drain as blanket drains (i.e. a horizontal lateral drain) for drainage of pavement sections, either placed directly beneath the pavement, beneath the base course, or within the subgrade, as shown in Figure 3.9. A geocomposite drain placed beneath the base layer, as shown in Figure 3.9a, provides two significant advantages:

1. Permeability requirement for base is much reduced (water only needs to flow through the thickness of the base rather than to the edge of the road), and
2. The vertical flow gradient through the base is much higher than the horizontal gradient.

This system can be 100 to 1000 times more effective than conventional drainage if the geocomposite drain is designed to deliver the required flow rates.

The middle figure (Figure 3.9b) shows the geocomposite drain replacing a drainable base layer directly beneath the pavement (concrete or full-depth asphalt). This detail may allow the pavement slab to be placed directly on subgrade or used for rehab pavements over crack and seal or rubblized sections. The geocomposite drain placed beneath asphalt concrete overlays offers the additional benefits of stress relief and reduced reflection cracking.

Figure 3.9c illustrates the use of a geocomposite drain to intercept capillary water that feeds ice lenses. This application can reduce the frost susceptibility of soils and avoid the need to import expensive non-frost-susceptible materials.

For the applications in Figure 3.9, the geocomposite drain must have the stiffness required to support traffic without significant deformation under cyclic traffic loading. At the same time, the drain must have a required flow capacity to rapidly drain the pavement section and prevent saturation of the base.

The following are construction considerations specific to the installation of geocomposite drains:

1. As with all geosynthetic applications, care should be taken during storage and placement to avoid damage to the material.
2. The placement of the backfill directly against the geotextile filter must be closely monitored, and compaction of the soil with equipment directly against the geocomposite drain should be avoided. Otherwise, the filter could be damaged or the drain could be crushed. Use of clean granular backfill reduces the compaction energy requirements.
3. At joints, where the sheets or strips of geocomposite drain butt together, the geotextile filter must be carefully overlapped and

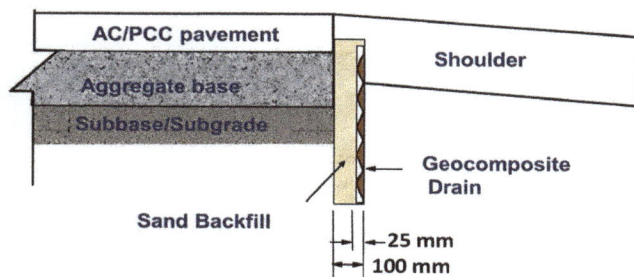

Figure 3.8 Retrofit Pavement Edge Drain (after Koerner et al, 1994)

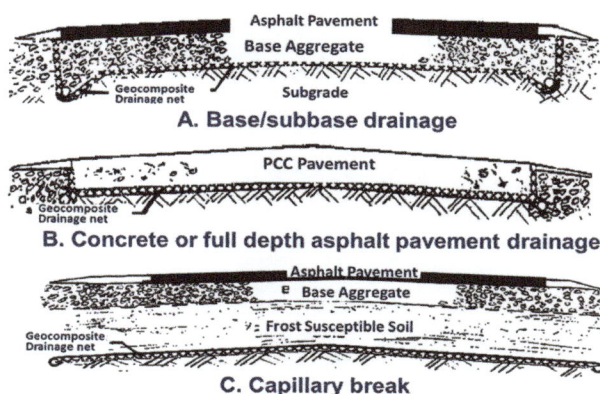

A. Base/subbase drainage

B. Concrete or full depth asphalt pavement drainage

C. Capillary break

Figure 3.9 Geocomposite blanket drains in pavement sections (Holz et al, 2008)

stitched or thermally welded to prevent soil infiltration. The geotextile should extend beyond the ends of the drain to prevent soil from entering at the edges.

4. Details must be provided on how the prefabricated drains tie into the collector drainage systems.

Key installation issues specific to the performance of geocomposite edge drain systems include:

> maintaining the verticality of the drain panel in the trench,
> proper positioning of the drain panel within the trench,
> as previously shown in Figure 3.8, placement of clean sand backfill between the aggregate and the geocomposite drain to maintain intimate contact between the base layer and the filter geotextile,
> timely installation of outlet fittings and pipes, and
> The use of outlet pipes with adequate pipe stiffness.

For installation of horizontal geocomposite drains in a roadway, the product roll width should be placed perpendicular to the roadway centerline on the prepared subgrade. The aggregate subbase should be placed on top of the geocomposite drain, spread evenly with a bulldozer, and compacted. To avoid damage to the geocomposite drain, the first lift above the drain should be initially compacted with the dozer and then with a smooth-drum roller, with the vibrator turned off.

3.3.3 Land Mass Dewatering/Consolidation

There are several applications where geosynthetic drains can be used for dewatering and consolidation of a land mass. This need is common for developed land adjacent to or surrounded by waterways. Geocomposite drains can be placed horizontally above fine-grain in situ or backfill soils, and a surcharge can be placed above the drain to allow consolidation of the soils under the surcharge load. An alternative surcharge method involves placing a geomembrane above the drain, sealed into the underlying soil, in a trench along the edge of the area to be improved. A vacuum is then applied to the geocomposite drain. In either method, for extremely soft soils, relatively narrow geocomposite vertical drains (also known as prefabricated vertical drains (PVDs); see Figure 1d) can be pushed, vibrated, or driven into the soil at relatively close spacing (1 to two meters apart) to reduce the drainage path length and accelerate consolidation. As with other geocomposite drains, PVDs consist of a geotextile filter material surrounding a plastic core that allows water flow. While

PVDs have a band-shaped, rectangular cross-section, the cross-section is significantly smaller than that of most other geocomposite drains, typically measuring only 100 mm wide and 3 to 9 mm thick. The geotextile filter is most often a heat-bonded nonwoven geotextile to reduce sidewall friction and thus drag during installation penetration.

PVDs are typically installed using a steel mandrel that is pushed or vibrated into the ground, as shown in Figure 3.10a. The PVD is inserted into the mandrel, which then acts as a guide and protective sleeve during installation. Once the mandrel reaches the desired depth, it is withdrawn, leaving the PVD embedded in the soil. The vertical drains are then connected to the horizontal drainage layer at the surface (again placed beneath the surcharge load) as shown in Figure 3.10b. As previously indicated in Section 3.3, geosynthetic drains can also be placed between layers in marginal fills to dissipate porewater pressure during compaction and placement of fill above the drainage layers. This application has been utilized in geosynthetic-reinforced soil walls and slopes.

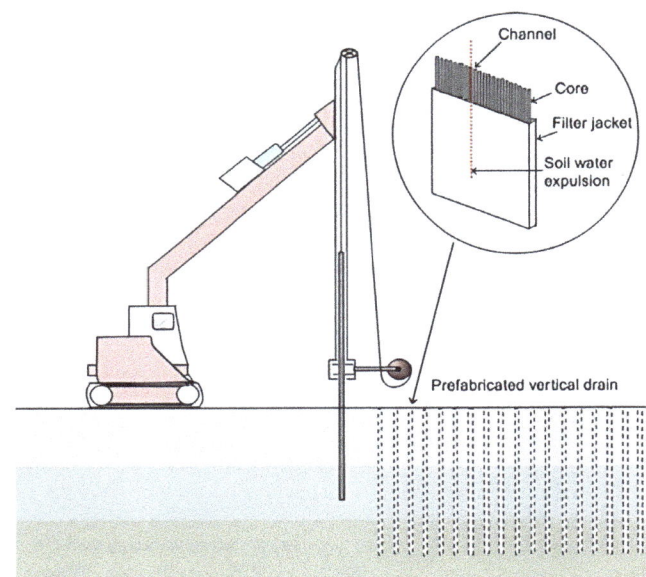

Figure 3.10a PVD installation with mandrel and guide sleeve (from Ocean Engineering, 2019)

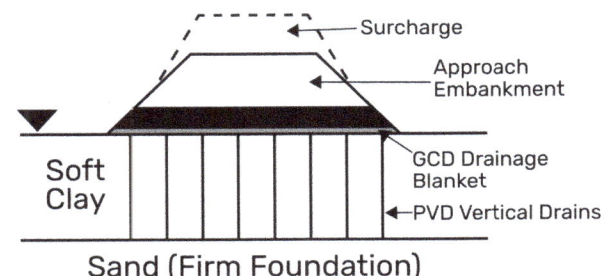

Figure 3.10b PVDs used in Surcharge Preloading (FHWA GEC 13)

3.3.4 Drainage for Preload Consolidation with and without Wick Drains

Consolidation is the reduction in soil volume resulting from water being squeezed out of the soil under load. A drainage blanket is required under the surcharge to relieve the water pressure and allow it to drain. As indicated in the previous section, the surcharge load can consist of either backfill soil or a geomembrane, sealed at the edges, with a vacuum applied to the drain. For this application, the groundwater will move upward as it is squeezed out of the saturated soil. The rate of consolidation is dependent on the permeability of the soil; however, it is also directly proportional to the square of the drainage distance to the drainage layer. For deep deposits of soil, this will be the most significant factor. For example, if the length the water has to travel, i.e. its drainage path, is 10 m versus 5 m, it will take 4 times longer, and if it is 20 m versus 5 m, it will take 16 times longer. Therefore, if we shorten the water path, we significantly decrease the time it takes to move the water. The method to decrease the consolidation time is to install vertical drains to shorten the drainage path. The installation of vertical drains changes the direction of water flow from vertical to horizontal, resulting in a significantly shorter path for water to travel, thereby accelerating the consolidation process. For example, vertical drains spaced 5 m apart will result in a drainage path of only 2.5 m in length.

In order to use PVDs, the subsoil should have moderate to high compressibility, low permeability, be fully saturated, and have low to moderate shear strength. There should also be no concern for secondary consolidation (e.g. if organics are high, PVDs may not be the correct solution). Soil thickness is typically limited to less than about 50 meters. The advantages include significant reduction in construction time, low cost, no spoil to be removed, durability, and extensive experience. The PVD also only requires a small material storage area.

One significant advantage of the PVDs over sand column vertical drainage systems is the simplicity of field control. Inspection primarily involves recording the depths and locations of each drain, observing the splices and verticality of equipment, taking occasional material samples for inspection and testing, and noting any significant deviations from the procedure. Of course, the drains must be installed in accordance with the project plans and specifications. Field inspection personnel must be aware of the procedures and potential consequences of any deviations from these procedures. The drainage blanket material quality and embankment placement and compaction must also be monitored, as covered in section 3.3 for horizontal geocomposite drain systems.

3.3.5 Underground Detention Systems / Subsurface Water Reservoirs

Optimizing real estate space is challenging. Large-scale infrastructure developments can lead to runoff issues, including flooding, erosion, and contaminated water. Stormwater management is required and is usually controlled with above-ground runoff basins. These basins occupy significant land area, are often unattractive, tend to collect pollutants, and attract mosquitoes over time.

Subsurface geosynthetic water reservoirs are underground structures that store water and significantly reduce these issues. Rain gardens above these underground reservoirs use appropriate vegetation to recycle storm runoff. Infiltration drain fields promote stormwater retention, detention, and infiltration into subsoils, and are increasingly recognized as best practices in managing stormwater runoff. Vegetation in the drain field absorbs contaminants in the water through a process called phytoremediation, converting them to less harmful chemicals within the plants (USEPA 2012). This provides a cost-effective and sustainable alternative to traditional water treatment methods that rely on chemicals and other artificial means. Underground reservoirs are even more suitable in areas where surface water is scarce or unreliable. These reservoirs serve various purposes, including irrigation, drinking water supply, and industrial processes, and, as a result, are a significant sustainable feature.

There are various geosynthetic systems, including polymeric boxes and arches, concrete boxes, block walls with concrete arches, corrugated pipes, and large stone and boulder beds (e.g. see Figure 3.11). Some synthetic cellular crate-type systems can hold up to 90% water in the excavated volume, which reduces the excavation requirements for the same volume of storage as other systems (DiLoreto, 2024). The construction of subsurface water reservoirs involves the use of geosynthetics to improve soil stability, drainage, and filtration. They are usually wrapped with a geotextile, which provides separation and filtration of the surrounding soil, and in some cases, reinforcement for barrier walls. Additionally, they may be coated with asphalt or a geomembrane liquid and gas barrier. Geosynthetic drains are also used to prevent the occurrence of pore pressure, accelerate the consolidation of soils during construction, remove uplift pressure, prevent an increase in porewater pressure, and prevent pore pressure from occurring in the case of unsatisfactory work of the sloping drains. A geocomposite drain may be placed immediately above the reservoir system, directly below a porous pavement or pavement block systems, along the adjacent areas, and/or in runoff channels in order to improve drainage into the reservoir.

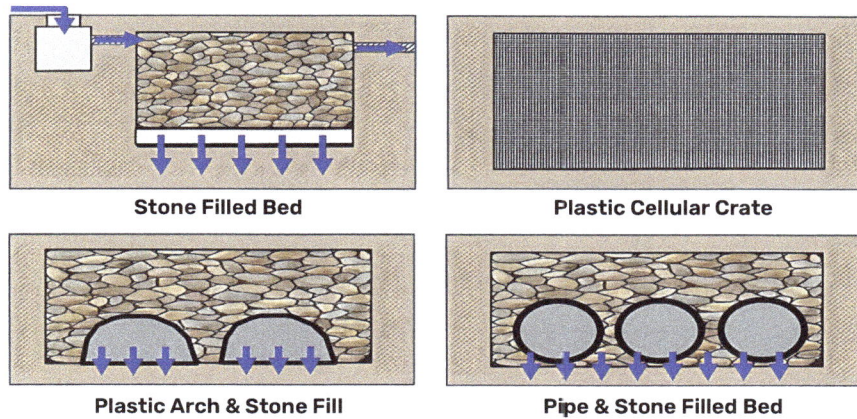

Figure 3.11 Examples of Subsurface water reservoirs (after DiLoreto. 2024)

The reservoir must be able to safely support the static and dynamic loads placed on the structures both during and after construction. Excessive deflection must be prevented, or the surfacing materials (i.e. asphalt, concrete, or block paving) will be damaged. Figure 3.12 shows an example of a subsurface stormwater retention reservoir under construction.

Standard calculations (e.g. analytical pavement design and Boussinesq solutions) should be performed to determine the stress at the top of the reservoir and compare the stress to the allowable load-carrying capacity and deformation serviceability limit of the reservoir system. Stress/strain data should be available for the specific reservoir system based on both laboratory and field measurements.

The following references provide more information on subsurface water reservoirs:

> Groundwater by Encyclopedia Britannica
> Water (Hydrologic) Cycle by Biology Libre Texts
> The Hydrologic Cycle by Engineering Libre Texts
> General Facts and Concepts About Groundwater by USGS Publications Warehouse

3.3.6 Soil Bioengineering Applications

These applications include the use of geosynthetic drains in sports fields, subsurface irrigation of plants, and structural root boxes. In sports fields, geosynthetic drains are used to replace traditional pipe and granular graded filter stone beneath the grass. Geosynthetic drains provide efficient drainage and help avert water ponding on the field surface. In addition to drainage, the geosynthetic also provides improved irrigation by directing water from hoses and sprinklers to be more uniformly distributed to the plant's roots. Geosynthetic drains are used for golf greens for the same purpose and in bunkers for drainage and separation of the sand from the subsurface soil.

For irrigation, geosynthetic drains can be placed on the surface between rows of plants to distribute the moisture from rainfall and reduce weed growth, as shown in Figure 3.13. They can also be used as irrigation mats beneath the subsurface, allowing for a more uniform distribution of water from irrigation pipes directly to the root structure of the plants. Geosynthetic drains with hydrophilic fibers (wicking fabrics) can pull water from the source and transport it to the plants, thereby

Figure 3.12 An example of a geosynthetic subsurface detention basin installation in Pennsylvania, USA

Figure 3.13 Geotextile use for irrigation distribution (from Significance of Geotextile Fabric in Modern Agriculture | Shankar Techx Pvt. Ltd.)

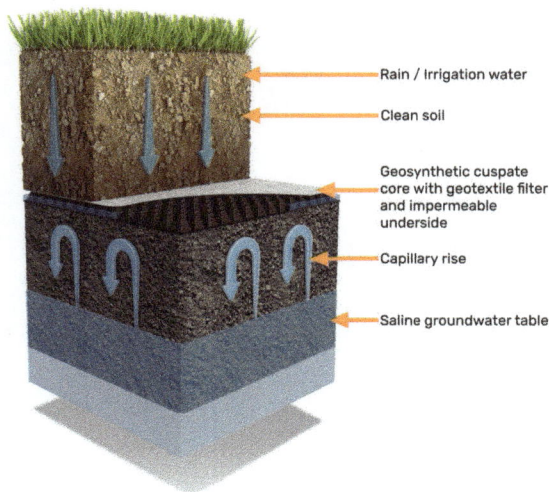

Figure 3.14 Geocomposite Drain used as a salt barrier (modified from ABG Ltd.)

minimizing the need for pumping. Additional information on geosynthetics in both agriculture and aquaculture applications can be found in Hsieh (2016.

Structural root boxes and root bags are a type of underground structure that is used to provide support to trees in urban environments and at nursery facilities. The use of geosynthetics in either structural root boxes or root bags can help to improve soil quality by reducing soil compaction and promoting healthy tree growth by providing irrigation directly to the root structure. One of the key requirements for the geosynthetic is to be strong enough to inhibit root growth, but not too strong so that the feeder roots can penetrate the system. This can be determined by attaching a load cell to a nail-sized probe to measure the force required to penetrate the geosynthetic drain.

An additional application for geosynthetic drains is the prevention of upward movement of saline groundwater through capillary action, thereby protecting topsoil and landscaping from the harmful effects of salt, as shown in Figure 3.14.

These geosynthetics typically consist of a geotextile filter bonded to a drainage core, such as a double cuspated core, creating a barrier that separates the "sweet" soil zone from contaminating saline water sources. In this way, they disrupt the natural capillary action, which would otherwise draw contaminating saline water upwards from the water table. The drain function serves to relieve uplift pressure from beneath and drains off rainwater from above while filtering the two separated soils. The use of these geosynthetic drains replaces the traditional 300 mm of crushed stone granular layer, reducing the environmental impact of manufacturing and transporting aggregate.

All of these geosynthetic drain applications provide a sustainable solution to conserve one of our most valuable resources, water.

3.4 Summary

Geosynthetic drains offer numerous advantages in civil and geotechnical engineering applications, making them a preferred alternative to conventional drainage systems. These materials - such as geotextiles, geocomposite drains, and prefabricated vertical drains - provide efficient, reliable water management by facilitating rapid drainage and reducing porewater pressures, thereby improving the stability of soil structures. Their lightweight and flexible nature simplifies transportation and installation, reducing construction time, improving safety and reducing costs. Geosynthetic drains can be engineered for specific flow rates and filtration characteristics, ensuring compatibility with a wide range of soil types and hydraulic conditions. Their use also minimizes the need for natural aggregate, leading to substantial environmental and economic benefits through reduced excavation and material consumption. Overall, geosynthetic drains contribute to more sustainable, cost-effective, and technically robust drainage solutions across infrastructure, environmental, and geotechnical projects.

3.5 References

AASHTO (2022), *Standard Specification for Geosynthetic Specification for Highway Applications*, Standard Specifications for Transportation Materials and Methods of Sampling and Testing, American Association of State Transportation and Highway Officials, Washington, D.C..

ABG Geosynthetics (2025). https://abg-geosynthetics.com/products/geocomposite-drainage/salt-break/, Retrieved online July 31, 2025.

ASTM International (2024), *Annual Book of Standards*, Vol. 4.13, Geosynthetics, ASTM International, Philadelphia, PA, USA.

Berg, R.R., Christopher, B.R. and Samtani, N. (2009). Mechanically Stabilized Earth Walls and Reinforced Soil Slopes, Design and Construction Guidelines. U.S. Department of Transportation, Federal Highway Administration, Washington, DC, FHWA-NHI-09-083 and FHWA GEC011, 668 p.

Cedergren, H.R. (1989). *Seepage, Drainage, and Flow Nets*, Third Edition, John Wiley and Sons, New York, 465 p.

Chai, J.C. (2016). "Geotextiles Used in Drainage," Chapter 13, editor: Koerner, R.M., *Geotextiles from Design to Application*, The Textile Institute, Elsevier Ltd., 2016, pp. 277–303.

Christopher, B.R., Zornberg, J.G. and Mitchell, J.K. (1989). Design Guidance for Reinforced Soil Structures with Marginal Soil Backfills, Proceedings of the Sixth International Conference on Geosynthetics, Atlanta, pp. 797–804.

City of Philadelphia, Pennsylvania. https://water.phila.gov/development/stormwater-plan-review/manual/chapter-4/4-8-subsurface-detention. Retrieved online July 31, 2025.

DiLoreto, R. (2024). Personal communication and presentation on subsurface water storage, ACF,.

Federal Highway Administration (2016). FHWA Geotechnical Engineering Circular (GEC) 13, Ground Modification Methods Reference Manual Vol. 1 and 2, authors Schaefer, V.R., Berg, R.R., Collin, J.G., Christopher, B.R., DiMaggio, J.A., Filz, G.M., Bruce, D.A., and Ayala, D., U.S. Department of Transportation, Federal Highway Administration, Washington, DC, FHWA-NHI-16-027 and FHWA-NHI-16-028, 351 p. and 489 p.

Geosynthetic Institute (2012). GRI GN2 and GRI GC13, *Joining and Attaching Geonets and Drainage Geocomposites*, originally published Sept. 25, 2012.

Giroud, J.P. (1980). "Introduction to Geotextiles and Their Applications," *Proceedings of the First Canadian Symposium on Geotextiles*, Calgary, Alberta, pp. 3–31.

Giroud, J.P. (1996). "Granular Filters and Geotextile Filters", *Proceedings of GeoFilters '96*, Lafleur, J. & Rollin, A.L., Editors, Montréal, Canada, pp. 565 – 680.

Guo Jun, Weng Fei, Ph.D. Zhang Xiong, Ph.D., Han Jie, Ph.D. "Quantifying Water Removal Rate of a Wicking Geotextile under Controlled Temperature and Relative Humidity", Publication: Journal of Materials in Civil Engineering, Volume 29, Issue 1

Heritage, R. and Shercliff, D. (2022). "Plastic or Concrete? The use of drainage geosynthetics to replace unnecessary environmentally damaging alternatives", *Proceedings of EuroGeo7*, Warsaw, Poland. International Geosynthetics Society.

Hsieh, C.W. (2016). Geotextiles in Agriculture and Aquaculture, Chapter 23, editor: Koerner, R.M., *Geotextiles from Design to Application*, The Textile Institute, Elsevier Ltd., pp. 511–530.

International Geosynthetics Society (2000). *Recommended Descriptions of Geosynthetics Functions, Geosynthetics Terminology, Mathematical and Graphical Symbols*, 4th edition, pp. 17.

ISO, International Organization for Standardization, https://www.iso.org/standards.html.

Koerner, R.M. (2016, Editor). *Geotextiles from Design to Application*, The Textile Institute, Elsevier Ltd., 2016, 617 p.

Koerner, R.M. (2012). *Designing with Geosynthetics*, 6th Edition, Vol. 1 and Vol. 2, Xlibris.

Koerner, R.M., Koerner, G.R., Fahim, A.K. and Wilson-Fahmy, R.F. (1994). *Long Term Performance of Geosynthetics in Drainage Applications*, National Cooperative Highway Research Program Report No. 367, 54 p.

Koerner, R. M. (1999). *Geosynthetic Reinforced and Geocomposite Drained Retaining Walls Utilizing Low Permeability Backfill Soils*, GRI Report No. 24, GSI, Folsom, PA, 140 pp.

Mohajerani, Dean, Munro, Soong (2019). "An overview of the behaviour of iron ore fines cargoes, and some recommended solutions for the reduction of shifting incidents during marine transportation", *Ocean Engineering*, Volume 182, pp. 451 – 474.

Shankar Packagings Limited. https://www.shankartechx.com/news-detail/an-overview-of-geotextile-types-and-uses. Retrieved online July 31, 2025.

Terzaghi, K., Peck, R.B., and Mesri, G. (1986). *Soil Mechanics in Engineering Practice*, Third Edition, John Wiley & Sons, New York, pp 330–332.

United States Environmental Protection Agency (1999), *Storm Water Technology Fact Sheet Infiltration Drainfields*, EPA 832-F-99-018, Office of Water, Washington, D.C.

United States Environmental Protection Agency (2012), *A Citizen's Guide to Bioremediation*, EPA 542-F-12-003, Office of Solid Waste and Emergency Response, Washington, D.C.

Wayne, M.H., K.W. Petrasic, Wilcosky, E., and Rafter, T.J. (1996). "An Innovative Use of a Nonwoven Geotextile in the Repair of Pennsylvania State Route 54," *Proceedings Geofilters '96*, Montréal pp. 513–521.

Chapter 4

Geosynthetics in Erosion and Sediment Control

4.0 Overview

Traditionally, natural materials such as straw fibers, burlap mats, brush piles, or stacked rock have been components commonly used in erosion and sediment control systems. However, the exclusive use of natural materials in site-built erosion and sediment control systems has shown performance limitations due to material and installation inconsistencies and susceptibility to progressive degradation. The inclusion of geosynthetics in erosion and sediment control systems has been proven to provide significant advantages when used in place of, or in combination with natural materials.

Typically, geosynthetics are buried and perform traditional functions such as separation and filtration. Yet, in most erosion and sediment control applications geosynthetics are used on the soil surface. As a result, they are expected to perform unique functions that contribute to controlling the dislodgement of soil, i.e. erosion control, as well as intercepting and removing dislodged soil out of runoff, i.e. sediment control. These unique functions have been described in ASTM D5819 "Standard Guide for Selecting Test Methods for Experimental Evaluation of Geosynthetic Durability" as follows:

> *Containment* - A geosynthetic provides <u>containment</u> when it encapsulates or surrounds materials such as sand, rocks, straw, brush, and fresh concrete.
> *Dynamic Filtration* - A geosynthetic performs the function of <u>dynamic filtration</u> when the equilibrium geotextiletosoil system allows for adequate liquid flow with limited soil loss across the plane of the geotextile over a service lifetime compatible with dynamic flows.
> *Screening* - A geosynthetic, placed across the path of a flowing fluid (ground water, surface water, wind) carrying particles in suspension, provides <u>screening</u> when it retains some or all fine soil particles while allowing the fluid to pass through. After some period of time, particles accumulate against the screen, which requires that the screen be able to withstand pressures

generated by the accumulated particles and the increasing pressure from accumulated fluid.
> *Surface Stabilization* - A geosynthetic, placed on a soil surface, provides <u>surface stabilization</u> when it restricts movement and prevents dispersion of surface soil particles subjected to erosion actions (rain, wind), often while promoting vegetative growth.
> *Vegetative Reinforcement* - A geosynthetic provides <u>vegetative reinforcement</u> when it extends the erosion control limits and performance of vegetation.

Erosion and sediment control systems can be categorized as either temporary/degradable or long-term/nondegradable. Temporary/degradable geosynthetic-enhanced erosion control systems commonly consist of a lightly stabilized geosynthetic and/or biodegradable components, such as biodegradable fibers or yarns. These systems are referred to as Rolled Erosion Control Products (RECPs) and perform a surface stabilization function. RECPs are further categorized as erosion control nettings (ECN), open weave meshes (OWM), or erosion control blankets (ECB), depending on their construction. These materials have a reported lifetime of as short as 6 months and as long as 3 or 4 years. These time periods correspond well to germination and initial growth periods for moderate and arid climate vegetation, respectively.

Sediment control systems generally fall into the temporary/degradable category, as well. These systems, also referred to as sediment retention devices (SRDs), are strategically deployed upstream of a sediment pond on construction sites to minimize the migration of suspended sediments to and through the pond. The SRDs are then removed once the site has been sufficiently stabilized either through revegetation or paving. These SRDs may be mostly natural materials encapsulated in geosynthetic fabrics, meshes, or nettings, known as wattles, filter logs, and compost socks, or they may be geotextile-based systems specifically designed to be deployed in or around storm drains.

Long-term/nondegradable erosion control systems are expected to provide extended performance. Traditional long-term systems include sod, riprap, and precast concrete blocks, and the performance of all these systems can be enhanced with geotextile filters. Still other geosynthetic-enhanced long-term systems include turf reinforcement mats (TRM) – another type of RECP, as well as engineered synthetic turf (EST), geocellular confinement systems (GCS), anchored geosynthetic systems (AGS), geosynthetic cementitious composite mats (GCCM), fabric formed concrete revetments (FFCR), and geotextile-encapsulated sand elements. The higher cost of these geosynthetic-enhanced long-term systems is often justified by the efficiency of their installation and their extended performance limits.

4.1 RECP Applications

4.1.1 Temporary Construction Site Erosion and Sediment Control

Having geosynthetic components enables rolled erosion control products (RECPs) and sediment retention devices (SRDs) to be manufactured under controlled conditions and to be more dependably "engineered" for specific project conditions. Temporary RECPs may be erosion control nettings (ECN), open weave meshes (OWM), or erosion control blankets (ECB). Following are descriptions of each of these geosynthetic-enhanced temporary RECPs.

Erosion Control Netting (ECN) - Erosion control netting is typically a plastic mesh. ECNs are used for anchoring loose fiber mulches. They can be rolled out over a seeded and mulched area and stapled or staked in place. Much more common in recent years is for the netting to be a component of an erosion control blanket which facilitates the installation of the mulch and netting together.

Open Weave Meshes (OWM) - Open weave meshes are woven of jute or coir or polyolefin yarns. Organic OWMs typically are 0.25 to 0.50 in. (6 to 12 mm) thick and have 1 in. (25 mm) or larger square uniform openings. Polyolefin meshes are considerably thinner with smaller openings. All meshes are very flexible, promoting intimate ground cover, though they do not provide full ground coverage. Organic meshes also absorb water, which can help maintain soil moisture.

Erosion Control Blankets (ECB) - Geosynthetic erosion control blankets are organic fiber filled "blankets" consisting of straw, wood (excelsior), or coconut fibers sewn to or between synthetic nettings. ECBs provide a thick (up to 0.5 in. (12 mm)) full coverage of mulch which better absorbs rainfall impact and retains moisture. The

nettings add strength to help ECBs resist erosive forces. Their useful life is limited to durability of the organic fibers. Geosynthetic ECBs can be further segmented into single net (1N), double net (2N), or even triple net (3N) and filled with straw (S), coconut (C), a combination of straw and coconut (SC), excelsior (X), or polymer fibers (FF), though polymer fiber-filled blankets are more typically considered long-term and are referred to as turf reinforcement mats.

4.1.2 Long-term RECP: Turf Reinforcement Mat

In more demanding runoff erosion control applications, such as concentrated pipe discharges and deep, turbulent channel flows rip-rap or concrete has traditionally been used, but increasingly these hard armor systems are being replaced by geosynthetic-enhanced vegetated erosion control systems for primary slope or channel lining. The geosynthetic in this application is commonly referred to as a Turf Reinforcement Mat, or TRM. A TRM is a rolled erosion control product composed of non-degradable synthetic fibers, filaments, nets, wire mesh and/or other elements, processed into a permanent, three-dimensional matrix of sufficient thickness. TRMs, which may be supplemented with degradable components, are designed to impart immediate erosion protection, enhance vegetation establishment and provide long term functionality by providing root and stem reinforcement during and after vegetation maturation. By reinforcing vegetation, TRMs allow for improved erosion control performance and maintain all the same ecological and environmental benefits of unreinforced vegetation. Following are widely recognized benefits of TRM-reinforced vegetation:

> Extended erosion performance
> Greater safety for public
> Reduced economical impact
> Improved water quality
> Wildlife habitat (though entrapment may be a problem)
> Improved aesthetics
> Ease of installation
> Reduced flow velocities
> Relief of pore water pressure buildup
> Improved flexibility against heaving and settlement
> Improved resiliency of vegetated, nature based solution
> Improved infiltration/groundwater recharge

4.2 Quantifying RECP Properties for Quality and Performance

An effective RECP is one which, in the short-term, absorbs the kinetic energy of rain, slows runoff, promotes water infiltration, and provides the microclimate needed for the germination of seeds and the subsequent establishment of a self-sustaining vegetative cover to provide, in the long-term, permanent erosion protection. To accomplish this combination of functions, the RECP must be selected based on its consistent material properties and associated performance characteristics. Quantifying RECP quality, consistency, and performance requires regular manufacturing quality checks and large-scale evaluations that simulate actual erosive forces.

Basic index tests are typically needed to assure manufacturing quality control of RECPs and to "benchmark" the performance results to specific material properties. As robust quality control is generally lacking in the manufacture of temporary RECPs, users more often rely on independent quality assurance testing such as the American Association of State Highway and Transportation Officials' (AASHTO's) Product Evaluation and Audit Solutions (PEAS) program which tests RECPs on a regular 3-year cycle in North America. This program is formerly known as the National Transportation Product Evaluation Program – NTPEP. The program includes both index tests and bench-scale "indexed performance" tests. Additionally, PEAS offers independently verified large-scale performance testing. The test methods used by PEAS are described in 4.2.1, and the multi-year database of test results is the basis for the product type characterizations provided in Table 1.

4.2.1 Index, Bench-scale, and Performance Testing of RECPS

Standardized index and bench-scale tests for both temporary and long-term RECPs have been developed to facilitate manufacturing quality control, purchasing quality assurance, and designer specification efforts. Following are the commonly used index test methods:

> Mass per Unit Area (Temporary RECPs): ASTM D6475, "Standard Test Method for Measuring Mass per Unit Area of Erosion Control Blankets".
> Mass per Unit Area (TRMs): ASTM D6566, "Standard Test Method for Measuring Mass per Unit Area of Turf Reinforcement Mats".
> Thickness: ASTM D6525, "Standard Test Method for Measuring Nominal Thickness of Permanent Rolled Erosion Control Products".

> Tensile Strength: ASTM D6818, "Standard Test Method for Ultimate Tensile Properties of Turf Reinforcement Mats".
> Light Penetration: ASTM D6567, "Standard Test Method for Measuring the Light Penetration of a Rolled Erosion Control Product (RECP)".
> Water Absorption: ASTM D8263, "Determining the Change in Mass of Rolled Erosion Control Products When Submerged in Water" (Formerly referred to as "Absorption").

Additionally, because of the high cost of full-scale testing, the following bench-scale "indexed" performance tests are routinely run on RECPs to confirm their performance characteristics:

> Slope Erosion and Runoff Reduction: ASTM D7101, "Standard Index Test Method for Determination of Unvegetated Rolled Erosion Control Product (RECP) Ability to Protect Soil from Rain Splash and Associated Runoff under Bench-Scale Conditions".
> Permissible Shear and Channel Erosion: ASTM D7207, "Standard Test Method for Determination of Unvegetated Rolled Erosion Control Product (RECP) Ability to Protect Sand from Hydraulically-Induced Shear Stresses under Bench-Scale Conditions".
> Germination/Vegetation Growth: ASTM D7322, "Standard Test Method for Determination of Rolled Erosion Control Product (RECP) Ability to Encourage Seed Germination and Plant Growth under Bench-Scale Conditions".

Still, large-scale performance tests are necessary to characterize actual RECP performance under expected field conditions, including recommended installation. Following are the internationally recognized standard test methods commonly used for full-scale evaluations of RECPs:

> ASTM D6459, "Standard Test Method for Determination of Rolled Erosion Control Product (RECP) Performance in Protecting Hillslopes from Rainfall-Induced Erosion"
> ASTM D6460, "Standard Test Method for Determination of Rolled Erosion Control Product (RECP) Performance in Protecting Earthen Channels from Stormwater-Induced Erosion"

Data from D6459 and D6460 is very useful for characterizing and differentiating between various RECP types. Dozens of large-scale slope and channel tests have been performed over the past 30 years demonstrating quite convincingly that there is a hierarchy of

Table 4.1 Property Characterizations for Geosynthetic RECPs

Property	ASTM Test	Unit	RECP Type							
			OWM	1NX, 1NS	2NX, 2NS	2NSC	2NC	2NFF	3N all fibers	3D Woven
Functional Longevity	n/a	Est.	3-6 mo.	3-6 mo.	6-12 mo.	12-24 mo.	> 24 mos.	Perm.	Perm.	Perm.
C-Factor[1]	D6459	–	≤ 0.10	≤ 0.08	≤ 0.06	≤ 0.04	≤ 0.02	≤ 0.10	≤ 0.10	≤ 0.10
Permissible Shear[2] – Unvegetated	D6460	psf	≥ 1.25	≥ 1.5	≥ 1.75	≥ 2.0	≥ 2.25	≥ 2.0	≥ 2.25	≥ 2.5
		Pa	≥ 55	≥ 70	≥ 85	≥ 100	≥ 110	≥ 100	≥ 110	≥ 120
Permissible Shear[2] – Vegetated	D6460	psf	n/a	n/a	n/a	n/a	n/a	≥ 9	≥ 10	≥ 11
		Pa	n/a	n/a	n/a	n/a	n/a	≥ 430	≥ 480	≥ 530
Tensile Strength (MD)	D6818	lb/in	≥ 20	≥ 4	≥ 8	≥ 12	≥ 16	≥ 16	≥ 20	≥ 240
		N/m	≥ 2.25	≥ 0.45	≥ 0.90	≥ 1.35	≥ 1.80	≥ 1.80	≥ 2.25	≥ 27.0
Tensile Strength (XD)	D6818	lb/in	≥ 20	≥ 2	≥ 4	≥ 6	≥ 8	≥ 8	≥ 10	≥ 120
		N/m	≥ 2.25	≥ 0.25	≥ 0.45	≥ 0.70	≥ 0.90	≥ 0.90	≥ 1.15	≥ 13.5
Mass / Unit Area	D6475 D6566	oz/yd^2	≥ 10	≥ 7	≥ 7	≥ 7	≥ 8	≥ 8	≥ 10	≥ 12
		g/m^2	≥ 340	≥ 240	≥ 240	≥ 240	≥ 270	≥ 270	≥ 340	≥ 410
Light Penetration	D6567	%	≤ 35	≤ 35	≤ 35	≤ 35	≤ 35	≤ 35	≤ 35	≤ 35
Bench-scale Slope	D7101	Soil Loss Ratio	≥ 4.0	≥ 5.0	≥ 6.0	≥ 8.0	≥ 10.0	≥ 5.0	≥ 5.0	≥ 5.0
Bench-scale Shear	D7207	psf	≥ 1.25	≥ 1.5	≥ 1.75	≥ 2.0	≥ 2.25	≥ 2.5	≥ 3.0	≥ 3.5
		Pa	≥ 50	≥ 60	≥ 70	≥ 80	≥ 120	≥ 125	≥ 145	≥ 165
Germination, % Improvement	D7322	%	≥ 150	≥ 150	≥ 150	≥ 150	≥ 150	≥ 150	≥ 150	≥ 150
UV Stability, % Retained 1000 hrs	D4355 D6818	%	n/a	n/a	n/a	n/a	n/a	≥ 80	≥ 80	≥ 80

1. See Section 4.2.3; 2. See Section 4.2.4

Figure 4.1 RECP Slope Protection (ECTC)

Figure 4.2 TRM Channel Lining (ECTC)

performance among the commonly available product types as shown in Table 1.

It should be noted that while ISO standards exist for other geosynthetics, they do not yet address erosion control products, sediment retention devices or their unique testing requirements.

4.2.2 Specifying RECPS

Table 4.1 presents product type characterizations based on the PEAS program described above. The values in Table 1 are generally set approximately 20% below (or above in the case of C-Factor and light penetration) the average value of a large data set to account for product and test variability. These threshold values are useful tools for establishing generic specifications for RECPs.

4.2.3 RECPs on Slopes – Design and Specification

On slopes, an RECP reduces soil loss_caused by rain and immediate runoff while creating an environment conducive to seed germination and initial plant growth, known as mulching. In humid/temperate regions, only short-term mulching is needed, so the RECP need only have a short functional longevity, the period during which an RECP adequately serves its primary function(s). Greater functional longevity is required in more

arid regions or where a longer vegetation establishment period wil be needed.

Once the designer has made general determinations related to RECP cost-benefit tradeoff, desired functional longevity, and applicable regulatory requirements, a common approach used to design and select the most appropriate type of RECP based on erosion control performance is based on Equation 4.1, the Revised Universal Soil Loss Equation (RUSLE).

$$A = R \cdot K \cdot LS \cdot C \cdot P \qquad (4.1)$$

Where:

> A = estimated average soil loss (tons per acre per year)
> R = rainfall-runoff erosivity factor (hundreds of foot·ton·inch / acre·hour·year)
> K = soil erodibility factor (ton·acre·hours / hundred acre·foot·ton·inch)
> L = slope length factor
> S = slope steepness factor
> C = cover factor
> P = support practice factor

The RUSLE2 software along with guidance for determining site-specific values for R, K, and LS can be downloaded at www.ars.usda.gov. The C-Factor is the improvement (reduction in soil loss) provided by

Table 4.2 Reduction in Soil Loss via C-Factors and P-Factors for Various Surface Treatments

Treatment		C-Factor	P-Factor
Bare Soil - Packed and smooth		1.00	1.00
Bare Soil - Freshly disked or rough, irregular surface		1.00	0.90
Sediment Barrier Systems (Sediment Retention Devices)		1.00	0.01-0.90
Bale or Sandbag Barriers		1.00	0.90
Rock (Diameter = 1–2 in. (25-50 mm)) Barriers at Sump Location		1.00	0.80
Asphalt/Concrete Pavement		0.01	1.00
Gravel (Diameter = 2.5–16 in. (60-400 mm)) at 300 tonnes/ha		0.05	1.00
Established Vegetation		0.01 – 0.25	1.00
Sod Grass		0.01	1.00
Temporary Vegetation/Cover Crop		0.45	1.00
Hydraulic Mulch at 4.5 tonnes/ha		0.10	1.00
Soil Sealant		0.10 - 0.60	1.00
Rolled Erosion Control Products		0.01 - 0.20	1.00
Hay or Straw Dry Mulch Applied at 4.5 tonnes/ha and anchored Assumes planting of grass seed has occurred before application, otherwise C-factor = 1.00.	Slope (%)		
	1 to 10	0.05	1.00
	11 to 20	0.10	1.00
	21 to 30	0.15	1.00
	> 30	0.20	1.00

Table 4.3 Typical CFactors for Various Slope Conditions and Growing Periods

Treatment	Dry Mulch Rate kg/m²	Slope %	CFactor For Growing Period**			
			< 6 Wks	1.5 6 Mos.	6 12 Mos.	Annualized*
No mulching or seeding	all		1.00	1.00	1.00	1.00
Seeded grasses	none	all	.70	.10	.05	.15
	0.22	< 10	.20	.07	.03	.07
	0.34	< 10	.12	.05	.02	.05
	0.45	< 10	.06	.05	.02	.04
	0.45	1115	.07	.05	.02	.04
	0.45	1620	.11	.05	.02	.04
	0.45	2125	.14	.05	.02	.05
	0.45	2633	.17	.05	.02	.05
	0.45	3450	.20	.05	.02	.05
Second Year Grass		all	.01	.01	.01	.01
Erosion Control Blanket		all	.07	.01	.005	.02
Composite Turf Reinf. Mat		all	.07	.01	.005	.02
Turf Reinforcement Mat		all	.14	.02	.005	.03
Fully Vegetated RECP		all	.005	.005	.005	.005

* annualized = (<6 wks value x 6/52) + (1.56 mos. value x 20/52) + (612 mos value x 26/52);
** approximate time periods for humid climates; <u>Conversion</u>: kg/m² x 4.46 = ton/acre

Table 4.4 Typical Hydraulic Roughness Values for Various Vegetated Conditions

Facility	Lining Minimum		Hydraulic Roughness (Manning's n) Values	
			Maximum	
Open Channel	Bare Soil	Unlined	0.020	0.023
	RECP Lined Soil	Erosion Control Blanket	0.025	0.045
		Turf Reinforcement Mat	0.021	0.036
	Grass Lined, < 8 in. (200 mm) flow	6 in. (150 mm) length	0.05	0.09
		12 in. (300 mm) length	0.09	0.18
	Grass Lined, > 8 in. (200 mm) flow	6 in. (150 mm) length	0.04	0.06
		12 in. (300 mm) length	0.07	0.12

the RECP and is obtained from large-scale testing, and the P-Factor is the improvement provided by a sediment retention device (SRD) and is also obtained from large-scale testing. SRD design/specification will be addressed in 4.3. Table 4.2 provides some "ballpark" values for various surface treatments, while Table 4.3 provides more detailed guidance on considering RECP type and degree of vegetation establishment.

4.2.4 RECPs in Channels – Design and Specification

In channel linings, the RECP protects against hydraulic shear stress on the sides and bottom of the channel caused by flowing water. As noted earlier, specific RECP selection will always include the necessity for judgments, including material durability, or functional longevity, but the key to designing and specifying an appropriate type of RECP for a given channel and the associated maximum flow event is to make sure that the permissible shear stress of the channel liner is much greater than the actual shear stress on that liner from the design flow. The permissible shear stress, $\tau_{permissible}$, is determined from large-scale testing of the candidate channel lining and compared to the actual shear stress computed from the site-specific design storm event. The actual shear stress is calculated from the expected channel hydraulics which are typically defined using the Mannings equation (Equation 4.2).

$$Q = (Z / n)\ A\ R^{2/3}\ S_f^{1/2} \qquad (4.2)$$

where: Q = Flow in m³/sec or cfs

Z = 1.0 for metric measurement units, and 1.486 for English

n = Manning's roughness coefficient

A = Cross-section area (m² or ft²)

R = hydraulic radius (m or ft)

S = slope of the energy gradeline (m/m or ft/ft)

Mannings roughness coefficient describes the resistance to water flow and will vary depending on the characteristics of the channel lining exposed to the flow. Some typical values are shown in Table 4.4.

4.2.5 Calculating Actual Shear Stress

Shear Stress at maximum flow depth can be calculated from the following equation (Equation 4.3):

$$\tau = \gamma \times D \times S_f \qquad (4.3)$$

where: τ = Shear stress at maximum flow depth: Pascal (lbs/ft²)

γ = Unit weight of water: 62.4 lbs/ft³ (9810 N/m²)

D = Flow depth: ft (m)

S_f = Friction slope or bed slope of channel: ft/ft (m/m)

But first, we need determine the maximum flow depth, D, using Mannings equation after selecting an appropriate Manning's Roughness Coefficient, n. Manning's "n" for RECPs can vary significantly with material type, amount of vegetation, and flow depth but typically ranges from 0.02 to 0.04

4.2.6 Permissible Shear Design Approach

If the permissible shear stress is greater than the computed shear, the lining is considered acceptable. The permissible shear stress, $\tau_{permissible}$, is determined from large-scale testing of the candidate channel lining and compared to the actual flow-induced shear stress, τ_{actual}. When TRMs are being evaluated the testing is done

after the vegetation has matured in order to assess long-term performance. It should also be noted that the hydraulic roughness values for vegetated TRMs are quite different than for TRMs alone. Table 4.1 presents the permissible shear values from large-scale testing of a wide range of RECPs.

Additionally, a designer will commonly include a factor of safety in the determination of permissible shear stress to account for soil type, soil compaction, vegetation type, and/or vegetation density different from that simulated in the testing.

4.2.7 RECP Installation in Slopes and Channels

Satisfactory performance of an RECP relies, more than anything else, on proper installation. The following sections highlight the techniques and considerations that are key to a satisfactory installation.

4.2.7.1 Soil Surface Preparation

Proper seedbed preparation includes loosening and spreading the surface soils to obtain a soil surface that will both "capture" the seed and allow the covering RECP to have intimate contact with the soil. Commonly this will include roughening and then lightly smoothing the soil surface prior to spreading/spraying seed. It is advantageous to incorporate any needed soil amendments, such as lime and fertilizer, at this time, as well.

4.2.7.2 Seeding and Backfilling

Seeding is commonly done before RECP deployment. If an RECP is installed prior to seeding, which is sometimes done with TRMs, it should have a very open structure so that the seed can fall through to the soil or seeding on top of the TRM should be followed with a thin layer of backfill. Following are key considerations to assure a satisfactory RECP installation:

> Seeding. For most RECPs, the recommended seed mixture shall be seeded onto the prepared soil surface. Every effort should be taken to obtain a uniform distribution and intimate contact between seed and soil surface over the entire seeded area.

> *Backfilling of RECPs.* A thin layer of fine soil or hydromulch should be spread/sprayed uniformly on top of the RECP so as to completely cover the seed.

> *Erosion of Backfill.* If soil backfilling, also called in-filling, is used, it may be necessary to prevent erosion of the in-fill when frequent and/or heavy precipitation is expected during the germination and early vegetation growth periods. Traditional mulch or, preferably, an additional light-weight temporary RECP can be deployed over the in-fill.

> *Maintenance of Seeded Areas.* Proper care of seeded areas is especially important during the period of vegetation establishment. Temporary irrigation may be necessary.

4.2.7.3 Vegetation Selection Information

Appropriate seed types and soil amendments, as well as recommended application rates for successful construction site revegetation are very specific to local growing conditions, the expected life of the vegetation, the expected roadside maintenance efforts, and the time of year. Local and regional agencies are good resources for seed mix and soil amendment recommendations.

4.2.7.4 Field Joining of Adjacent Rolls

When unrolling the materials over the prepared and seeded soil surface, it is recommended to overlap or "shingle" adjacent rolls in the direction of flow. For both slopes and channels, this means that the upslope RECP should shingle over the downslope RECP. All overlaps must be sufficient to accommodate a row of anchors, such as staples, pins, or other type of ground anchor.

4.2.7.5 Securing of the RECP

A satisfactory RECP installation will limit the extent of runoff flowing below the RECP and prevent it from being lifted off the surface by emerging vegetation. Anchor trenches at the top and bottom of slopes, as well as, frequent use of staples or pins will assure that the deployed RECP maintains uniform ground contact. Uniform securing is accomplished by "patterned" staking, pinning, or stapling of positioned RECPs. The pattern depends on the steepness of the slope or channel. It is important to follow the manufacturer's recommended securing pattern and frequency. Extra wide rolls are often used for installation efficiency and should be secured in the same manner as individual rolls. It should be noted that temporary RECPs may be secured using degradable anchors.

4.2.7.6 Penetrations and Repairs

Additional securing and overlapping material should be provided at all penetrations and structure interferences as they are notoriously prone to concentration of erosive forces. Likewise, if a repair is required because the RECP has been accidentally damaged, a patch of the same base RECP type should be cut to fit over and sufficiently beyond the damaged area to permit joining to or securing through the parent RECP.

4.3 Temporary Applications – Sediment Retention Devices

Sediment retention devices (SRDs) are deployed on construction sites to intercept and slow stormwater flows and, in so doing, to minimize the migration of suspended sediments. The SRDs are then removed once the site has been sufficiently stabilized either through revegetation or paving. The most common geosynthetic SRD is silt fence – geotextile stretched between posts and toed into the ground along the base of slopes. Other SRDs may be natural or man-made materials encapsulated in natural or geosynthetic fabrics, meshes, or nettings, known as fiber rolls. Still, other SRDs are 100% geosynthetic-based systems such as many inlet protection devices or floating turbidity curtains to isolate in-water or near-water construction. Typically, thicker or 3-dimensional systems are better able to retain sediment while maintaining seepage flows. Following are more detailed descriptions of each geosynthetic-enhanced system.

Silt Fence - A temporary vertical sediment barrier made of porous fabric having its bottom edge buried in the ground. It is held up by fence posts driven into the ground. It is relatively inexpensive and relatively easy to remove (Figure 4.3).

Fiber Roll – Fiber rolls are tube-shaped sediment retention devices filled with straw, flax, rice, coconut fiber material or composted material. The tube-shaped wrap may be a 100 percent biodegradable material, such as burlap, jute or coir, but if greater functional longevity is needed, the wrap is commonly a geosynthetic and the fill may be soil, stone, or recycled plastic or rubber. Also referred to as wattles, filter logs, and compost socks (Figure 4.4).

Inlet Protection – Inlet protection can be either internal or external to a stormwater inlet (manhole). Internal controls consist of a filter insert or filter bag. External controls intercept the path of the stormwater as it reaches the inlet using some type of 2- or 3-dimensional permeable filter barrier. These are often proprietary devices (Figure 4.5).

Turbidity or Silt Curtain - A flexible, geosynthetic barrier used to trap sediment in water bodies. This curtain is generally weighted at the bottom to ensure that sediment does not travel under the curtain which is supported at the top through a flotation system. Staked curtains are available for applications with very limited exposure to water flow or wave action (Figure 4.6).

4.3.1 Quantifying SRD Performance

Both bench-scale and large-scale performance tests have been developed to characterize the filtration component only and the "as installed" performance of SRDs, respectively. Products are installed per the product manufacturer's published installation recommendations.

Figure 4.3 – Silt Fence (Sprague and Sprague, 2016)

Figure 4.4 - Fiber Roll in Channel (Sprague, GRI 25, 2013)

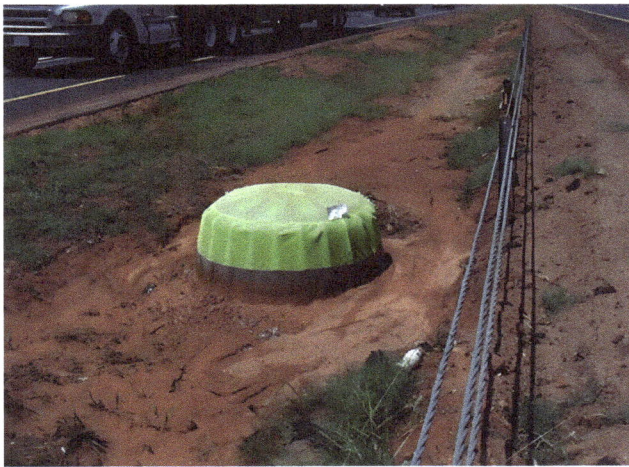

Figure 4.5 – SRD Protecting Inlet (Sprague and Sprague, 2016)

Figure 4.6 - Turbidity Curtain (Sprague and Sprague, 2016)

The results of the bench-scale test are used primarily for product acceptance, while the large-scale results are reasonably indicative of field performance and are acceptable for use in performance specifications and, often, in design calculations.

The most unique thing about SRD's is that, typically, for them to be very effective in retaining sediment they must also impound most of the runoff. Conversely, for them to freely pass runoff, they have to pass a significant amount of sediment. Thus, the test methods used for SRDs are used to quantify the "balance" of retention and flow provided under standard conditions. As shown in Table 2 (the P Factor column) SRD effectiveness in retaining sediments can be as low as 10% or as high as 99%. Following are the standards commonly used for performance evaluations of SRDs:

> ASTM D5141, "Standard Test Method for Determining Filtering Efficiency and Flow Rate of the Filtration Component of a Sediment Retention Device" (Figure 4.1);

> ASTM D7351, "Standard Test Method for Determination of Sediment Retention Device Effectiveness in Sheet Flow Applications" (Figure 4.2).

4.3.2 Sediment Retention Devices – Design and Specification

The nearly impossible dream of designers of SRDs is to remove all the sediment in the runoff while allowing only clear water to pass without any backup or ponding. This is especially true during peak flow events. Thus, designers typically opt for maximizing the retention of sediment, accepting the need for ponding to allow sedimentation and the gradual seepage of relatively clear water through the SRD. Thus, SRDs are expected to: 1) initially screens silt and sand particles from runoff; 2) develop a soil filter upstream face of the SRD which also reduces the ability of water to flow through the SRD; and, thus, 3) creating a pond behind the device which serves as a sedimentation basin to collect suspended soils from runoff water. To facilitate these steps, the SRD must have properly sized openings and/or porosity

Figure 4.7 ASTM D5141 Test Setup (TRI Environmental)

Figure 4.8 ASTM D7351 Full-Scale Test (TRI Environmental)

to screen the sediments that subsequently form the soil filter while providing sufficient storage capacity behind the device to contain the volume of water and sediment anticipated during a major storm. Thus, filtering efficiency, flow rate and structural capacity are especially important properties of SRDs. Also, because of long-term outdoor exposure typical of SRDs, UV resistance may also be a critical parameter.

Having geosynthetic components enables SRDs to be more dependably engineered for specific project conditions. Additionally, when properly designed and specified, SRDs will provide the following benefits over traditional sediment controls. These benefits include the minimal labor required for installation, lower cost, higher efficiency in removing sediment, and durability and reusability.

It is important to note that while SRDs are generally used to intercept sheet-flow, they can also be useful in reducing sediment migration in stormwater channels. Intermediate check structures have been used in channels to slow concentrated flows making them less erosive until the associated channel can vegetate

sufficiently to resist soil loss during concentrated flow events. Critical elements of this protection are the ability of the temporary check structure to: slow and/or pond runoff to encourage sedimentation, thereby reducing soil particle transport downstream, trap soil particles upstream of a structure, and decrease soil erosion.

4.3.3 "Designing" Sediment Retention Devices

The design of silt fence and other SRDs begins with defining the storage capacity required of the SRD and is a function of the drainage area it serves, the design rain storm, and the anticipated soil loss during the storm. Thus, a rigorous design must properly size the SRD for both water storage and sediment loading. The quantity of water that can be expected to pond behind the silt fence can be estimated using the Rational Method, which incorporates the drainage area served by the SRD, a runoff coefficient for the soil cover conditions, and a design event rainfall. The impoundment volume behind the SRD must also be adequate to contain the sedimentation that will occur in the impounded pool. Each storm will deposit a volume of sediment behind the fence until the sediments fill the pool and render the fence useless, at which time the sediment must either be removed or a new silt fence constructed immediately downstream from the existing one. The quantity of sediments resulting from the chosen design storm(s) can be calculated using the Revised Universal Soil Loss Equation (RUSLE) as described in 2.1.2. Conveniently, the RUSLE includes the P-Factor (see Table 2) as a means to include the benefit provided by an SRD. The appropriate P-Factor is obtained from large-scale performance testing of the chosen SRD as described in 2.2.1.

4.3.4 Simplified Generic SRD Specifications for Routine Applications

"The Standard Specifications for the Construction of Roads and Bridges on Federal Highway Projects" is a public-domain source (available online) for material specifications related to constructing roads and bridges in the US. These specifications are cited as "FP-14", contain both United States Customary and Metric units of measure, and apply to routine applications.

4.3.5 SRD Installation and Maintenance Considerations

Equally as important as material properties are the installation details and periodic maintenance that assure performance of SRDs. Each device must be installed in such a way that sediment laden runoff passes through it rather than undermining or bypassing, and accumulated sediments must be periodically removed to maintain SRD capacity.

A field study conducted in the United States found a disappointingly large percentage of silt fence installations to be inadequate at retaining stormwater and associated sediments. The study of 56 randomly selected sites in 12 states, reflecting different soil types, different hydraulic conditions, and two different installation methods – trenching and slicing. Of the 56 sites, 30 were traditional trenched installations and 26 were sliced installations. Both installation methods are acceptable practices according to ASTM D6462, "Standard Practice for Silt Fence Installation and Maintenance." In general, the study found the slicing method provided more consistent results than trenching. This consistency was manifested in more uniform silt fence fabric burial depth, substantially higher level of compaction of the soil adjacent to the buried fabric, and greater frequency of satisfactory performance. Conversely, the trenched installations exhibited frequent inadequacy of both burial depth and soil compaction and, often, indications of being undermined by runoff. The study clearly showed that installation matters and, all too often, it is inadequate. In addition to ASTM D6462, the AASHTO M288 specification has a well-regarded specification for silt fence materials and installation. In both, the specifications emphasize the importance of both proper installation and routine maintenance.

4.3.6 Sediment Ponds and (Geosynthetic) Floating Pond Skimmers

A discussion of sediment retention devices would be incomplete without referring to what is still the most common approach to construction site sediment control – the sediment pond. The purpose of a sediment pond is to detain runoff waters and trap sediment from erodible areas to protect properties and drainage ways below the installation from damage by excessive sedimentation and debris. The water is temporarily stored, and the bulk of the sediment carried by the water drops out to be retained in the basin while the water is gradually released.

Still, critical to the optimal performance of a sediment pond is sufficient retention time of the runoff to allow for extensive sedimentation to take place in the pond, and to facilitate the clearest water being discharged from the pond. Retention time within the sediment pond is regulated by the pond discharge system, and the traditional discharge system has been a perforated riser pipe that discharges from the bottom of the pond as the pond fills. More recently, a more effective geosynthetic discharge system has emerged – the floating pond skimmer.

The floating pond skimmer operates at the surface

Figure 4.9 Floating Pond Skimmer Installed (Rymar Waterworks)

Figure 4.10 Floating Pond Skimmer Installed (Rymar Waterworks)

of the ponded water and will not withdraw sediment from the submerged volume of the basin. A floating pond skimmer "skims" water at a controlled rate of flow from the surface of the sediment pond where sediment concentrations are at a minimum in the water column, instead of draining from the bottom where sediment concentrations are their highest. A well-designed skimmer improves the performance of a sediment pond or basin by reducing retention time associated with meeting a desired water quality standard, discharging cleaner water, and providing consistent, predictable draw down times. A typical floating pond skimmer consists of three main components: a flexible coupling, a rigid tube that serves as the inlet, and floating headworks that serves to support the inlet at or near the surface of the impounded water.

Floating Pond Skimmer Design Criteria. If the volume of the sediment pond and the required number of hours/days to drain the basin is known, then the candidate skimmer discharge rate can be checked to verify that pond dewatering will be completed in the appropriate time interval. This is a product-specific verification, because each pond skimmer product (and each product size) has a unique design, including the associated hydraulics that are affected by the floatation, inlet, and connecting tube/coupling designs chosen. Thus, the discharge rate is dependent on the specific product/size and can only be determined through product-specific, full scale, "as installed" testing. ASTM D8107 – "Standard Practice for Measurement of Floating Pond Skimmer Flow Rate" is the test method used to characterize pond skimmer flow rate.

4.4 Long-term Applications – Erosion Resistant Hydraulic Structures

In highly erosive hydraulic applications, such as swift moving currents or high energy waves, large rock or precast concrete units that interlock or are cabled together have traditionally been used as the primary "energy absorbing" component of erosion resistance hydraulic structures. Yet, under these extreme conditions, the traditional rock and concrete solutions have proven to provide only short-term protection. This is because of the material and installation inconsistencies and susceptibility of the structures to undermining. Thus, at a minimum, geotextile filters are now routinely being placed under riprap and concrete units to protect them from being undermined by the scouring of the underlying subgrade. And, increasingly, hard armor is being replaced by geosynthetic-enhanced erosion resistant engineered systems for shoreline and coastal defense. Long-term geosynthetic-enhanced erosion control systems provide durable, flexible, and permeable erosion protection for:

> Coastal shorelines
> Earthen dams
> Storm channels and ditches
> Bulkhead
> Pipeline crossings
> Lakes and reservoirs
> Rivers, streams and bayous
> Dikes and levees
> Bridge abutments
> Water control structures
> Ponds and holding basins
> Sand dunes
> Embankments
> Spillways
> Wildlife channel crossings
> Boat launch ramps
> Berthing structures

Along with improved hard armor installations that incorporate a geotextile filter system, long-term/non-degradable geosynthetic-enhanced erosion resistance

hydraulic structures are routinely being used in challenging hydraulic conditions. These geosynthetic-enhanced systems include engineered synthetic turf systems, geocellular confinement systems, anchored geotextiles, geosynthetic cementitious composite mats, fabric formed concrete revetments, and geotextile-encapsulated sand elements. Following are brief descriptions of each type of geosynthetic-enhanced system, and Table 4.5 provides some advantages and disadvantages of each system. Table 4.6 compares the relative performance limits associated with the various erosion resistant systems when exposed to channel flows.

Geotextile (GTX) Filter Systems - A properly designed geotextile performs the function of dynamic filtration when placed on the soil surface beneath a hard armor system. The geotextile provides support for the armor layer over the seepage-induced softened subgrade as the water surface rises, and when the water surface recedes, seepage from the subgrade cannot carry soil particles with it which could cause undermining of the armor layer.

Engineered Synthetic Turf (EST) - Engineered synthetic turf is fabricated by attaching synthetic fibers, manufactured to resemble natural grass, to a backing material to create a simulated turf surface. The fibers are typically made from nylon, polypropylene or polyethylene and the backing material is typically a durable geotextile. The synthetic turf is then deployed in a system that may include an underlying structured geomembrane for drainage and impermeability and in-fill to protect the turf and provide ballast to the system.

Geocellular Confinement Systems (GCS) - Geocellular confinement systems, often called geocells, are made of strips of polymer sheet or geotextile connected at staggered points so that when the strips are pulled apart, a large honeycomb mat is formed that can be filled with soil, rock, or concrete. Geocell thickness (depth) typically ranges from 2 to 12 in. (5 to 30 cm).

Anchored Geosynthetic System (AGS) - An anchored geosynthetic system is a soil slope stabilization technique aimed at shallow slope instability problem areas. It is an in-situ technique in which a geosynthetic material is placed on the unstable or questionable slope and anchored to it with high-capacity anchors at discretely reinforced nodes, typically 3 to 6 ft (1–2 m) apart. To provide slope stability in addition to surface erosion control, these anchors must be long enough to penetrate the actual or potential failure surface.

Geosynthetic Cementitious Composite Mats (GCCM) - Geosynthetic Cementitious Composite Mats are factory assembled geosynthetic composites consisting of

Table 4.5 - Some Potential Advantages and Disadvantages of Longterm/Nondegradable Geosynthetic-enhanced Systems

Geosynthetic System	Potential Advantages	Potential Disadvantages
Geotextile Filters	Low cost addition to hard armor systems; Longterm (indefinite); All flows.	Can be damaged during installation; High material and environmental cost of hard armor.
Engineered Synthetic Turf	Impermeability; Vegetation Alternative; Aesthetics; Landfill gas capture when utilized with welded membrane	High; infill replacement and maintenance; no vegetation establishment; service life
Geocellular Confinement Systems (GCS)	Longterm (indefinite); Low to Moderate flows (soil filled); Moderate to High flows (concrete filled); Low to moderate sediment yields; Durable & Low maintenance.	Moderate to high costs; When filled with concrete, no vegetation establishment and infiltration ; Special deployment and equipment requirements
Anchored Geosynthetic System (AGS)	Increased surficial slope stability; densification of soil surface; allows for vegetation establishment	Moderate to high cost; Maintenance required to re-drive or tighten anchors if pretensioning is lost due to high-water content cohesive soils
Fabric Formed Concrete Revetments (FFCR)	Longterm (indefinite); Moderate to high flows; Low to moderate waves; Durable & Low maintenance; Low sediment yields.	High to very high costs; No vegetation establishment for some styles; Some styles prevent infiltration; Special deployment/equipment req'ts
Geotextile-encapsulated sand elements	Longterm (indefinite); High to very high currents and waves; Durable & Low maintenance.	High to very high costs; Limited vegetation establishment; Special deployment/equipment requirements

Table 4.6 Typical Performance Limits of Various Systems in Channel Flows

Erosion Control System	Limiting Shear (psf)	Limiting Shear (Pa)
Erosion Control Blankets	1.5 - 3.0	75 - 150
Natural Vegetation	up to 2	up to 100
Unvegetated TRM	**< 3**	**< 150**
Fully Vegetated TRM	**> 10**	**> 500**
Geocellular Confinement[#,*]	4 – 20+	190 – 950+
Fabric Formed Revetments*	4 – 20+	–190 - 960+
Riprap*	4 – 15+	–190 – 720+
Gabions[#,*]	12 – 20+	570 - 960+
Articulating Block Mats*	6 - 25	290 - 1200

[#] Depending on fill material; [*]Depending on size and thickness of units.

a cementitious layer contained within a layer or layers of geosynthetic materials that harden after hydration. GCCM's provide the utility of rolled geosynthetic systems for transport and deployment, and the strength, stiffness, and impermeability of concrete once hardened, or cured.

Fabric Formed Concrete Revetments (FFCR) - Fabric formed concrete revetments take advantage of low-cost, durable, (typically) synthetic fabrics to produce 3dimensional forms for casting concrete slabs. By pumping a very fluid fineaggregate grout into a fabric envelope consisting of 2 layers connected by tiechords or by interweaving at points, a concrete mattress can be constructed in minutes. FFCRs provide the durability of rigid linings, such as castinplace concrete or asphaltic concrete, and the flexibility and/or water permeability of protective rock systems such as riprap or gabions.

Geotextile-Encapsulated Sand Elements (GeoSystems) - Geotextile-encapsulated sand elements are three-dimensional systems manufactured from geotextiles and filled with sand. There are four types of elements: geotextile bags, geotextile mattresses, geotextile tubes, and geotextile containers.

4.4.1 Geosynthetic-enhanced Riprap and other Hard Armor: Geotextile Filter Systems

Protective rock systems such as riprap, precast concrete units, and rock filled baskets (gabions) are commonly used to resist the erosive forces of concentrated flowing water at the end of culverts, at the toe of riverbanks, or even to line steep channels. Generally these systems are cost effective only in areas where rock is readily available and when the site is easily accessible to heavy equipment. Still, these systems have their limitations.

It is well understood that a properly constructed armor system includes a filter layer placed between the bank soil and the armor to prevent piping. Typical installations are shown in Figures 4.11 and 4.12. Traditional filter layers have been graded sand and aggregate layers. These graded filters are very costly to construct because they are constructed of select graded materials. Also, the filter layer must be a controlled thickness. On a steep slope, it can be very difficult to properly construct. For these two reasons, aggregate filter layers are often inadequate or missing. Geotextiles overcome the drawbacks of graded sand and aggregate filters. First,

they can be easily selected with the correct hydraulic properties to complement the soil that needs protection. Secondly, they can be installed with ease on slopes – even under water. In addition to these benefits, geotextiles have a smaller carbon footprint than granular filters.

4.4.1.1 Simplified Generic Specifications for Routine Applications

Geotextile filters are discussed in detail in Chapter 3. It is important to note that in erosion control applications, filters are generally exposed to dynamic flow conditions. Additionally, filter specification should take into account both construction survivability and in-service performance. For example, mechanical properties of the geotextile may need to protect against damage from placement of the armor layer from excessive drop heights while the filtration properties will need to provide for sufficient clogging resistance and permeability.

4.4.1.2 Site Prep and Geotextile Deployment

The slope or channel should be graded smoothly and compacted, if possible. Unroll the geotextile on the prepared soil. The geotextile should be placed parallel to small ditch and stream alignments and perpendicular to lake or ocean shores. This alignment minimizes the exposure of the geotextile to current or wave uplift. Overlap the geotextile a minimum of 1.5 ft (0.5 m) on dry ground and 3.0 ft (1.0 m) in submerged installations in order to provide continuous erosion protection. Secure the geotextile in place using pins or staples, fill material or rocks. The spacing of pins or staples should be based on the soil and anchor type, but typically is no more than 18 in. (45 cm).

4.5 Engineered Synthetic Turf (EST)

Among the most difficult erosion challenges is establishing a stable, dense vegetative cover on the steep side-slopes associated with closing and capping a landfill. It is difficult enough to place and compact a geotechnically stable soil cover while accounting for interface friction, drainage, and gas uplift issues, but it can be equally as challenging to obtain and add a high quality top soil layer and to nurture a vegetative cover on those steep slopes while being subject to the erosive forces associated with heavy rain and high winds. Thankfully, a geosynthetic system has been developed to address this and similar large-expanse erosion challenges – engineered synthetic turf (EST).

Engineered synthetic turf is fabricated by attaching synthetic fibers, manufactured to resemble natural grass, to a backing material to create a stable simulated turf surface. The fibers are typically made from nylon, polypropylene or polyethylene and the backing material is typically a durable textile. The synthetic turf is then

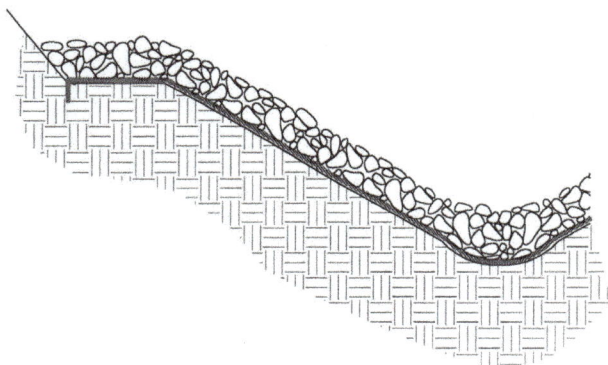

Figure 4.11 Geotextile Filter under Channel Armor

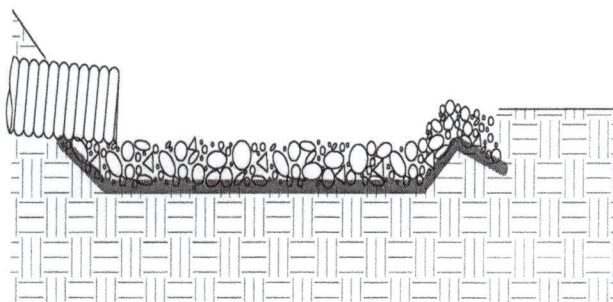

Figure 4.12 Geotextile Filter under Pipe Scour Armor

Figure 4.13 Engineered Synthetic Turf Landfill Cap (WatershedGeo)

Figure 4.14 Engineered Synthetic Turf Downshoot (WatershedGeo)

deployed in a system that may include an underlying structured geomembrane for drainage and impermeability and in-fill to protect the turf and provide ballast to the system. Engineered synthetic turf systems can be installed rapidly with much less environmental impact than traditional construction. EST systems provide exceptional erosion resistance, even under high flows such as landfill down chutes, and at much less cost than hard armor systems. These systems require minimal long-term maintenance and provide extended functional longevity. Thus, they provide the aesthetic benefits of vegetation without the maintenance of vegetation.

4.5.1 Designing and Specifying

Designing and specifying EST systems are typically done in collaboration with system suppliers, considering site conditions and regulatory requirements.

4.6 Cellular Confinement Systems (CCS) - a.k.a. "Geocells"

Cellular Confinement Systems (CCS), or simply Geocells, are honeycomb products manufactured by joining polymeric strips or geotextile strips by welding, gluing or stitching to create a three-dimensional, compartmentalized, polymeric structure having discrete cells that are formed by expanding the structure, that is subsequently filled with soil, aggregate, concrete, or other in-fill material. Geocells provide lateral confinement to the soil or other material in-filled into the cells (or act as formwork for concrete in-fill). Geocells can be used on arid slopes, when a thick topsoil layer is required for allowing vegetation growth. Geocells have many practical uses, including:

> Geocells filled with topsoil could be deployed on steep slopes, berms, levees, chutes, aprons and spillways to create an erosion-resistant vegetation layer.
> Geocells filled with structural fill and vegetation could be deployed to create vegetated slopes that could carry traffic loads
> Geocells filled with sand and other granular material could be deployed on gradual slopes to create stable erosion-resistant slope faces.
> Geocells filled with coarse aggregate could be deployed in channels and on slopes, except for severe grades, to create erosion-resistant faces under moderate sheet flow.
> Geocells filled with concrete could be deployed around bridges, on severe slopes, and in high flow rate channels, spillways and chutes to created erosion-resistant surfaces.

Geocell systems used to create an erosion-resistant surface for inclined slopes must be designed to resist downslope sliding forces (caused by infill weight) through friction of the geocell/infill composite on the slope face and an array of anchors or tendons, or combinations thereof, anchored along the crest of the slope or berms along the slope. The configuration of the geocell and the required anchoring method to the slope should ensure a sufficient factor of safety against downslope sliding

Figure 4.15 Geocells in Slope Erosion Control (Geo Products)

Figure 4.17 Tensioned anchor holding spider netting (Geosynthetic Institute)

Figure 4.16 Geocells in Channel Erosion Control (Geo Products)

Figure 4.18 Earth anchors (Gripple Inc.)

4.6.1 Design and Specification

Designing with and specifying geocells are typically done in collaboration with system suppliers, considering site conditions and expected hydraulics. Design of geocell systems used to create erosion-resistant surfaces on in channels subject to high concentrated flows depends on the infill-geocell-flow interaction and should be evaluated through performance testing. Large-scale erosion testing as described in 1.2.1 is an example of appropriate performance testing. Limited experimental data has suggested that erosion resistance (maximum permissible flow velocity) of the geocell system is clearly dependent on the in-fill material used. In general, coarse aggregate in-fill can be used up to 10 ft/s (3 m/s); vegetated soil up to 20 ft/s (6 m/s); and concrete at flows higher than 20 ft/s (6 m/s).

4.7 Anchored Geosynthetic System (AGS), a.k.a. Anchored Spider Netting

An anchored geosynthetic system (AGS) is a soil slope stabilization technique aimed at shallow slope instability problem areas. It is an in situ technique in which a geosynthetic material (generally a geotextile or HPTRM) is placed on the unstable or questionable slope and anchored to it with high capacity anchors at discretely reinforced nodes, typically 3–6 ft (1-2 m) apart. To provide surficial slope stability in addition to erosion control, these anchors must be long enough to penetrate the actual or potential failure surface. Anchored spider netting is an example of a slope stability AGS (Figure 4.17). When the anchors are properly tensioned, the surface geosynthetic is pulled against the soil placing the geosynthetic in tension and the contained soil in compression. A variety of anchor types can be used, including cables with earth anchors (Figure 4.18).

Depending on the site conditions, the slope can be seeded either before or after the placement of the geosynthetic, although seeding before is generally preferred. Growth of vegetation through the geosynthetic enhances the long-term surface stabilization of the slope. For long-term slope stabilization, particularly with high water-content cohesive soils (silts, clays, and their respective mixtures) it is necessary to return to the slope periodically to redrive/retension the anchors. This is required because of the long-term consolidation characteristics of high water-content fine-grained soils, as described previously. This aspect of the system must be carefully tuned to the local conditions but will result in a greatly stabilized site.

4.8 Geosynthetic Cementitious Composite Mat (GCCM)

ASTM and ISO define a GCCM as 'a factory-assembled geosynthetic composite consisting of a cementitious layer contained within a layer or layers of geosynthetic materials that becomes hardened when hydrated'. A GCCM provides unique advantages when a hardened protective surface is required, but conventional concrete is difficult or impossible to obtain or install. A GCCM combines the ease of installation of a geosynthetic during installation with the durability of hardened concrete long-term. It is a flexible cement-impregnated

Figure 4.19 GCCM deployed as slope protection (Concrete Canvas)

Figure 4.20 GCCM deployed as a channel liner (Concrete Canvas)

composite that hardens when hydrated to form a thin, durable concrete layer.

A GCCM can be especially useful if faced with one or more of these project conditions:

> Your location is remote, or difficult to access with standard concrete equipment.
> You have limited time, and need a simple, quick installation that does not require a return visit to the site to remove forms.
> You have limited labor or equipment and need a solution that requires only a small crew and basic equipment.
> You cannot stop the water flow, so the solution can be installed underwater.

4.8.1 Material Selection

Appropriate GCCM thickness will depend on site-specific conditions, including the expected flow rates, traffic, subgrade support, and durability requirements. Thin GCCMs, typically less than 0.32 in. (8 mm) in thickness, may be adequate for noncritical erosion protection or remediation and non-trafficked slope protection while thicker GCCMs may be used for more robust protection applications. ASTM D8364, Standard Specification for Geosynthetic Cementitious Composite Mat (GCCM) Materials covers the requirements and properties for GCCMs in various applications.

4.8.2 Handling and Installation Considerations

Another international standard, ASTM D8173-18 Standard Guide for Site Preparation, Layout, Installation, and Hydration of GCCMs, provides important handling and installation guidance associated with the use of GCCMs and is the basis for this section.

It is important to recognize that GCCMs harden when exposed to water and therefore should be properly stored to prevent exposure to water and moisture prior to the desired installation. Also, once the manufacturer's original sealed packing is opened, the material should be deployed within 24 hours and, if there is leftover material, the leftover material should be rewrapped to prevent moisture contact prior to subsequent use. Another issue with handling and deploying large GCCM rolls is that they may require heavy equipment and a spreader bar. Alternately, large rolls may be cut into smaller sections or rolls on-site that may be handled manually, depending on weight and local lifting regulations.

Prior to actual GCCM installation, the subgrade should be graded and compacted so that it is firm and free from debris, sharp or protruding rocks, and vegetation, including roots. Typically, GCCMs should be deployed in a specific orientation to allow for hydration. It is also important that the GCCM lie smooth and flat and be in intimate contact with the subgrade and be properly joined to minimize the potential for undermining of the GCCM, especially in channel armoring applications. All unjointed edges should be properly installed with an anchor trench, or alternatively secured in such a way that liquid, wind, or both cannot get under the installed mat. Installation of the GCCM in water conveyance channels usually begins at the lowest elevation, with successive segments installed as installation proceeds upstream.

As noted above, adjacent panel segments of GCCM should be jointed (shingled or butted) to minimize the potential for water seepage between adjacent layers. Segments are usually placed in a shingled joint - upstream edge over downstream edge. A simple shingled joint is often sufficient to minimize seepage if continuous contact between the segments (no gaps or ripples) is maintained. Water may still be able to seep through the joint in the event of hydrostatic pressure buildup beneath or ponding above the GCCM, but this can be advantageous when accommodating rising and falling water tables. To make joints of this type soil tight, an underlayment such as a nonwoven or other geosynthetic material may be used to reduce the loss of fines through the joint in the event of seepage. Seepage-resistant joints can be made by incorporating corrosion-resistant fasteners and adhesive sealants together in the segment overlaps. It may also be acceptable to use butt joints where adjacent segments meet as long as the GCCM bottom layer extends beyond the rest of the mat to provide a "flap" that can extend under the adjacent panel to provide segment-to-segment continuity when deployed. Segments should be closely touching when forming a butt joint and adhesive or other bonding technique should be used between the bottom layer flap and the bottom layer of the adjacent piece of GCCM.

4.9 Fabric Formed Concrete Revetment (FFCR)

Conventional rigid linings, such as cast-in-place concrete, asphaltic concrete, grouted riprap, stone masonry, and soil cement, can all be considered nonerodible. However, these treatments are usually expensive, and tend to progressively fail when a portion of the lining is damaged. Once a rigid lining deteriorates, it is very susceptible to erosion damage because large, flat, broken slabs are easily moved by channel flow. Lining deterioration and structural instability usually develop as a result of poor subgrade conditions, such as settlement, swelling soils, embankment slumping, frost heave, or hydrostatic uplift. Repair of rigid linings is often expensive and time-consuming. In response to

the questionable economics and durability of conventional rigid linings, fabric formed concrete revetment (FFCR) systems have emerged. FFCR systems combine the durability of conventional rigid linings with the more flexible and/or water-permeable properties of rock systems. When facing conditions such as limited access, remote region construction, or the need for quick installation using unskilled labor, even under water, FFCRs are often the optimal solution. FFCRs are made by pumping high strength fine aggregate concrete into fabric forms to form blocks within a network of fabric cells.

4.9.1 FFCR Hydraulic Properties and System Selection

FFCR hydraulic properties include a modest roughness that provides greater hydraulic efficiency than rock systems. By offering less resistance to flow, FFCRs are self-cleaning, require minimum maintenance, and provide a clean and environmentally attractive appearance which is far less hazardous to pedestrians and animals than stone riprap. The result is a less expensive erosion

Figure 4.21 FFCR deployed as slope protection (Huesker)

Figure 4.22 FFCR deployed as a channel lining (Huesker)

Table 4.7 Fabric Formed Concrete Revetment Selection Considerations

Fabric Form Type	Flow Velocity	Bedload and Ice Formations	Subgrade Support	Roughness coefficient "n"	Wave Action	Underwater Placement	Allows Seepage and Drainage
FPM	Low	Light	Well compacted	0.025-0.030	Light	Yes	Yes
USM	Low	Light	Well compacted	0.015	Light	Yes	If weep tubes added
ABM	Moderate to High	Light to Heavy	Can tolerate moderate deformation	0.045-0.050	Light to Heavy	Yes	Yes

control system which conveys stormwater much more efficiently than stone riprap linings. Thus, large areas can be drained more efficiently, quickly, and at a lower installed cost.

FFCRs common in many styles to accommodate various design conditions, but the three most common are: Filter Point Mat (FPM), Uniform Section Mat (USM), and Articulating Block Mat (ABM). Regardless of the style there are basic design concepts and methodologies for proper selection of mat style and thickness to successfully resist erosive forces associated with waves and currents. Generally, the design uses common analysis techniques for channel flow or current and wave impact, but selection of the most appropriate FFCR system will depend on many considerations such as shown in Table 4.7.

4.9.2 FFCR Construction

There are three basic steps to be taken in any fabric formed concrete revetment installation:

1. Site preparation
2. Panel placement and field assembly
3. Structural grout pumping

Slope grading equipment, such as backhoes or drag lines, is used to excavate to required depths, contour the slopes to the specified slope ratio, and form the anchor, toe, and terminal trenches around the periphery of the installation. The graded slope should be free of rock, brush, roots, or large soil clods. The area above the anchor trench should be graded so that surface water will run off and not saturate the slope.

A geotextile filter is typically deployed over the prepared slope, followed by the fabric forms. The fabric form panels can be quite large to minimize joining on the site, so their deployment can require both equipment and manpower. The panels are positioned according to a carefully prepared drawing. The top of the panels is typically positioned before unrolling the panel down the slope and then further unfolding and positioning side-to-side by a work crew. The panels should be positioned loosely along the slope. Wrinkles and loose fabric should be expected as they are necessary to compensate for form contraction. Once positioned, the upper edge of the panel is folded into the anchor trench atop the slope. The extra fabric provided for contraction during pumping should be accumulated and held at the top of the slope and gradually released as the form is filled.

Adjacent panels are joined by field sewing or zippering the double-layer fabric forms, bottom edge to bottom edge, and top edge to top edge. All field sewn seams, zipper connections, and lap joints must be carefully inspected. Colored thread contrasting to the color of the fabric is advised for all field sewn seams to facilitate inspection. As the first two panels are being joined, the third is positioned so that the seaming crew can start on it, upon completion of the first seam. This procedure is continued until all field seams are made. No more material than can be pumped in one day should be unrolled and positioned. The upper edge of the fabric form panel which has been placed into the anchor trench should be pumped first, thus forming an anchor to prevent the remainder of the form from sliding down the slope as it is pumped with structural grout. The pumping crew should then inject grout into the lower mat area, proceeding gradually up the slope until the fabric form has been filled.

Structural grout is injected into the fabric form by inserting a 2 in. (5 cm) diameter grout injection pipe through a small slit cut in the upper layer of fabric. A grout tight seal is formed by wrapping the injection pipe with burlap, which is held in place by a laborer as grout is being injected. When the pipe is withdrawn, the burlap is stuffed in the hole where it remains until the grout stiffens to a point that it is no longer fluid. The burlap is then removed and the concrete surface at the hole smoothed by hand.

4.9.3 Concrete Filled Fabric Bags

Fabric formed concrete is not limited to mats. Fabric bags can be filled with concrete and used as 'building blocks' for numerous applications. Freshly filled bags of concrete are easily joined by 'stabbing' a smooth pointed reinforcing dowel into a filled bag and inserting the exposed end through the fabric of the succeeding bag. Two bags may be joined side-by-side using re-bar 'staples'. Bars of full cages may be inserted in the bags through openings in the fabric, the opening being closed before filling by means of a portable sewing machine or zipper. Cages are centered in the bag by suspending them from the upper surface of the bag

4.10 Geotextile-encapsulated Sand Elements

Small geotextile containers, such as sandbags, have been used for decades primarily as temporary construction devices. More recently, the design and construction of very large geotextile containers has gained popularity because of their simple placement and construction, cost effectiveness and minimum impact on the environment. The containers may be large hand-filled bags, prefabricated hydraulically-filled tubes, or site-fabricated mechanically-filled containers. A variety of fill material types, including concrete, gravel, sand, and, most recently, fine-grained materials having very high water content have been used. Still, for constructing large-scale, cost-effective erosion resistant structures, the fill of choice tends to be sand because it is easily transported and facilitates a dependable long-term geometry, and the finished units can be used as engineered elements in a "geosystem". Thus, the finished units are commonly referred to as geotextile-encapsulated sand elements.

According to Bezuijen and Vastenburg (2013), geotextile-encapsulated sand elements are three-dimensional systems manufactured from textile materials that are filled with sand and used as elements in erosion-resistant hydraulic engineering structures. These elements include: Geotextile bags, Sand Filled Mattresses, Geotextile tubes and Geocontainers. Further, the use of these elements has the advantage that local materials can be utilized and that no stone needs to be extracted and transported from a quarry to the site, potentially adding considerable operational advantages to the execution of hydraulic works, plus better financial and environmental aspects.

One question related to building with sand in particular is the potential for liquefaction. This is the physical phenomenon that the soil loses its strength when a shaking or other rapid loading causes the particles to spread, losing confinement/compaction. The mass of soil then behaves as a liquid. Once contained within a geotextile the sand particles are prevented from lateral spreading. Over time, sand becomes so tightly packed that no excess hydrostatic pressure is generated to cause liquefaction.

4.10.1 Geotextile-Encapsulated Sand Elements – Materials and Manufacture

Geotextile-encapsulated sand elements will typically be a somewhat flattened tube, or oval, in cross-section with the ends typically taking on the shape of a pillow when filled with sand unless restricted by an adjacent structure or tube. Hydraulic (dredge) filling under pressure or dry filling while placed in a large form enable greater height in the filled unit. The necessary material properties of the encapsulating geotextile are based on both short-term deployment and filling and long-term in-service stresses. Additionally, successful deployment and filling is very much dependent on proper manufacturing.

4.10.1.1 Materials

Initial loss of fine particles during filling process is acceptable and common, however there must be no loss of fill material over time. Thus, sufficient knowledge of the fill material (often the result of dredging) to assure compatibility with the geotextile components is essential. Of primary importance in characterizing the fill material is its permeability and particle size distribution.

Geotextile component compatibility involves matching fabric opening size, permeability, and tensile strength to the filling material and to the deployment and filling operation. For example, though the sand fill generally has a high relative density enabling efficient settling , the sand is often relatively fine-grained and therefore the fabric must have openings sufficiently small to retain small soil particles but large enough to allow efficient dewatering. Additionally, either higher filling pressures or vertical sidewall forms are necessary to increase the height of the cross-section, thus higher tensile strengths (especially at seams and fill ports) are required to achieve higher profiles. Strength is also an important consideration in surviving construction handling. The in-service conditions add additional considerations, including the geotextile must:

> retain the sand fill under dynamic hydraulic forces, such as waves;
> resist erosive forces that might undermine unit support;
> resist hydraulic abrasion by suspended sediments or mechanical abrasion;
> resist puncture and tearing by debris; and
> resist ultraviolet light degradation.

4.10.1.2 Manufacturing

The simplest geotextile containers involve two sheets of geotextile sewn along the edges with an open end, such as with bags. Larger, more sophisticated systems, such as mattresses, tubes, and site-fabricated containers will involve multiple geotextile panels assembled with heavy-duty seams and specially designed inlets or ports. These larger systems are commonly manufactured for geotextile panels having widths of 16 ft (5 m) or more and lengths only limited by project requirements and handling limitations. Permeable and impermeable liners have been used as an additional inner layer on some systems. Impermeable liners increase the velocity in the containers for material transport over long distances when hydraulically filling tubes. The manufactured units are either folded and placed on pallets or rolled for ease of transport and handling.

The most critical elements of manufacturing are the seams and the fill ports. The seam strength is normally the weakest link in the design and may be only half of the fabric ultimate strength ,depending on the seaming technique. Inlet/outlet ports are points of maximum pressure, and thus, must be carefully manufactured. Ports are normally spaced along the length of the tube at interva s suitable for the sedimentation characteristics and angle of repose of the dredged materials. Generally, port spacing is less than 50 ft (15 m).

4.10.2 Geotextile Mattresses

Geotextile mattresses can be uniquely appropriate for river and stream applications such as bank protection or scour hole repair in areas where sand is plentiful and a protective surface layer of sand and/or vegetation is possible. Geotextile mattresses are comprised of two interconnected layers of geotextile where the space between is filled with sand and, in localized areas, concrete. Ce ls, chambers or tubes form compartments within the mattress, which facilitates an even distribution of the fill material in the geotextile mattress and maintains its shape and combats movement of the fill material during use. It's important that the mattress geotextile be secured in place with anchor trenches and steel anchors along the length of the slope. They can

Figure 4.23 Geotextile mattress installed on a slope (Solmax)

Figure 4.24 Deployment of prefilled mattress (Solmax)

Figure 4.25 Smaller geotextile bags deployed (Solmax)

Figure 4.26 Special handling of large bags (Solmax)

be filled in place or filled separately and transported to the slope.

4.10.3 Geotextile Bags

Geotextile bags have been used for small coastal and river applications. Smaller bags are typically used for emergency repairs, while larger bags, made from durable geotextiles provide longer-term solutions, especially when covered with a protective layer of soil or rock.

4.10.4 Geotextile Tubes

A geotextile tube can be defined as "a large tube [greater than 7.5 ft (2.3 m) in circumference] fabricated from high strength, woven geotextile, in lengths greater than 20 ft (6.1 m)", according to GRI Test Method GT11: Standard Practice for "Installation of Geotextile Tubes used as Coastal and Riverine Structures." Dredged material filled geotextile tubes are constructed by hydraulic filling of the tubular geotextile envelope with a water-soil mixture using a cutter suction pipeline dredge. The dredged material filled geotextile tubes can also be pre-filled and placed using a "cradle" bucket on a barge-mounted crane or they can be installed using a continuous position-and-fill procedure. Geotextile tubes are better suited for larger coastal applications requiring large, lengthy erosion-resistant structures and are commonly used to assist in dike and groin construction. Tubes can be used as the core of these structures then covered by a durable armor system to provide resistance to degradation from debris impact, abrasion, or UV exposure. If the tube outer layer is durable enough, it can remain uncovered.

4.10.5 Geocontainers

For very large underwater scour problems, very large containers of sand, or Geocontainers, can provide an economical, practical solution for construction of both coastal and river erosion resistant structures. Geocontainers may be mechanically or hydraulically filled inside of bottom dump hopper scows, moored into place, and dumped. A geocontainer is a large geotextile-encapsulated sand element containing 130 to 1050 ft³ (100 to 800 m³) of sand and is dropped through water from a split bottom barge. The available barge determines the size of the container which, consequently, leads the

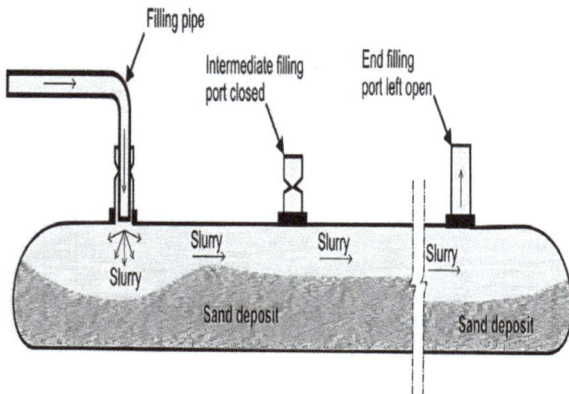

Figure 4.27 Geotextile tube filling schematic (Solmax)

Figure 4.29 Geocontainer in split-bottom barge (Solmax)

Figure 4.28 Geotextile tube deployed (Solmax)

Figure 4.30 Geocontainer dropped from split-bottom barge (Solmax)

final design. Geocontainers are commonly used when water depth is deeper than 10 ft (3 m).

4.10.6 Applications

As the technology associated with manufacturing, deploying and filling of geotextile-encapsulated sand elements has been demonstrated, additional innovative uses have been proposed, including a wide range of river, estuary, and shoreline/coastal geosystem projects.

4.10.6.1 River Applications.

Geotextile-encapsulated sand elements, because of their simplicity, flexibility, and resulting stability against erosive forces, provide an innovative, cost-effective alternative to the traditional techniques used in the construction of typical current-guiding and scour-preventing structures in rivers.

Current-Guiding Structures. Usually, river-guiding structures are constructed with natural rock. As an alternative solution, a structure can be built up by one or more geotextile-encapsulated sand elements. By varying size, number and composition of the elements, any structure can be realized. Current-guiding structures such as revetments, groins, and longitudinal dikes are used to prevent localized erosion, regulate the river bed to concentrate the river into one channel, and to normalize the river bed to fix the horizontal profile

Bottom Scour-Preventing Structures. Bottom scour can be prevented by securing the river bed with an armoring system. Usually, this bottom protection consists of a primary armor rock layer with one or more filter layers underneath to prevent sediment from passing through the rock protection. By filling in and covering a depression in the river bed with geotextile-encapsulated sand elements the same result can be achieved. The elements have to be positioned closely together to prevent current access and resulting sediment passage.

4.10.6.2 Estuary Applications.

An estuary is that general area at the mouth of a river into which the tide flows. Extensive wetlands (i.e. marshes and swamps) are characteristic of estuaries and are frequently threatened by erosion and subsequent salt-water intrusion. Efforts to protect, reclaim, or increase these areas can be enhanced by using longitudinal dike structures composed of geotextile-encapsulated sand elements in these areas where traditional construction techniques are ill-suited or cost-prohibitive. Dredged material can be used to both fill the dike elements and to fill in behind the dike structure to reclaim or increase wetlands, or to simply contain dredge spoils resulting from waterway dredging and forming an artificial island.

4.10.6.3 Shoreline/Coastal Applications

Beach renourishment projects often incorporate groins, breakwaters or sill structures as effective tools for trapping littoral drifting sediments which stabilizes the shoreline. Because of their flexibility, structural integrity and relatively large mass, geotextile-encapsulated sand elements are very suitable to be used as groins, breakwaters, or sills. Additionally, gullies, caused by tidal currents, can be interrupted or filled using geotextile-encapsulated sand elements.

Groins. Where there is an abundant supply of littoral drift, the function of a groin is to build or to widen a beach by trapping the sediment movement along the shore. Moreover, properly designed groins will reduce the long-shore transport of sediment. This is achieved by reorienting the compartmented shoreline, so that it will be in closer equilibrium to the predominant wave direction. Sand filled elements can be used as the core of these structures then covered by a rock armor or, depending on how durable is the fabrication material, the geotextile container can remain uncovered.

Offshore Breakwaters. Offshore breakwaters can be used to reduce the force and vary the direction of waves striking the shore, and thereby reducing shore erosion without harming the recreational aspects of the beach. Breakwaters are located away from the shore, and can either be submerged or the crest can be elevated above sea level during all tides. Breakwaters tend to reduce littoral transport along the shoreline side of the structure. Sand filled elements can be used as the core of these structures then covered by a durable armor system to provide resistance to degradation from debris impact, abrasion, or UV exposure. If the tube outer layer is durable enough, it can remain uncovered.

Sill Structures. A wide protective (perched) beach or shallow offshore can be retained by means of terracing with the construction of beach retaining sills. Wave energy is dissipated while propagating over this shallow region by breaking and bottom friction. Hence, waves have a reduced erosive effect upon the shoreline. Sand filled elements present positive interaction with the environment and can be used as a low sill solution as well as providing a beneficial reuse of dredged sediments.

Land Reclamation and Artificial Islands. Artificial islands or land reclamation, constructed with dredged material filled fabric tube systems present an additional use for both underwater construction techniques and shore protection schemes. Advantages for using the dredged material filled fabric tube systems include the reduction of required fill materials, speed of construction and relatively low costs. Removal is also easily accomplished as is often required for temporary energy exploration structures.

Gully Repair. Gullies are caused by tidal currents. In the vicinity of coastal structures they can cause geotechnical instability of these structures or provide a breach point from which the structure progressively erodes.

4.10.7 Design Guidance

Geotextile-encapsulated sand elements are relatively new and promising development in hydraulic engineering and are, thus, somewhat hampered by the lack of explicit design rules. In fact, most of the design rules are empirical in nature, making it especially challenging for an inexperienced designer to incorporate geotextile-encapsulated sand elements into new structures. Fortunately, some of the pioneers in the field have documented their expertise in these authoritative publications:

> "Geosynthetics and Geosystems in Hydraulic and Coastal Engineering"by Kristian W. Pilarczyk
> "Design Rules and Applications" by A. Bezuijen & E.W. Vastenburg

The design considerations and detailed approaches covered in the authoritative publications include:

> Geometry and installation issues that must be determined including the required structure geometry; key topographic information; hydraulic conditions such as water depth, currents, tides, and waves; construction accessibility and available equipment.
> Factors of safety will be incorporated into the project design relating to local and global stability; certainty of hydraulic conditions; variability of filling material; and, long-term durability of the encapsulating geotextile.
> Design experience and the tools and procedures that are available to thoroughly design a project. Experience is critical for identifying important material, manufacturing, and installation details and can be enhanced by working closely with experienced designers, manufacturers, and installers.

4.5 References

AASHTO (2023), *M288 – Geosynthetic Specification for Highway Applications*, American Association of Highway and Transportation Officials, Washington, DC.

Bezuijen, A. and Vastenburg, E.W. (2013), *Geosystems. Design Rules and Applications*, CRC Press/Balkema, The Netherlands.

FHWA (2014), "Standard Specifications for Construction of Roads and Bridges on Federal Highway Projects", FP-14 (U.S. Customary Units), U.S. DEPARTMENT OF TRANSPORTATION, Federal Highway Administration.

Fowler, J., and C. J. Sprague (1993), "Dredged Material Filled Geotextile Containers," Proceedings from Coastal Zone '93, ASCE, pp. 24152428.

Koerner, R. M. (2012), *Designing with Geosynthetics*, 6th Edition, Xlibris.

PEAS (formerly NTPEP) (2024), "DataMine," Product Evaluation and Audit Solutions, AASHTO, https://transportation.org/product-evaluation-and-audit-solutions.

Pilarczyk, K. (2000), *Geosynthetics and Geosystems in Hydraulic and Coastal Engineering*, AA Balkema, The Netherlands

Sprague, C.J. (1997), "Geotextile Containers for Erosion Control A Literature Review", Proceedings of Geosynthetics '97, IFAI, Long Beach, CA, pp. 135145.

Sprague, C.J. (1999a), "Assuring the Effectiveness of Silt Fences and Other Sediment Barriers", Proceedings of Conference XXX, International Erosion Control Association, Nashville, pp. 133-154.

Sprague, C.J. (1999b), "Green Engineering Part 1 – Optimizing Erosion Control with Vegetation and RECPs", *CE News*, February, pp. 54-58.

Sprague, C.J. (1999c), "Green Engineering Part 2 – Design Principles and Applications using Rolled Erosion Control Products", *CE News*, March, pp. 76-81.

Sprague, C.J. (2001), "Dredged-Material Filled Geotextile Tubes: Design and Construction", *GFR*, March, pp. 46-51.

Sprague, C.J. (2003), "A Field Study of Silt Fence Installations Around the United States", Conf. XXXIV, International Erosion Control Assoc., Las Vegas, pp. 359-368.

Sprague, C.J (2013), "The Evolution of Geosynthetics in Erosion and Sediment Control", GRI-25, Geosynthetic Institute Conference, Long Beach, CA.

Sprague, C.J., Allen, S.R. and Sprague, J.E., (2018), "Performance Evaluations of Erosion and Sediment Control BMPs using Independent Full-scale Simulations", *Proc. of the 11th International Conference on Geosynthetics*, IGS, Seoul, Korea.

Sprague, C.J., and Koutsourais, M.M. (1992), "Erosion Control Using Fabric Formed Concrete Revetment Systems," *Geotextiles and Geomembranes*, Vol. 11, Nos. 46, pp. 587609.

Sprague, C.J. and Sprague, J.E. (2015), "Testing and Specifying Sediment Retention Devices", *Proc. of Geosynthetics 2015*, IFAI, Portland, OR.

Sprague, C.J. and Sprague, J.E. (2016a), "Testing and Specifying Erosion Control Products", 2016 Conference of International Erosion Control Association, San Antonio, TX.

Sprague, C.J. and Sprague, J.E. (2016b), "Chapter 24: Geosynthetics in Erosion and Sediment Control", *Geotextiles, From Design to Applications* (1st Edition), Edited by R. Koerner, Elsevier (Woodhead), ISBN-9780081002216.

Sprague, C.J. and Sprague, J.E. (2023), "Hydraulic Roughness Testing of Geosynthetics ", *Proc. of Geosynthetics 2023*, IFAI, Kansas City, MO.

Sprague, J.E., and Sprague, C.J. (2017), "Evaluation of Anchor Systems Through Full Scale Pullout Testing", 2017 Conference of International Erosion Control Association, Atlanta, GA.

Sprague, J.E., and Sprague, C.J. (2020), "Comparative Analysis of the Flow and Filtration Capabilities of Sediment Retention Devices (SRDs)", Conf. L, International Erosion Control Association, Raleigh, NC.

Sprague, J.E., Sprague, C.J., and Ruzowicz, B. (2015), "Evaluating Floating Surface Skimmers", 2015 Conference of International Erosion Control Association, Portland, OR.

Chapter 5

Geosynthetics in Reinforced Soil Systems

5.0 Overview

5.0.1 Geosynthetic reinforcements

Specific geosynthetics can be used for reinforced soil applications. Figure 5.1 shows the fundamental concept of reinforced soil as it relates to a basal reinforced embankment on a soft foundation; but this same concept applies to other reinforced soil applications as well. Here, the geosynthetic reinforcement intersects potential failure surfaces that pass through the base of the embankment and through the geosynthetic reinforcement. In doing so, the geosynthetic reinforcement improves the stability of the embankment through its tensile behaviour. For basal reinforced embankments, the major form of bond resistance is through friction between the reinforcement surfaces and the adjacent soil.

The geosynthetic reinforcements must exhibit the following attributes in order for them to behave as reinforcements in the soil;

> Tensile Strength: The geosynthetic reinforcements must have the required tensile strength to resist the out-of-balance forces causing instability.
> Tensile Stiffness: The geosynthetic reinforcements must generate the tensile loads at tensile strains that are consistent with serviceability.
> Durability: The geosynthetic reinforcements must have the required in-ground durability for them to support the tensile loads and strains over the required design life of the structure.
> Temperature Resistance: The geosynthetic reinforcements must have the required strength, strain and durability behaviour at ambient in-ground temperatures encountered in the soil environment.
> Bond Strength: The geosynthetic reinforcements must form an efficient bond with the adjacent soil so that the tensile loads are readily redistributed into the surrounding soil.

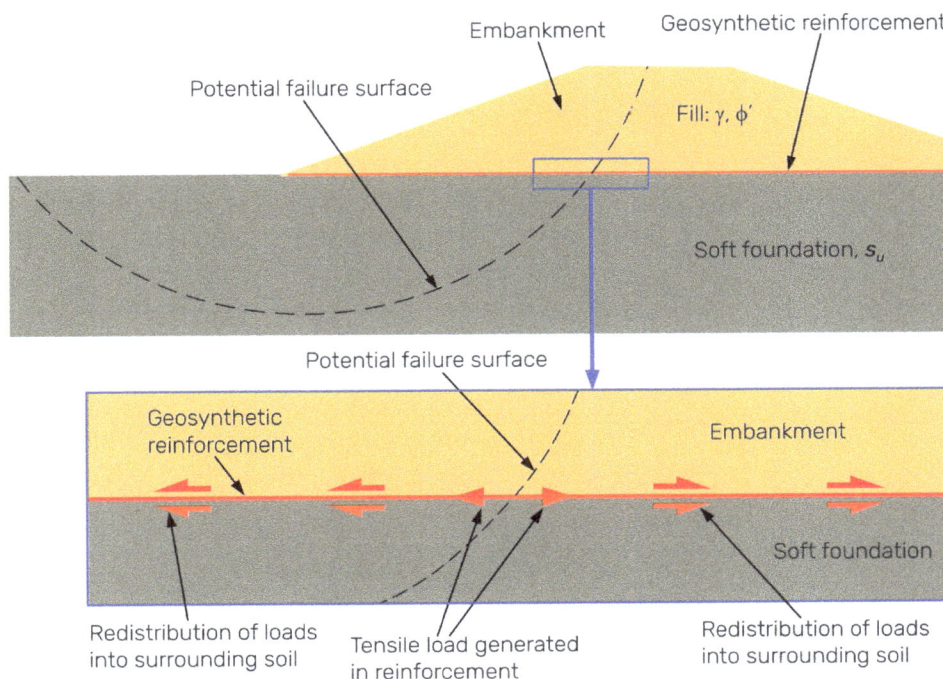

Figure 5.1 Fundamental concept of reinforced soil for a basal reinforced embankment on a soft foundation.

For certain reinforced soil applications, e.g. retaining walls, geosynthetic reinforcements are normally combined with other components, such as facing units, etc., to form a "system", where the system overall provides the stability for the structure (e.g. see Section 5.4). In these cases, reinforced soil is commonly referred to as a system.

5.1.2 Types of geosynthetic reinforcements

Chapter 1 of this IGS Handbook provides an overall description of the different types of geosynthetics materials, however, as stated above, geosynthetic reinforcements must carry tensile loads at defined strains over different required design lives at ambient in-ground temperatures. This fundamental requirement restricts the types of geosynthetics that can be used as geosynthetic reinforcements.

The element of time, i.e. design life, is a very important factor for geosynthetic reinforcements, because the attributes of strength, strain and in-ground durability are all related to time. Table 5.1 lists the typical required reinforcement design lives for different reinforced soil applications according to BS8006:2010 (which is a reinforced fill code of practice used throughout much of the English-speaking world), with the required reinforcement design lives ranging from 1 to 2 years to 120 years depending on the application. Thus, to be used for these different applications, the geosynthetic reinforcements must perform over these different design lives. For example, in the United States the Federal Highway Administration has also listed required design lives for

geosynthetic reinforcements according to its state highway requirements which range from 3 years for temporary structures to 100 years for permanent retailing walls. This again shows the wide range of required reinforcement design lives for different structures.

To fulfil the reinforcement function, the load carrying elements of the geosynthetic reinforcements must be able to sustain the applied tensile loads, at defined strains, over different required design lives, in an in-ground environment, and in a cost-effective manner. For this reason, the polymer load carrying elements that are used in geosynthetic reinforcements are limited to high modulus polyester (HMPET) yarns and polyester (PET) strips, high modulus polypropylene (HMPP) yarns, high density polyethylene (HDPE) and polypropylene (PP) extruded polymer sheets, polyvinyl alcohol (PVA) yarns and polyaramid (PA) yarns.

The typical tensile behaviour of these polymer load carrying elements is shown in Figure 5.2. It is observed that these different elements provide a wide range of tensile behaviour ranging from very strong and stiff reinforcements (e.g. PA yarns) to lower stiffness and strength reinforcements (e.g. extruded HDPE sheets). (What is interesting is that PA yarns have tensile behaviour similar to prestressing steel tendons.) What is to be remembered is that all these load carrying elements are suitable for geosynthetic reinforcements, it is just that a greater cross-sectional area may be required to achieve a required tensile strength.

There are three families of geosynthetics that are readily used for reinforced soil applications. These are woven and knitted geotextiles, geogrids and geostrip reinforcements, and are summarized in Chapter 1 of this handbook. These geosynthetic reinforcements all utilize the different polymer load carrying elements shown in Figure 5.2 to varying extents. Further, these three families of geosynthetic reinforcements have three beneficial structural characteristics in common. First, each of the geosynthetic reinforcement families have their polymer load carrying elements distributed in a linear, flat array within their structure and this enables them to directly reproduce the tensile loads and strains of the component polymer load carrying elements in an efficient manner. Second, the surface construction of these materials is designed to form an efficient bond with the surrounding soil. Third, if the polymer load carrying elements used are durable in soils over long periods of time, then the resulting geosynthetic reinforcements can carry the tensile loads at defined strains over long-term design lives.

Table 5.2 lists the basic structures of the three families of geosynthetic reinforcements along with the types of polymer load carrying elements used in each. The combination of different structure and different load carrying element gives the wide variety of geosynthetic reinforcement properties that are available.

Table 5.1 Required geosynthetic reinforcement design lives for different reinforced soil structures according to BS8006:2010

Reinforced soil structure	Required design life
Temporary structures	Several months up to 3 yrs
Basal reinforced embankments on soft foundations	1 to 10 yrs
Basal reinforced piled embankments	60 to 120 yrs
Basal reinforced embankments spanning voids	60 to 120 yrs
Reinforced soil slopes	30 to 75 yrs
Reinforced soil retaining walls	30 to 75 yrs
Reinforced soil walls and slopes with highway structures	120 yrs

5.1.2.1 Woven and knitted geotextiles

Woven and knitted geotextile reinforcements are multi-functional materials in that they can reinforce, separate and filter all within a single geosynthetic structure. This makes them highly beneficial for reinforced soil applications. Their structure is textile-based where one set of polymer load carrying elements lie flat in the longitudinal direction and the other set lie flat in the cross direction. Woven and knitted geotextile reinforcements can utilize a range of polymer load carrying elements as listed in Table 5.2 and these control the mechanical behaviour of the geotextile reinforcement. While it is possible to have the geotextile reinforcements with similar tensile strengths in both the length and cross directions (termed b directional or biaxial products), it is more common to have significantly higher tensile strengths in the length direction (termed unidirectional or uniaxial products), as these result in more efficient reinforcements.

Table 5.3 shows the wide range of initial tensile properties (strengths and strains) available for woven and knitted geotextile reinforcements. This wide range is achieved by utilizing different polymer load carrying elements arrayed in different densities within the geotextile structure.

5.1.2.2 Geogrids

Geogrid geosynthetic reinforcements are singular-function materials providing reinforcement (only) to soils. They have open, grid-like structures, as described in Chapter 1 of this handbook, which enables an efficient bond to be developed with the adjacent soil as well as the ability to efficiently utilize different polymer load carrying elements for good initial tensile strength behaviour. Their structure may be either textile-based, composed of polymeric sheets, or composed of welded polymer strips, as listed in Table 5.2. Textile-based geogrid materials are available in a wide range of tensile strengths and maximum strains (see Table 5.3) depending on the polymeric load carry elements used. Geogrids based on extruded polymer sheets of HDPE and PP are more limited in their range of tensile properties available because of manufacturing of punching and drawing of these (same) polymer sheets (see Table 5.3). Geogrids based on bonded polymer strips are dependent on the tensile behaviour of the polymer strips used and their relative spacings in the geogrid structure (see Table 5.3).

Geogrids may have similar strength properties in both the length and cross directions (termed bidirectional or biaxial geogrids) however, it is more common for soil reinforcement applications to require the length

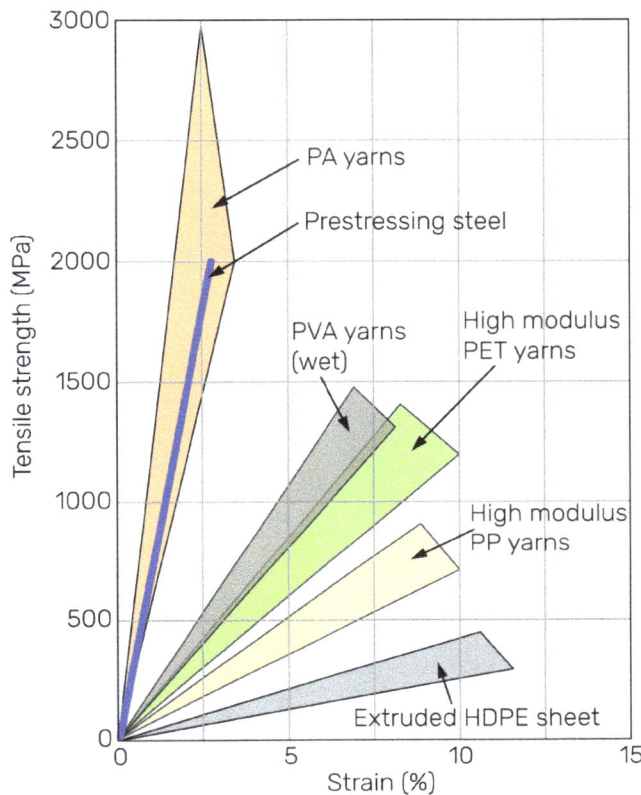

Figure 5.2 Tensile behaviour of different polymer load carrying elements used in geosynthetic reinforcements (after Lawson, 2022).

Table 5.2 Basic structures and polymer load carrying elements used for different geosynthetic reinforcements (After Lawson, 2022)

Geosynthetic reinforcement family	Basic structure	Polymer load carrying elements used
Woven and knitted geotextiles	Textile-based	Polyester (HMPET), Polyaramid (PA), Polyvinyl Alcohol (PVA) or Polypropylene (HMPP) yarns
Geogrids	Textile-based	Polyester (HMPET), Polyaramid (PA) or Polyvinyl Alcohol (PVA) yarns
	Extruded and drawn polymer sheets	Polyethylene (HDPE) or Polypropylene (PP) sheets
	Bonded polymer strips	Polyester (HMPET) or Polypropylene (PP) strips
Geostrips	Continuous yarns with polymer matrix or casing	Polyester (HMPET), Polyaramid (PA) or Polyvinyl Alcohol (PVA) yarns

Table 5.3 Initial tensile properties of current geosynthetic reinforcements (After Lawson, 2022)

Geosynthetic reinforcement type	Initial tensile strength (kN/m)	Strain at maximum load (%)
Woven and knitted geotextiles	80 - 5,000	3 - 12
Geogrids: Textile-based Extruded and drawn polymer sheets Bonded polymer strips	30 – 2,000 30 – 160 30 – 400	3 – 12 12 – 15 10 – 15
Geostrips	50 – 500 kN	3 – 12

direction significantly stronger than the cross direction (termed unidirectional or uniaxial geogrids). This results in a more efficient reinforcement layout for reinforced soil applications.

5.1.2.3 Geostrips

Geostrip reinforcements are also singular-function materials providing reinforcement (only) to soils. They are linear elements which consist of a polymer load carrying core surrounded by a polymer casing or coating.

The linear geostrips are used individually in certain reinforced soil applications, e.g. in proprietary reinforced fill retaining walls. The resulting tensile properties of geostrip reinforcements are dependent on the polymer load carrying elements used and their reinforcement packing density within the structure, see Table 5.3. Alternatively, these geostrip reinforcements may be bonded together to form 2-D planar reinforcements for basal reinforced embankment applications. Geostrip reinforcements have unidirectional tensile strengths only.

5.1.3 Reinforced soil applications

Geosynthetic reinforcements are used for a range of reinforced soil applications. In all these applications, the placed reinforcements intersect potential failure surfaces thereby enhancing the stability of the reinforced soil structure. Further, the tensile stiffness of the geosynthetic reinforcements limit strains and deformations in the reinforced soil structures to serviceable limits. Figure 5.3 shows the more common reinforced soil applications using geosynthetic reinforcements.

5.1.3.1 Basal reinforced embankments on soft foundations, Figure 5.3a

In this application a layer of geosynthetic reinforcement is placed on top of the soft foundation beneath the embankment to provide additional stability for the constructed embankment. The presence of the geosynthetic reinforcement enables the embankment to be constructed higher and/or with steeper side slopes than would be possible without the use of the basal reinforcement. This application is covered in further detail in Section 5.2.

5.1.3.2 Basal reinforced embankments on piles, Figure 5.3b

In this application a layer of geosynthetic reinforcement is placed at the base of an embankment over a pile foundation platform to improve stability and prevent settlements. This technique enables the embankment to be constructed to any height, at any rate, without stability and settlement problems. This application is also covered in further detail in Section 5.2.

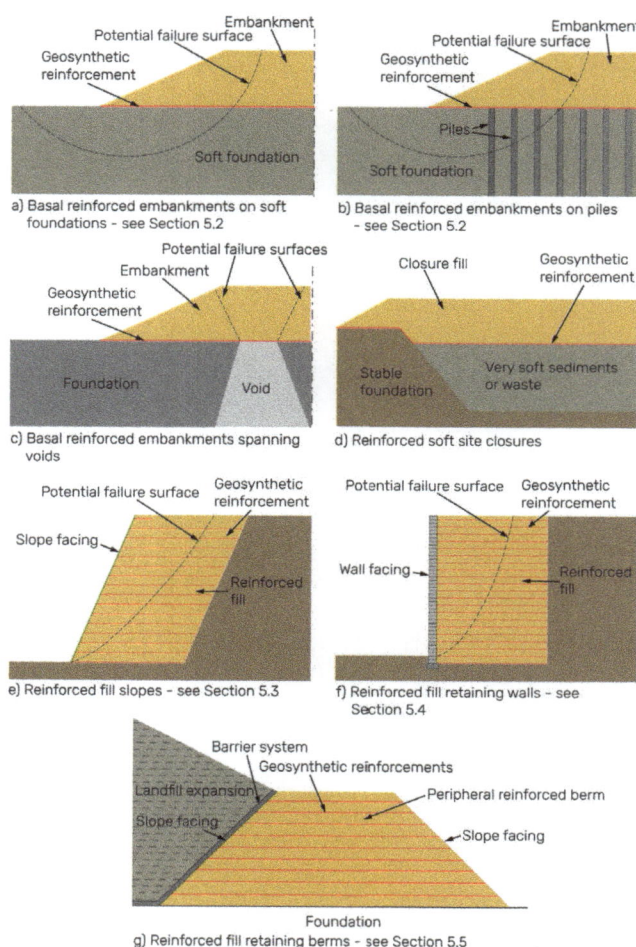

a) Basal reinforced embankments on soft foundations - see Section 5.2

b) Basal reinforced embankments on piles - see Section 5.2

c) Basal reinforced embankments spanning voids

d) Reinforced soft site closures

e) Reinforced fill slopes - see Section 5.3

f) Reinforced fill retaining walls - see Section 5.4

g) Reinforced fill retaining berms - see Section 5.5

Figure 5.3 Reinforced soil applications using geosynthetic reinforcements.

5.1.3.3 Basal reinforced embankments spanning voids, Figure 5.3c

In this application a layer of geosynthetic reinforcement is placed at the base of an embankment over a foundation that is prone to the formation of voids. Here, the geosynthetic reinforcement prevents embankment distress during the formation of foundation voids, thereby ensuring localized embankment stability and controlling localized embankment settlements.

5.1.3.4 Reinforced soft site closures, Figure 5.3d

In this application a layer of geosynthetic reinforcement is placed across the surface of very soft deposits prior to the placement of a fill closure. The presence of the geosynthetic reinforcement provides local stability, enabling a stable working platform to be constructed across the very soft deposit.

5.1.3.5 Reinforced fill slopes, Figure 5.3e

In this application multiple layers of geosynthetic reinforcement are placed in the compacted fill slope to provide additional stability and limit deformations. The presence of the geosynthetic reinforcement enables stable steep fill slopes to be constructed to any height and at any slope angle, at the same time utilizing a range of suitable slope facings. This application is covered in further detail in Section 5.3.

5.1.3.6 Reinforced fill retaining walls, Figure 5.3f

In this application multiple layers of geosynthetic reinforcement are placed in the compacted reinforced fill of the retaining wall to provide stability and limit deformations. The use of the layers of geosynthetic reinforcement enable stable retaining walls to be constructed to

a wide range of heights. This application is covered in further detail in Section 5.4.

5.1.3.7 Reinforced fill retaining berms, Figure 5.3g

In this application reinforced fills are used to construct stable peripheral retaining berms to enable the vertical expansion of engineered landfill structures. Here, the reinforced fill structure behaves as a reinforced fill slope or retaining wall depending on its geometry. This application is covered in further detail in Section 5.5.

5.2 Reinforced embankments over soft foundations

5.2.1 Introduction and brief historical overview

Soft foundation soils suffer from two fundamental problems when embankment fills are placed on top - low shear resistance and high compressibility. Low foundation shear strengths result in instability of the embankment, where the degree of instability is not only a function of the undrained shear strength of the soft foundation, but also a function of the embankment weight and geometry and the rate of application of the embankment load. High foundation compressibility results in embankment settlement, where it is not just the magnitude of settlement that is important, but also the rate of settlement occurring over time. Settlements are a function of the compressibility of the soft foundation, the foundation soil geometry, as well as the embankment weight applied.

Several techniques exist to design and construct embankments over soft foundation soils, where those

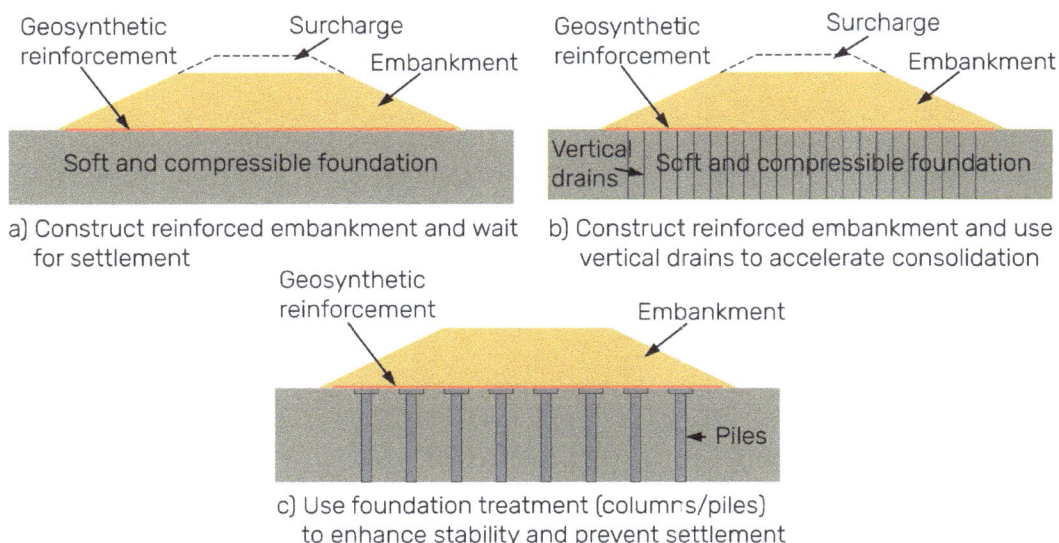

a) Construct reinforced embankment and wait for settlement

b) Construct reinforced embankment and use vertical drains to accelerate consolidation

c) Use foundation treatment (columns/piles) to enhance stability and prevent settlement

Figure 5.4 Geosynthetic reinforcement solutions to enhance stability and control settlements for embankments constructed on soft foundations.

showing the use of basal geosynthetic reinforcement are depicted in Figure 5.4. These three techniques that utilize basal geosynthetic reinforcement are:

> Basal reinforced embankments on soft foundations, Figure 5.4a. Here, the basal reinforcement is used to enhance the short-term stability of the embankment until such time as the foundation has consolidated and can support the full embankment weight. Surcharging can be used to accelerate the consolidation of the soft foundation. Here, the basal geosynthetic reinforcement is considered to only enhance the short-term stability of the embankment and does not affect the settlement of the embankment over the long-term. This subject is discussed in further detail in Section 5.2.2 with relation to the required performance of the basal geosynthetic reinforcement.

> Basal reinforced embankments on soft foundations with prefabricated vertical drains, Figure 5.4b. Here the basal reinforcement is used to enhance the stability of the embankment until such time as the accelerated consolidation of the soft foundation has occurred using the prefabricated vertical drains. The combination of surcharging and the use of prefabrication vertical drains can provide cost effective embankment solutions. Likewise (the same as for the previous case), the basal geosynthetic reinforcement is considered to only affect the short-term stability of the embankment and does not affect its settlement in the long-term, which is governed using the prefabricated vertical drains and any surcharge. This subject is discussed in further detail in Section 5.2.3 regarding the required performance of the basal reinforcement.

> Basal reinforced embankments on piles, Figure 5.4c. Here the basal reinforcement is used to span across the tops of the foundation columns or piles to transfer the embankment loads onto the caps. The basal geosynthetic reinforcement/piles combination enhances stability and mitigates settlements. This subject is discussed in further detail in Section 5.2.4 with regard to the required performance of the basal reinforcement.

Basal reinforced embankments on soft foundations (Figure 5.4a) were first used in The Netherlands in the early 1970's (Volman et al, 1977) using woven geotextiles. The technique then spread internationally and by the early 1980's it was being used in the United States by the US Corp of Engineers (e.g. Fowler, 1982) and in SE Asia (e.g. Risseeuw and Voskamp, 1993). Today, this

technique has not only become standard practice in many parts of the world but has also expanded to incorporate other soft foundation soil treatments such as prefabricated vertical drains (Figure 5.4b).

The use of basal reinforcement in combination with prefabricated vertical drains, shown in Figure 5.4b, began during the mid-1980's where it became evident that by combining accelerated foundation consolidation with enhanced embankment fill stability using basal reinforcement a cost-effective solution could be obtained, with the foundation consolidation occurred over a relatively short time period. Today, this is a very common technique driven by the benefits of having foundation consolidation occurring within the (relatively short) construction contract timeframe.

Primarily woven geotextiles are used for basal reinforced embankments on soft foundations. The reason for this is that this geosynthetic reinforcement type combines the multi-functional behaviour of the reinforcement, separation, filtration and reinforcement in a single material, which provides the ideal properties when dealing with soft soils.

Basal reinforced embankments constructed on piles were first developed in Sweden in the early 1970's, Holtz and Massarsch (1976). The technique adopted was a modern refinement of an earlier approach used in Scandinavia where pile groups with large pile caps were used to support embankments constructed over soft foundations since the late 1920's. The addition of the basal geosynthetic reinforcement enabled the pile caps to be reduced significantly in size with the basal reinforcement used to transfer the embankment loads directly onto the adjacent piles.

By the early 1980's this piled embankment technique had spread to the United Kingdom (e.g. Reid and Buchanan, 1984) and SE Asia (e.g. Tan et al, 1985). In the UK the geosynthetic reinforcement used consisted of bonded geostrip materials while in SE Asia it was woven geotextiles. Since that time the use of the technique has spread throughout Europe (e.g. Gartung and Verspohl, 1996; Briançon et al, 2008; van Eekelen et al, 2010), the USA (e.g. Collin et al, 2005) and to China (e.g. Liu et al, 2007) where woven geotextiles and geogrids have been commonly used.

Today, all three types of geosynthetic reinforcements—woven geotextiles, geogrids and bonded geostrip reinforcements—are used for basal reinforced embankments constructed on piles.

5.2.2 Basal reinforced embankments on soft foundations

Where geosynthetic reinforcement is used alone at the base of an embankment over a soft foundation, its role is to enhance the stability of the embankment until such time as the soft foundation consolidates and

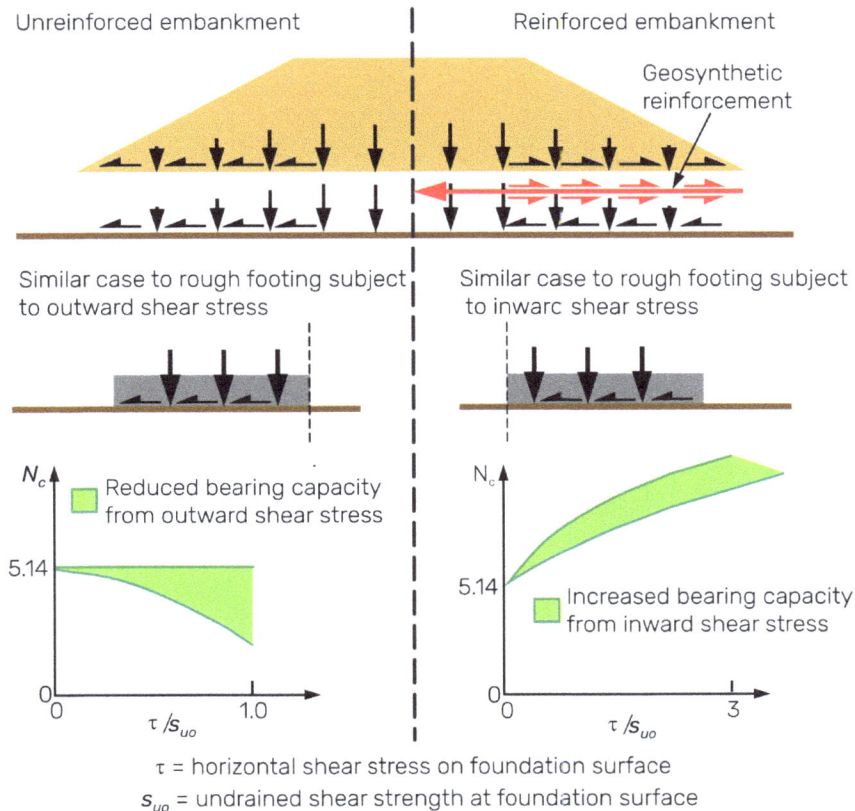

Figure 5.5 Basic mechanics of basal reinforced embankments on soft foundations (after Jewell, 1996).

supports the embankment loading by itself. In this role, the presence of the geosynthetic reinforcement is not considered to influence any settlements that may occur during the foundation consolidation process.

The basic mechanics of basal reinforced embankments on soft foundations can be described in terms of bearing capacity theory. Figure 5.5 shows the two cases of an unreinforced embankment (the left-hand side of Figure 5.5) and a basal reinforced embankment (the right-hand side of Figure 5.5).

In an unreinforced embankment (the left-hand side of Figure 5.5), the base of the embankment has vertical loading due to the weight of the embankment fill and beneath the side-slopes there are horizontal outward shear stresses acting. In terms of bearing capacity, the combination of vertical loading and horizontal outward shear stresses result in a lower bearing capacity (compared to vertical loading only). This is shown in terms of the bearing capacity factor N_c reducing below a value of 5.14 (for vertical loading only) by an amount related to the magnitude of the horizontal outward shear stress ratio t/s_{uo}.

Conversely, in a basal reinforced embankment (the right-hand side of Figure 5.5), the presence of the geosynthetic reinforcement absorbs the horizontal outward shear stresses at the base of the embankment and imparts horizontal inward shear stresses. In terms of bearing capacity, the combination of vertical loading (from the embankment weight) and horizontal inward

shear stresses result in a higher bearing capacity (compared to vertical loading only). This is shown in the bearing capacity factor N_c increasing above a value of 5.14 (for vertical loading only) by an amount related to the magnitude of the horizontal inward shear stress ratio t/s_{uo}.

Thus, the improvement in bearing capacity by using the geosynthetic reinforcement enables basal reinforced embankments to be constructed to a greater height and/or have steeper side-slopes than unreinforced embankments. While the use of basal reinforcement provides these two performance improvements, close attention is still required to the rate of application of the embankment fill to ensure stability is maintained during construction.

5.2.2.1 Potential failure modes

Figure 5.6 shows the three potential failure modes that must be analysed for basal reinforced embankments on soft foundations. These are:

> Side-slope stability, Figure 5.6a. Here, the required embankment long term side-slope stability is assessed under effective stress conditions.
> Rotational stability, Figure 5.6b. Here, the required reinforcement design strength is determined to ensure adequate stability exists through the base of the reinforced embankment. This is discussed in more detail below.

> Excessive strain in the basal reinforcement, Figure 5.6c. Here, the basal reinforcement is limited to a maximum strain level to ensure excessive deformations do not occur at the base of the embankment.

5.2.2.6 Load profile in the geosynthetic reinforcement over time

The tensile load generated in the basal geosynthetic reinforcement is not constant but varies over time, as shown in Figure 5.7. During embankment construction, the tensile load increases until it reaches a maximum T_r when the embankment has reached its maximum height.

Following this, the soft foundation soil consolidates under the loading of the embankment fill, and it gains in shear strength. As it gains in shear strength, the soft foundation can support more of the embankment loading and thus the tensile load in the basal reinforcement reduces. This process continues until the soft foundation has fully consolidated over time t_d and can fully support the embankment loading without need for the basal reinforcement. Thus, the basal reinforcement is required only for the (relatively short) time period t_d and this application is commonly referred to as a "short-term" reinforcement application, but in practice it may take up to 10 to 15 years for the soft foundation to fully consolidate.

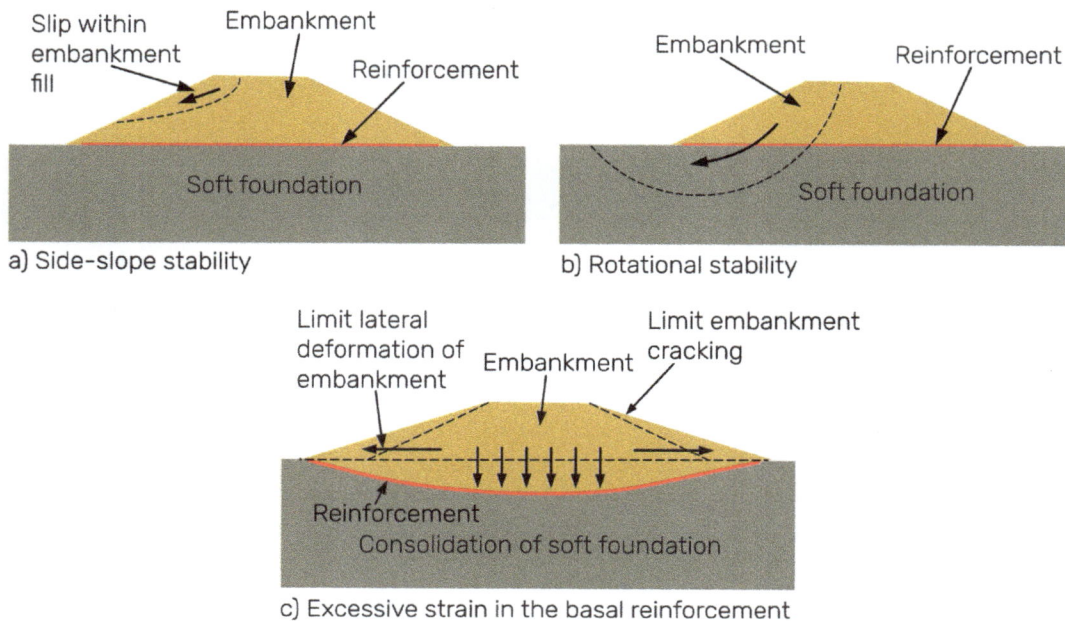

a) Side-slope stability

b) Rotational stability

c) Excessive strain in the basal reinforcement

Figure 5.6 Potential failure modes for basal reinforced embankments on soft foundations.

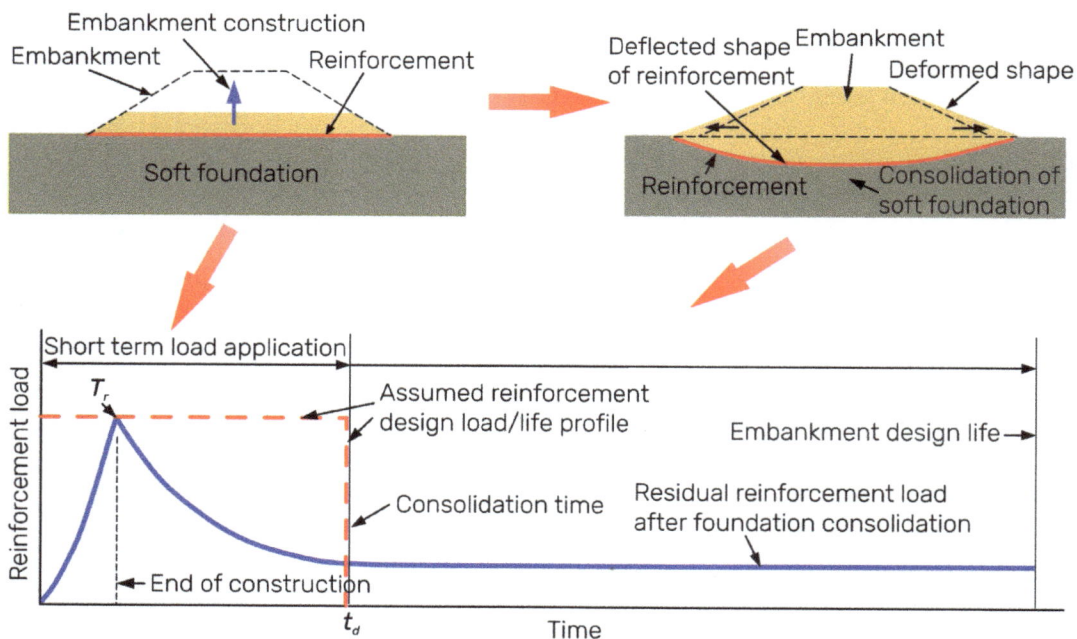

Figure 5.7 Load profile in geosynthetic reinforcement over time (after Lawson, 2022).

IGS Geosynthetics Handbook, 1st Edition

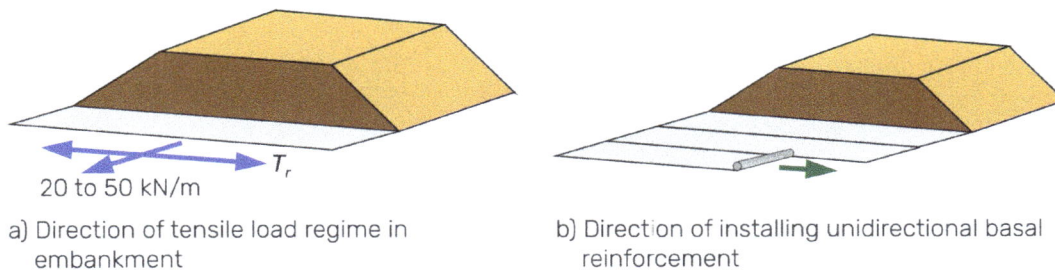

a) Direction of tensile load regime in embankment

b) Direction of installing unidirectional basal reinforcement

Figure 5.8 Layout of basal reinforcement on sit.

Following the foundation consolidation period t_d, the basal reinforcement load does not reduce to zero but normally retains a residual load locked in due to the deformation (settlements) of the embankment foundation during consolidation. The magnitude of this residual load is relatively small compared to T_r and it is considered to remain at this level for the remainder of the design life of the embankment.

From a reinforcement design perspective, the two critical parameters of importance are the maximum tensile load T_r and the time for consolidation to occur t_d as shown in Figure 5.7. It is difficult to accurately determine the actual variation of reinforcement load over this relatively short time period, so to simplify the design approach it is common to conservatively assume that the tensile load remains constant at magnitude T_r over the construction and up to the consolidation time period t_d (the red-dashed line shown in Figure 5.7).

A range of design/analysis methods are available to determine the maximum reinforcement load T_r. These are:

> Plasticity methods. Plasticity methods are based on bearing capacity theory (similar to that described in Figure 5.5) and provide a simple approach to determine the maximum reinforcement load T_r.
> Limit equilibrium methods. Limit equilibrium methods which are commonly used to assess the stability of geotechnical structures can be used to determine the maximum reinforcement load T_r.
> Continuum methods. Continuum methods consisting of finite element and finite difference numerical analyses can model the behaviour of the basal reinforced embankment in various geometries and determine its maximum tensile load T_r.

5.2.2.7 Layout of the geosynthetic reinforcement on site

Figure 5.8a shows the direction of the tensile loads in the geosynthetic reinforcement at the base of the embankment. The major tensile load T_r (determined by Figure 5.7) occurs in the direction across the width of the embankment, which is the direction of the primary tensile loading. In the direction along the length of the embankment there are relatively small tensile loads – 20 to 50 kN/m – depending on the control of the embankment filling process.

Prior to placement of the basal reinforcement the foundation surface should be prepared by removing any tall vegetation, e.g. trees and shrubs. Low-lying grass can be left in place and undisturbed because its root matter can help to bind the surface of the soft foundation layer. Tall grass should be removed, and any holes should be filled to make a more even surface.

The most efficient way of meeting the tensile load profile at the base of the embankment shown in Figure 5.8a is to use a single layer of unidirectional (uniaxial) geosynthetic reinforcement (with high strength in the longitudinal reinforcement direction) placed perpendicular to the embankment alignment, Figure 5.8b. The reinforcement layer should extend across the full width of the base of the embankment with no joins in this direction. Along the length of the embankment (where only relatively low strength is required), joins can be accepted by using either a geosynthetic overlap (0.5 m to 1.0 m depending on ground conditions) or by seaming on site. It should be recognized that sewn seams don't have the same strength as the original geosynthetic reinforcement material with typical sewn seam strength efficiencies ranging from 25% to 75% depending on the strength of the reinforcement, the type of seam being used, and the type of sewing yarn being used.

The initial layer of embankment fill should be placed carefully on top of the basal reinforcement layer. Mechanical equipment used to place and spread the fill should be limited in weight and not traverse directly on the reinforcement. Fill should be placed on the existing fill platform before it is spread over the exposed basal reinforcement by use of the lightweight equipment. Where the foundation is particularly soft a swamp dozer may be used for spreading the initial embankment fill ensuring only low vertical stresses are applied to the soft foundation.

The type of fill used immediately above the basal reinforcement should be granular (i.e. frictional) in nature and preferably be free draining. Also, there should be no large rocks in this initial fill layer, and the fill grading should be consistent with that assumed for design.

5.2.3 Basal reinforced embankments on soft foundations with prefabricated vertical drains

5.2.3.1 Prefabricated vertical drains

Prefabricated vertical drains (PVDs), also known as wick drains, vertical drains, band drains or consolidation drains, are used to accelerate the removal of excess pore water from consolidating soft foundation soils. The benefits of the use of PVDs in basal reinforced embankments are twofold. First, foundation consolidation can be made to occur relatively quickly, within the construction timeframe, thus minimizing later costly maintenance. Second, the undrained shear strength of the soft foundation increases more quickly (compared to not using PVDs) thus providing earlier improvement in foundation stability. Typical embankment foundation consolidation timeframes when using PVDs can range from 6 months to 2 years depending on PVD spacing and foundation hydraulic conductivity, versus the 10 to 20 years if no PVDs are used.

PVDs consist of a geotextile filter jacket encapsulating a plastic drainage spacer, see Figure 5.9a. The geotextile filter jacket acts as the filter between the soft foundation soil and the drainage spacer while the plastic drainage spacer transports the pore water along the length of the PVD to the ground surface. PVDs are typically 100 mm in width and 3 mm to 5 mm in thickness and are only required to transport relatively low volumes of (pore) water under varying confining pressures along their length. It is important that the structure of the drainage spacer maintains its drainage flow path even when large foundation settlements and possible PVD buckling occur.

The PVDs are inserted into the soft foundation by vibrating a vertical hollow-steel mandrel (with the PVD inside), Figure 5.9b. The mandrel, containing the PVD, is inserted into the soft foundation layer to the required depth and then the mandrel is withdrawn leaving the PVD in place. The PVD is then cut, and the installation machine moves to the next location where the PVD installation process is repeated. The installation process can be very efficient with ideal installation rates as high as 1,500 m/hour for a single installation machine. However, local ground conditions may limit installation rates to less than this ideal rate. It is easy to increase the rate of PVD installation by simply increasing the number of installation machines working on site, as shown in Figure 5.9b.

5.2.3.2 Load profile in the geosynthetic reinforcement over time

The combination of PVDs and basal geosynthetic reinforcement interact together to provide an efficient method of embankment construction over soft foundation soils, Figure 5.4b. Figure 5.10 shows this interaction and compares it to the case where no PVDs are used, as described in Section 5.2.2. For the case of the basal reinforcement used without PVDs the reinforcement tensile load profile is the same as that shown in Figure 5.7 and is also shown for comparative purposes in Figure 5.10.

When PVDs are used an identical embankment loading occurs, but the maximum tensile load generated in the basal reinforcement T_r does not reach the same level as that without PVDs, Figure 5.10. The reason for this is that the PVDs are already enabling foundation consolidation to occur during the embankment filling process and thus the maximum tensile load generated in the basal reinforcement is reduced (compared to the without PVDs case). Also, the presence of the PVDs enable a quicker rate of consolidation and thus a shorter time over which the basal reinforcement is required to support the embankment, so the design life over which the reinforcement is required is reduced (compared to the without PVD case). The magnitude

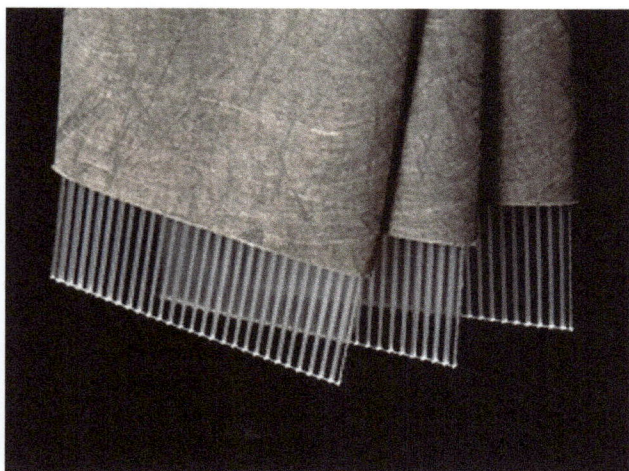

a) Common type of PVD

b) PVD installation equipment

Figure 5.9 Prefabricated vertical drains (PVDs). Photo courtesy of Solmax.

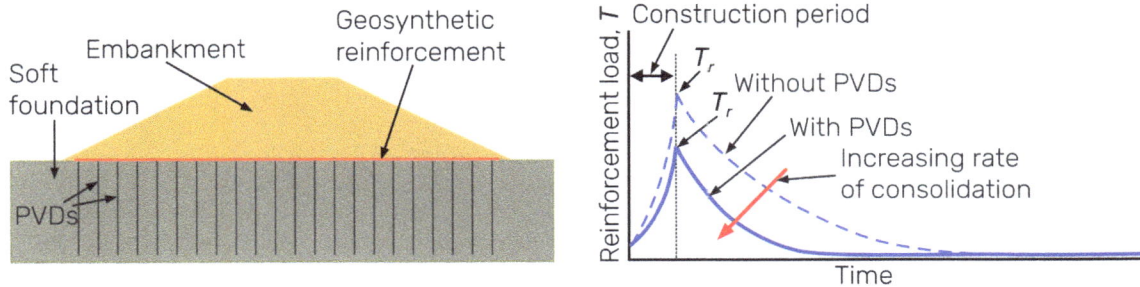

Figure 5.10 Load profile in the basal reinforcement over time with and without PVDs (after Lawson, 2022).

of this time reduction is dependent on the PVD spacing used and the hydraulic conductivity behaviour of the soft foundation. This combination of lower required maximum tensile load T_r and shorter required reinforcement design life t_d enables a more cost-effective basal reinforcement solution to be chosen compared to the without PVD case.

While the use of PVDs enables accelerated consolidation of the soft foundation, care is still required regarding the rate of embankment fill placement to ensure maintenance of embankment stability during construction.

5.2.3.3 Layout of the geosynthetic reinforcement on site

The direction of tensile loads in the basal reinforcement is the same as shown in Figure 5.8a where no PVDs are used. The major tensile load T_r (see Figure 5.10) occurs in the direction across the width of the embankment, while the minor tensile load occurs in the direction along the length of the embankment.

Large vegetation (trees, tall plants, etc.) should be cleared from the site but low-lying vegetation (grass, etc.) can be left in place. Next, a geotextile separator should be placed over the area followed by a granular layer of adequate thickness to construct a stable working platform to support the PVD installation machines (see Figure 5.9b). Following installation of the PVDs, further granular material is placed to complete the horizontal drainage blanket for the PVDs. Finally, the unidirectional (or uniaxial) geosynthetic reinforcement is placed across the top of the drainage blanket perpendicular to the direction of the embankment in the manner shown in Figure 5.8b. The embankment fill is then placed and spread on top of the geosynthetic reinforcement.

5.2.4 Basal reinforced embankments on piles

While the basal reinforced embankment techniques shown in Figures 5.4a and 5.4b, and described in Sections 5.2.2 and 5.2.3, provide enhanced short-term stability for the embankment they do not influence settlements over time. Additional foundation treatments must be applied to reduce these settlements.

One technique used is to pile the foundation where a combinaton of basal reinforcement and the piled foundation provides enhanced stability as well as reduced settlements.

The basal reinforced piled embankment technique shown in Figure 5.4c utilizes piles or columns to mitigate embankment settlement and enhance stability, and thus enables the embankment to be constructed at any rate and to any height without subsequent foundation stability and settlement problems. The basal reinforcement provides local stability across the base of the embankment fill enabling the embankment loads to arch across adjacent pile caps. This technique is sometimes referred to as a load transfer platform.

5.2.4.1 Applications

Basal reinforced embankments on piles are typically used where differential settlements and stability cause problems, namely:

> Transitions between piled and non-piled structures, Figure 5.11a. To prevent differential settlements at the junction of piled structures (e.g. bridges), it is common to use a basal reinforced piled foundation beneath the approach abutments. As the distance from the bridge structure increases, the embankment piles are shortened in length and spaced further apart to provide a smooth transition with the existing embankment.
> Embankment widening preventing differential settlements, Figure 5.11b. Where new embankments are constructed abutting an existing embankment a basal reinforced piled foundation can be used to prevent differential settlements from occurring between the old and new embankments.
> Embankments with vertical walls, Figure 5.11c. When embankments are constructed in confined areas it is common for them to have vertical faces. To ensure structural stability and prevent the retaining walls from rotating inwards and over-stressing the walls, a basal reinforced piled foundation is commonly used.

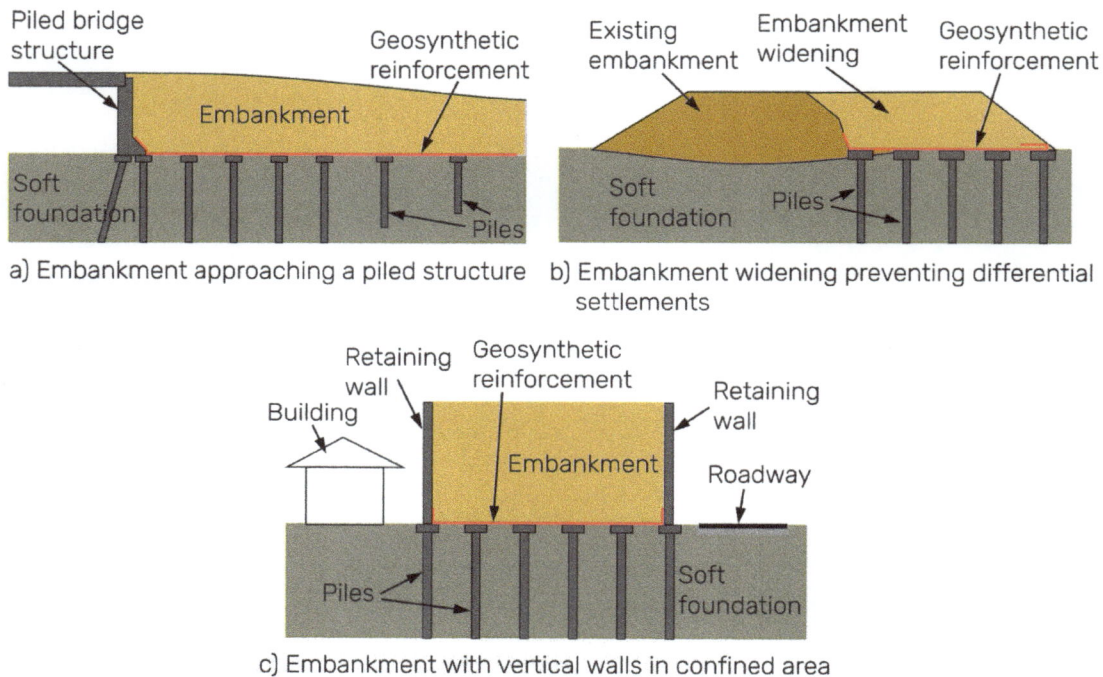

Figure 5.11 Three applications for basal reinforced embankments on piles.

> Preventing large embankment settlements in general. In some cases, it may be beneficial to prevent large embankment settlements from occurring, such as when constructing embankments over peat.

> Where speed of construction is paramount. Basal reinforced embankments on piles can be constructed at any rate, thus the technique is ideal where speed of embankment construction is of the essence.

5.2.4.2 Types of piles and columns used

A variety of different piles and columns have been used for this basal reinforced embankment technique. A summary is given in Table 5.4. Piles, which are relatively small in diameter, normally require pile caps to spread the embankment loadings onto the tops of the piles in an efficient manner. Columns, being larger in diameter, generally don't require caps, or have the caps built into the upper part of the column. It is common to design the piles and columns as end-bearing units, which means they should be extended down to a firm foundation stratum. In some cases, this has required the piles to be installed at over 30 m in depth.

Care should be taken when using pile caps to ensure there are no sharp edges around the periphery of the caps which may damage the basal reinforcement. The vicinity around the edges of the pile caps is a critical location because the maximum tensile stresses in the reinforcements occur here. Preferably, caps with smooth or bevelled edges should be used to minimize localized tensile stresses and any damage.

Potential failure modes in basal reinforced embankments on piles

Figure 5.12 shows the potential failure modes for basal reinforced embankments on piles. These are:

> Pile group capacity, Figure 5.12a. The structural and geotechnical capacity of the pile group must support the embankment loading.

> Pile group extent, Figure 5.12b. The lateral extent of the pile group must be such that embankment edge failures do not impact on the embankment crest.

> Embankment arching, Figure 5.12c. The basal reinforcement arches across the tops of the adjacent pile caps at the base of the embankment. This is described in further detail below.

> Lateral sliding, Figure 5.12d. The basal reinforcement must resist the lateral sliding of the embankment fill above the pile caps.

> Overall stability, Figure 5.12e. The overall stability should be assessed especially if the foundation is very soft.

> Other potential failure modes include excessive settlement of the pile group and excessive differential settlements of the embankment surface.

5.2.4.3 Load profile in the geosynthetic reinforcement over time

Tensile loads are generated in the basal geosynthetic reinforcement due to the vertical arching at the base of the embankment fill, Figure 5.12c, and the horizontal outward thrust of the embankment fill, Figure 5.12d. Both of these tensile loads are generated simultaneously.

Table 5.4 Types of piles and columns used in basal reinforced embankments on piles (After Lawson, 2022)

Type	Features
Piles:	
Pre-cast concrete	High load capacity which maximises pile spacings. Requires use of pile caps.
Bored concrete	Large diameter, high capacity. Relatively uncommon use.
Timber	Low load capacity. Used where consistent high groundwater levels occur. Normally, no pile caps used.
Columns:	
Concrete/cement	The head of the columns are 'belled out' so that caps are integrated with the columns.
Stone	Relatively low load capacity. No pile caps used.
Confined-stone	Used in very soft foundations. No pile caps used.
Semi-rigid	Column rigidity used as design criterion. Normally, no pile caps used.

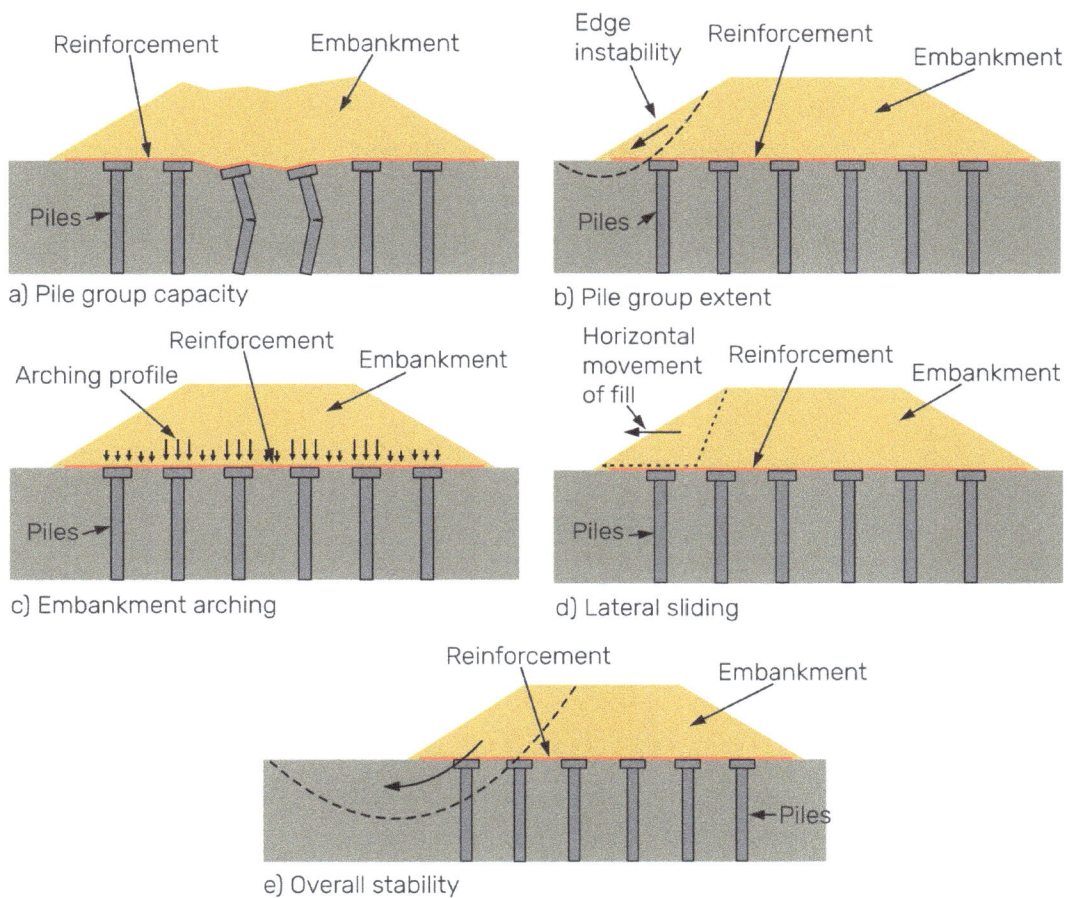

Figure 5.12 Potential failure modes for basal reinforced embankments on piles (after BS8006:2010).

The tensile load profile in the basal reinforcement due to embankment fill arching over time is shown in Figure 5.13. As the embankment fill height increases the tensile load in the basal reinforcement increases because of some arching at the base of the embankment. When the embankment construction is complete the tensile load continues to increase until such time as an equilibrium condition is reached between the fully arched embankment loads in the reinforcement, the consolidation of the soft foundation between the pile caps and the deformation of the basal reinforcement. At this point, the tensile load in the basal reinforcement is a maximum and continues at this level for the remaining design life of the piled embankment. In this case the reinforcement is required to perform for the full design life of the embankment.

The shape of the actual reinforcement load profile in Figure 5.13 is complex and difficult to analyse because

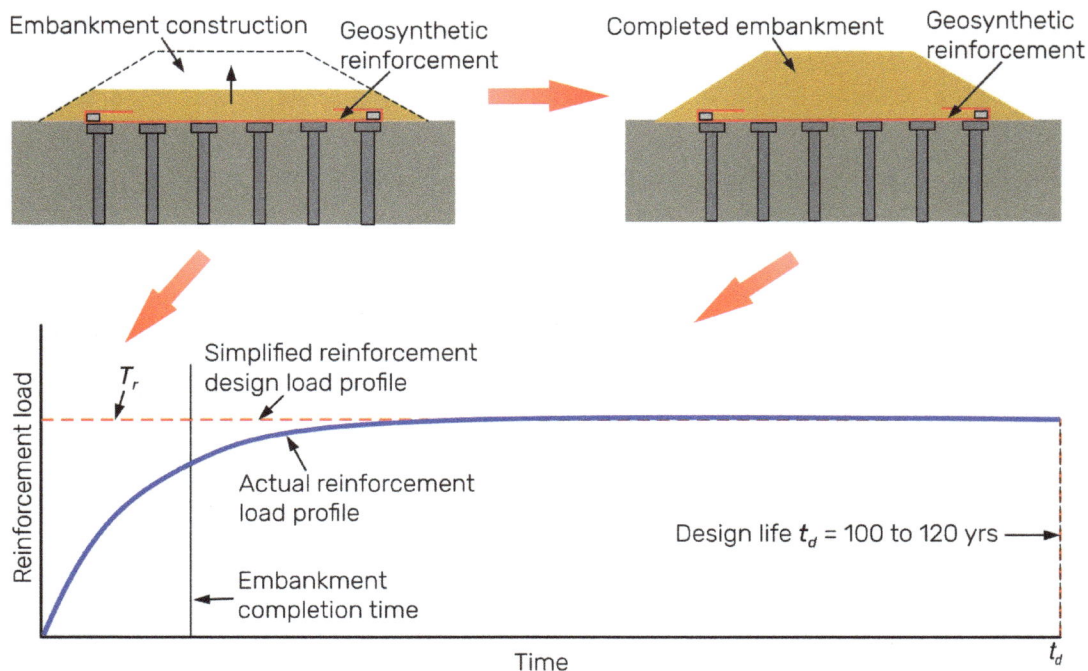

Figure 5.13 Load profile in geosynthetic reinforcement over time (after Lawson, 2022).

of the interaction between the different components. Consequently, from a design perspective a simplified approach is normally adopted where a constant maximum tensile load T_r is assumed to occur over the full design life of the embankment.

5.2.4.4 Arching across the base of the embankment

Different analytical models have been developed to enable the calculation of the basal reinforcement tensile load acting across the base of piled embankments, e.g. BS8006:1995, 2010, EBGEO (2011) and van Eekelen and Brugman (2016). The models range from simple (e.g. BS8006:1995) to complex (e.g. van Eekelen and Brugman, 2016).

5.2.4.5 Layout of the geosynthetic reinforcement on site

Figure 5.14 shows the bidirectional (biaxial) nature of the reinforcement loads beneath a basal reinforced embankment on piles. In the direction along the length of the embankment the embankment loads T_{rp}, are due to arching at the base of the embankment (see Figure 5.12c). In the direction across the width of the embankment the embankment loads are $T_{rp}+T_{ds}$, where T_{rp} is due to embankment arching and T_{ds} is the load generated due to the outward thrust of the embankment fill (see Figure 5.12d).

The most efficient geosynthetic reinforcement layout that satisfies this bidirectional tensile load profile at the base of piled embankments is to use two layers of unidirectional (or uniaxial) reinforcement (with the high strength in its longitudinal direction) laid orthogonally on top of each other, Figure 5.14. Across the width of the embankment, the geosynthetic reinforcement should be laid in one continuous sheet. Along the length of the

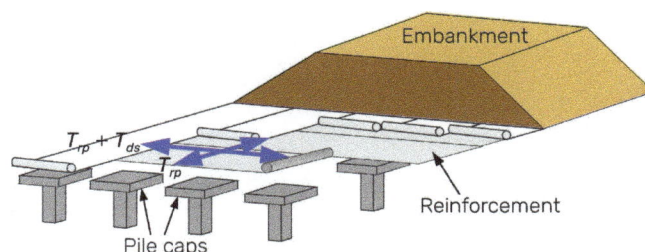

Figure 5.1.4 Bidirectional basal reinforcement loads and efficient reinforcement layout (after Lawson, 2022).

embankment, the geosynthetic reinforcement should be laid in long lengths, with any end-on-end joins overlapped by a two-pile spacing length. The two orthogonal geosynthetic layers can be placed immediately on top of one another.

5.3 Reinforced fill slopes

5.3.1 Introduction and brief historical overview

Geosynthetic reinforcements are used to enhance the stability of soil fill slopes by resisting the forces that would normally cause instability of the slope. The technique, shown in Figure 5.15, utilizes horizontal layers of geosynthetic reinforcement placed in the compacted reinforced fill at predetermined vertical spacings. The layers of geosynthetic reinforcement provide additional shear resistance in combination with that of the compacted reinforced fill, which results in a stable constructed fill slope.

The layers of geosynthetic reinforcement work by intersecting the critical failure plane in the reinforced fill

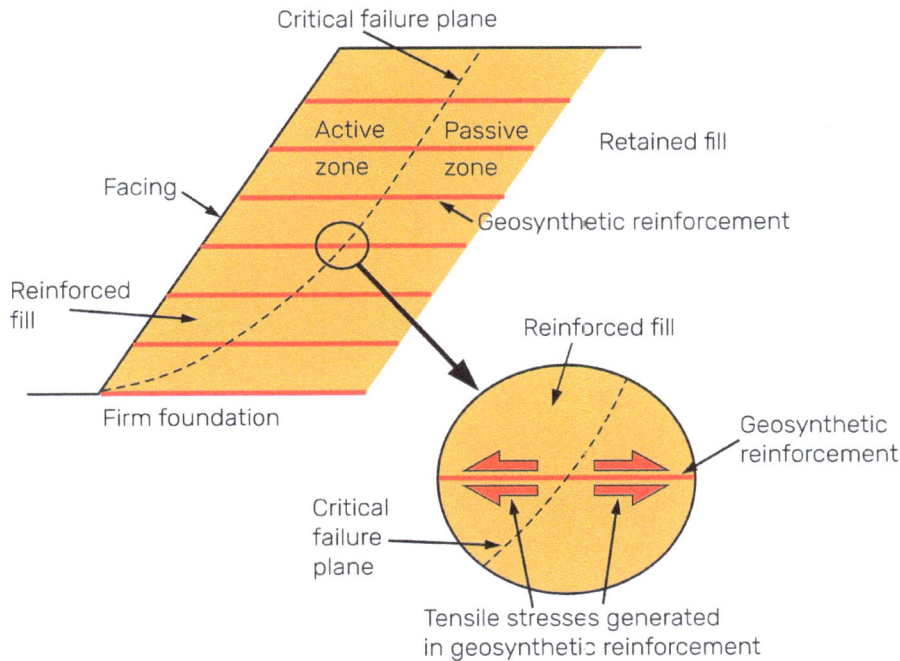

Figure 5.15 Basic features of a reinforced fill slope.

(see Figure 5.15) where tensile stresses are generated in the geosynthetic reinforcements. These tensile stresses are then dissipated back into the surrounding reinforced fill by the action of geosynthetic reinforcement/reinforced fill bond further away from the critical failure plane. This dissipation of reinforcement tensile stresses occurs in both the active zone (towards the slope face) and the passive zone (away from the slope face).

For reinforced fill slopes most of the shear resistance is provided by the compacted reinforced fill (can be as much as 90% of the total in some cases) while the minority is provided by the geosynthetic reinforcements (which may be as small as 10% of the total). Thus, for long-term slope stability it is very important that the behaviour of the reinforced fill remains constant and does not deteriorate over time. Achieving this can place constraints on the types of suitable reinforced fill used for long-term reinforced fill slopes.

Geosynthetic reinforcements first began to be used in reinforced fill slopes in the late 1970's in the United Kingdom. The reinforcements used at that time were woven geotextiles and HDPE sheet geogrids. During the 1980's the use of geosynthetic reinforced fill slopes expanded rapidly driven by the engineering economics of the technique and increased understanding of the role of the geosynthetic reinforcements in fill slopes. The main economics governing this application is the ability to use a relatively wide range of fill types to construct stable slopes using geosynthetic reinforcements. Lower quality fills require greater quantities of geosynthetic reinforcement to construct a stable slope, while better quality fills require less reinforcement.

Today, the geosynthetic reinforced fill slope technique is ubiquitous the world over. Its economics and long-term performance have been demonstrated repeatedly over many years. Geogrid reinforcements are by far the most common reinforcement type used in reinforced fill slopes, with woven and knitted geotextiles also used.

5.3.2 Classification of reinforced fill slopes

Reinforced fill slopes may be classified according to slope face angle which provides a convenient and descriptive approach, Figure 5.16. This simple classification divides reinforced fill slopes into two categories based on slope face angle – "shallow" reinforced slopes and "steep" reinforced slopes. Shallow reinforced slopes normally have slope face angles less than 45° and for these slope geometries any reinforcement tensions are dissipated before reaching the slope face. Consequently, for these slopes, only a "passive" (i.e. non-structural) slope facing is required, and this commonly takes the form of an erosion control mat to resist any surface water run-off.

Steep reinforced slopes have slope face angles greater than 45°. For these slopes, an "active" (i.e. structural) slope facing is required to provide local stability at the slope face as the reinforcement tensions are not fully dissipated on reaching the slope face. Further, most international design codes, e.g. BS8006:2010, allow for steep reinforced slopes with face angles ≥70° to be designed as retaining walls as does the U.S. Federal Highways Administration. Reinforced fill retaining walls are discussed in further detail in Section 5.4.

The combination of the slope facing, the geosynthetic reinforcement, the connection between the

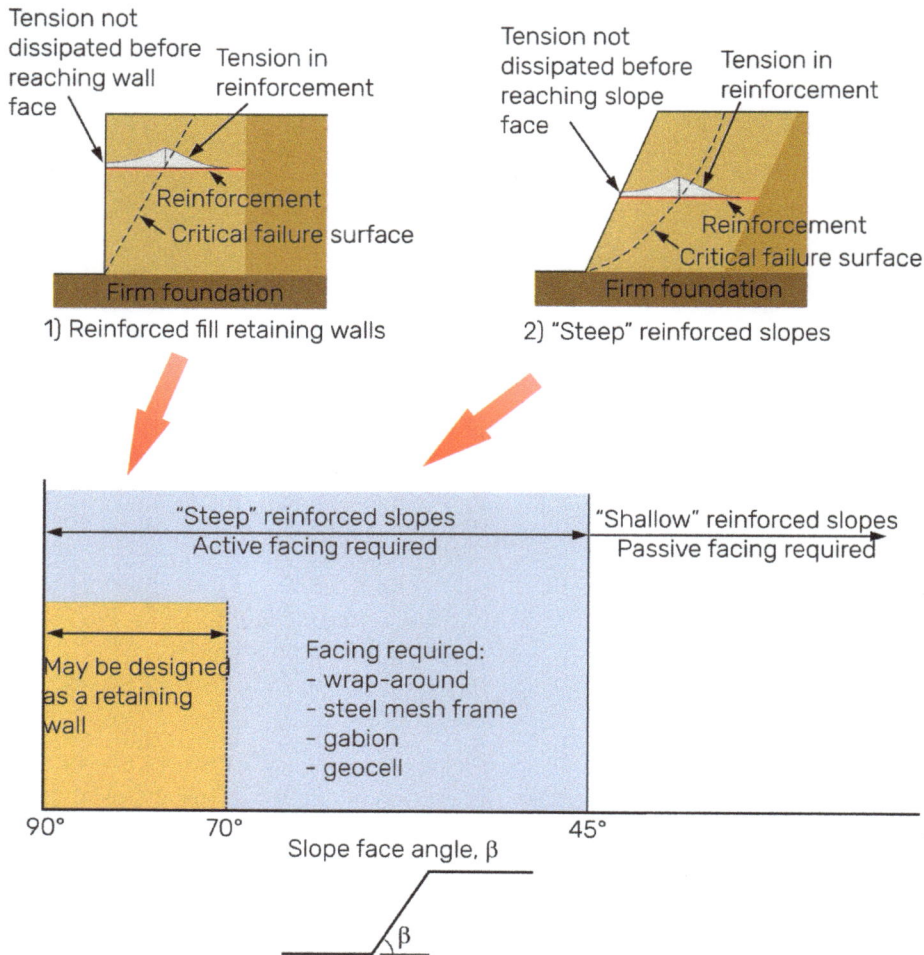

Figure 5.16 Classification of "steep" reinforced fill slopes according to slope face angle.

facing and the reinforcement and the reinforced fill all act together as a reinforced soil "system", where the overall system provides the necessary stability and deformation resistance for the steep slope. This is the same as for reinforced fill retaining walls where the reinforcement components also act as a system for these structures (see Section 5.4).

The design procedure for steep reinforced fill slopes is similar to that of reinforced fill retaining walls. The required stresses for equilibrium need to be determined, and a suitable geosynthetic reinforcement layout included in the slope to satisfy these required stresses. A steep reinforced fill slope has the advantage over a vertical reinforced fill retaining wall in that the facing arrangement and any deformations in the structure are much less critical – accurate alignments must be maintained if retaining walls are to have an aesthetic finish. Because deformations are less critical in steep reinforced fill slopes, it is easier to utilize a wider range of suitable reinforced fills than is the case for vertical reinforced fill retaining walls.

5.3.3 Facings for steep reinforced fill slopes

Slope facings for steep reinforced slopes are commonly durable, flexible and free-draining. The slope surface is erosion resistant, while the permeable nature of the facing allows groundwater flows to pass from the reinforced fill out through the slope surface without loss of the reinforced fill.

Figure 5.17 shows three different structural facings commonly used for steep reinforced fill slopes. These are soil-filled bags with wrap-around geosynthetic reinforcement facing, Figure 5.17a, angled steel mesh facing with retained top-soil, Figure 5.17b, and stone-filled gabion slope facing, Figure 5.17c. The reasons why these three techniques are popular are because they readily fulfil the performance requirements for steep slopes, are cost-effective and easy to construct, and can be easily adjusted to cater for different slope angles and alignments.

The soil-filled bags shown in Figure 5.17a are normally made from hessian and filled on site with topsoil and grass seed. These soil bags are placed on the slope face and shaped to the required slope and compacted shape by use of an excavator bucket. The reinforced fill is placed and compacted behind the soil bag facing up

a) Soil-filled bags with wrap-around geosynthetic facing

b) Angled steel mesh facing with retained top soil

c) Stone-filled gabion facing

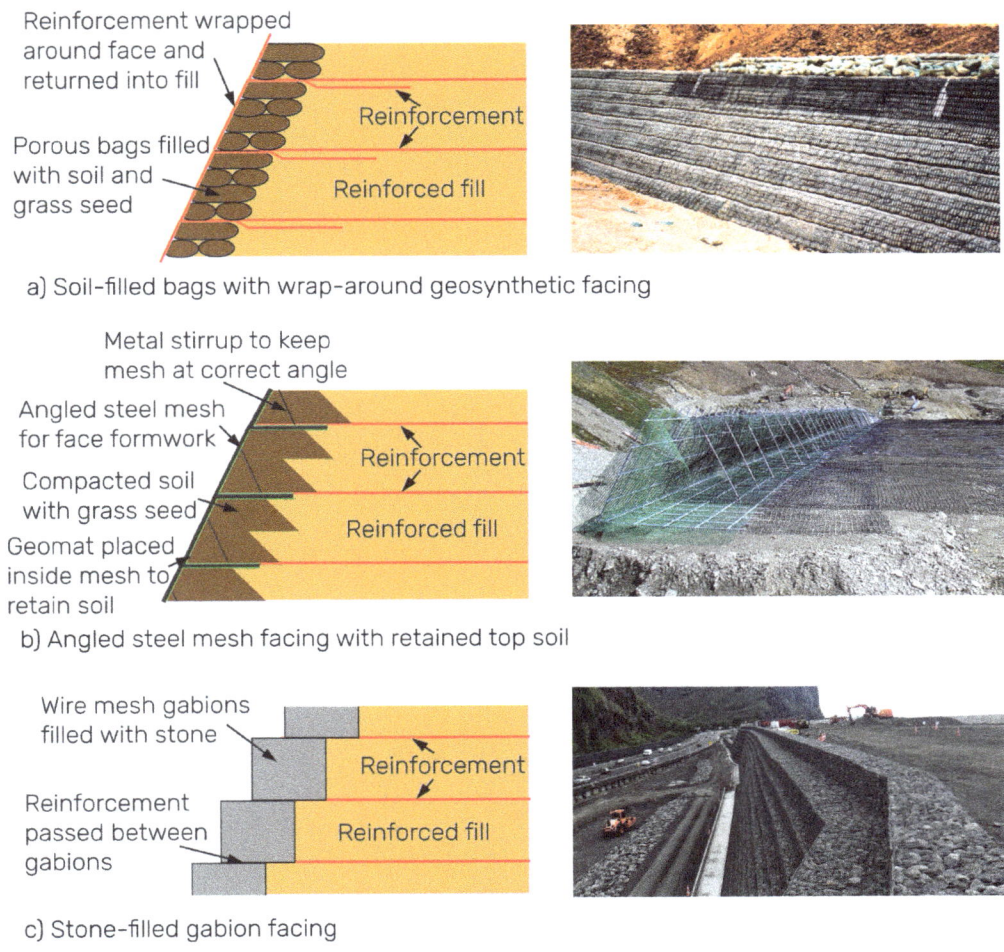

Figure 5.17 Three common structural facings used for steep reinforced fill slopes. Photo courtesy of Solmax.

to the next (upper) reinforcement level. The geosynthetic reinforcement is brought out through the slope face and wrapped upwards around the soil bag facing units and returned into the reinforced fill at the next upper reinforcement level. This technique provides a full positive connection between the geosynthetic reinforcement and the facing units. This process is repeated to construct the completed reinforced fill slope. After completion, the grass seed grows out through the slope face providing a vegetated covering for the reinforced slope. Figure 5.17a shows an example of a steep reinforced slope with a wrap-around soil bag facing being constructed.

The angled steel mesh facings shown in Figure 5.17b are bent according to the slope angle required. These facings are placed on the slope with the geosynthetic reinforcement placed onto the base of the mesh units forming a frictional connection between the facing units and the reinforced fill. Commonly, a geomat is placed on the inside of the angled mesh units prior to filling with compacted topsoil and grass seed. Once the topsoil has been placed behind the mesh units the reinforced fill is placed and compacted for the slope fill. This process is repeated to construct the complete reinforced slope. The seeded grass grows out through the slope face

providing a vegetated covering for the reinforced slope. Figure 5.17b shows an example of an angled steel mesh facing used for a steep reinforced fill slope.

Stone-filled gabion units shown in Figure 5.17c provide a relatively hard surface finish to steep reinforced fill slopes. The gabion facings are filled with stone creating a dense, hard, resistant surface layer. The slope facing angle is varied by adjusting the stacking geometry of the gabion units. The geosynthetic reinforcement layers are passed between each adjacent gabion layer providing a frictional bond connection between the facing units and the reinforcement. The reinforced fill is then placed and compacted behind the gabion units. A geotextile filter may be placed at the rear of the gabion units to prevent the reinforced fill from being eroded out through the facing. This process is repeated until the reinforced fill slope has been completed.

5.3.4 The reinforced fill in steep reinforced fill slopes

A wide range of fills can be used for steep reinforced fill slopes, with the addition of the geosynthetic reinforcements to provide the additional shear resistance to meet stability requirements. Poorer quality fills require

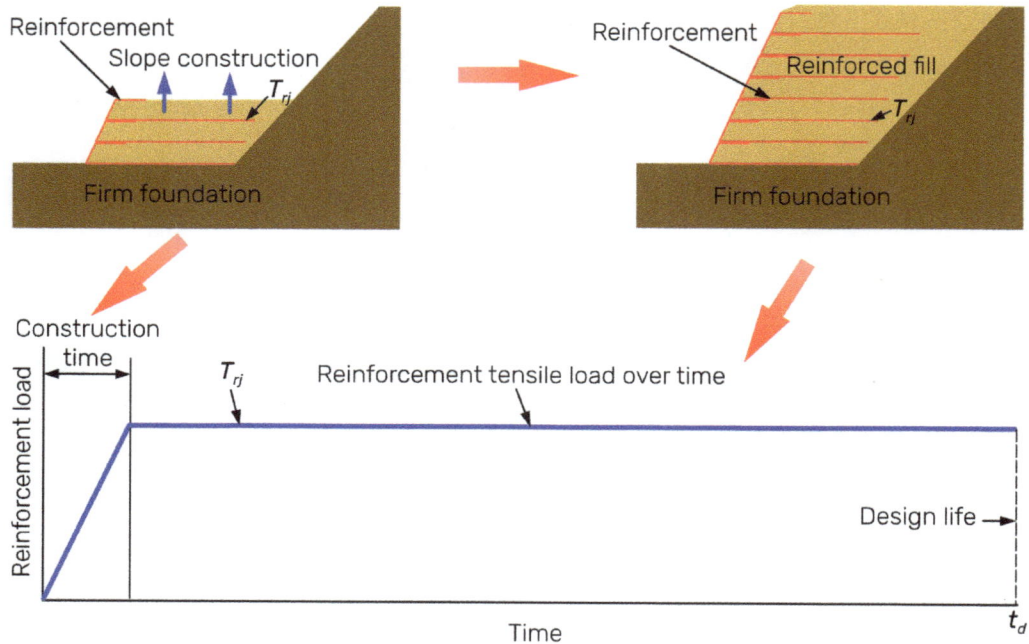

Figure 5.18 Tensile load in geosynthetic reinforcements over time for reinforced fill slopes (after Lawson, 2022).

greater quantities of geosynthetic reinforcement to achieve stability than do better quality fills. The amount of geosynthetic reinforcement required can be adjusted to achieve the required slope structural stability.

The most common type of reinforced fill used for steep reinforced fill slopes is cohesive-frictional fill whose typical gradation range includes that shown in Figure 5.29. This type of fill is commonly available naturally and can be accessed locally. The compacted peak shear resistance f$'_p$ of this fill type lies commonly in the range of 30° to 36° which makes it suitable as reinforced fill for steep slopes. For high, steep reinforced fill slopes good attention to drainage detailing is also important (see Section 5.3.7).

While Figure 5.29 refers to the grading limits of cohesive-frictional fills specifically used for retaining walls, it should be noted that greater flexibility can be applied to the grading limits for reinforced fill slopes. For example, the U.S. Federal Highway Administration (FHWA 2021) have recommended that the maximum amount of fines (% passing the No 200 sieve) should be limited to less than 50% for reinforced fill slopes covering its jurisdiction. Naturally, close attention to compaction and drainage details are required for reinforced fill slopes containing these high percentages of fines.

5.3.5 Tensile loads in the geosynthetic reinforcements over time

The tensile load profile over time in the geosynthetic reinforcements in steep reinforced fill slopes is shown in Figure 5.18. During slope construction the tensile loads increase in the layers of geosynthetic reinforcement until such time as the slope construction is completed.

Following this, the tensile loads remain constant with time (denoted by T_{rj}) for the remaining design life of the reinforced fill slope (denoted by t_d). Required design lives for reinforced fill slopes may range from several years to many years (75 to 120 years) depending on requirements. It is important for the geosynthetic reinforcement to carry these tensile loads over the required design life of the steep reinforced fill slope.

Where steep reinforced fill slopes are constructed for long design lives (75 to 120 years) it is important that the reinforced fill used be durable and provide the same, consistent shear resistance behaviour throughout its design life. If the shear resistance of the reinforced fill reduces with time, then the geosynthetic reinforcements have to support higher tensile loads to compensate for this. If this can't be done, then it can result in unexpected deformations or even failure of the steep reinforced fill slope.

5.3.6 "Standard" geometry versus "non-standard" geometry steep reinforced fill slopes

A distinction can be made between what is considered a "standard" slope geometry and a "non-standard" slope geometry. Figure 5.19 shows the major differences between these two different slope geometries. Standard slope geometries consist of a firm foundation with a horizontal base and crest, Figure 5.19a. Here, the presence of the firm foundation ensures no failure surfaces pass through or behind the base of the reinforced fill zone and consequently all critical failure surfaces exit at the toe of the slope. This geometry simplifies greatly the analysis methods used for the reinforced slope.

The standard geometry reinforced fill slope is by far

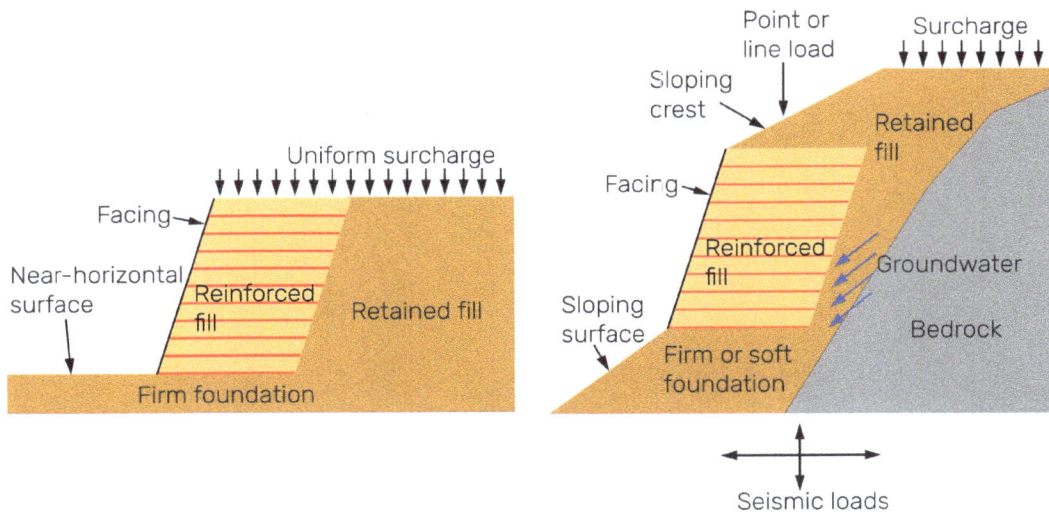

a) "Standard" reinforced fill slope geometry   b) "Non-standard" reinforced fill slope geometry

Figure 5.19 Typical "standard" versus "non–standard" reinforced fill slope geometries.

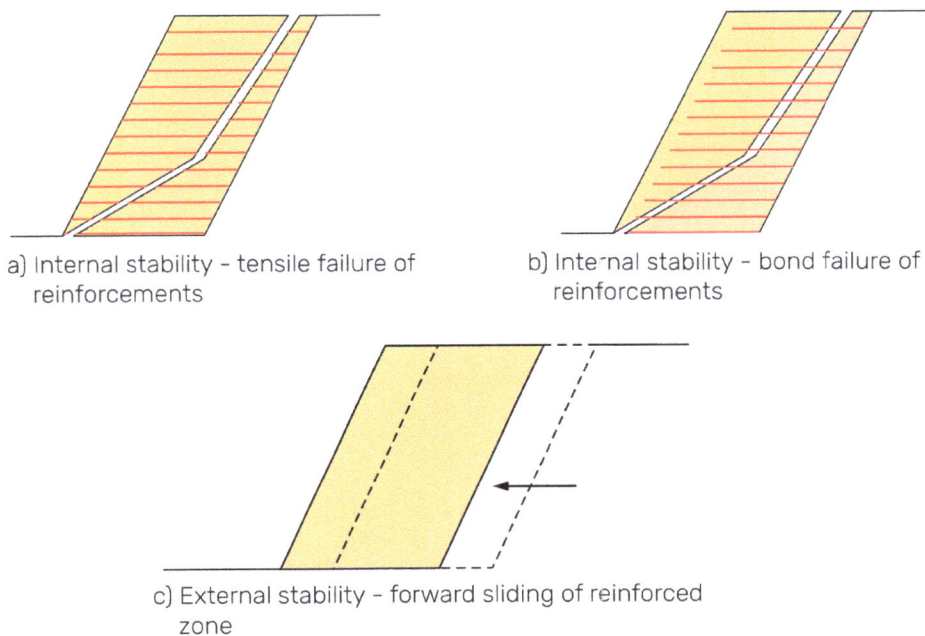

a) Internal stability - tensile failure of reinforcements

b) Internal stability - bond failure of reinforcements

c) External stability - forward sliding of reinforced zone

Figure 5.20 Potential failure modes for standard geometry reinforced fill slopes.

the most common form of slope geometry utilized. Figure 5.20 shows the three potential failure modes that are to be analysed for standard geometry reinforced fill slopes – internal stability for tensile load and bond failure of the reinforcements and external forward sliding of the reinforced fill zone across the surface of the firm foundation.

The methods of analysis used are normally simple limit equilibrium procedures. Several different limit equilibrium analysis methods are available to analyse the internal stability of reinforced fill slopes and these are shown in Figure 5.21. These range from circular slip analyses (e.g. Taylor, 1948), to two-part wedge analyses (e.g. Jewell, 1996), and log-spiral analyses (e.g. Leschinsky and Boedecker, 1989). Many computer software programs are readily available for the design/

analysis of these standard reinforced fill slope geometries using these different limit equilibrium analysis approaches.

Non-standard geometry reinforced fill slopes, as shown in Figure 5.19b, provide added complexity to the analysis method of steep reinforced fill slopes due to the magnitude of external loads present (e.g. seismic loads) and the geometry of the external soil surfaces. Further potential failure modes must be analysed in addition to the three standard geometry modes shown in Figure 5.20 and these modes are shown in Figure 5.22. The additional potential failure modes deal with compound stability and overall stability. The potential global failure mechanism becomes a major consideration for non-standard geometry slopes and any computer software used should be

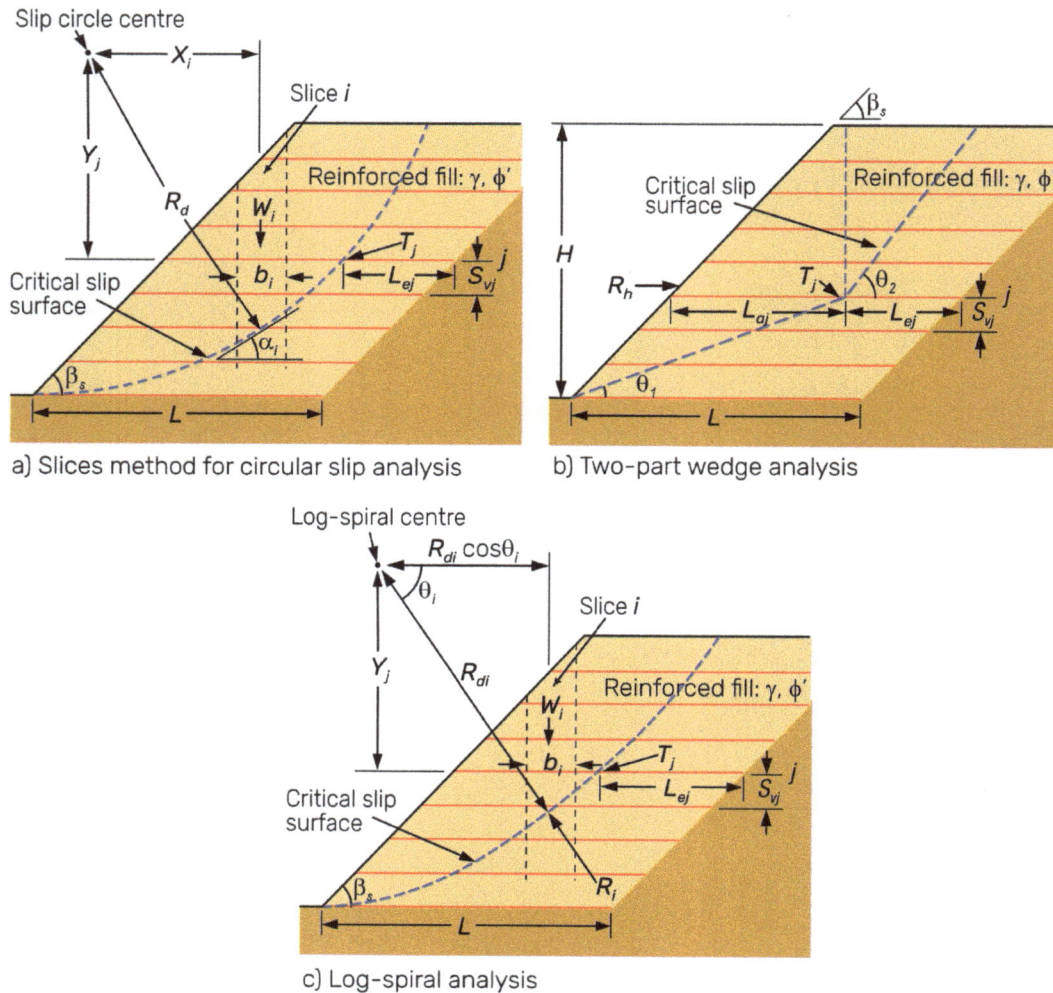

a) Slices method for circular slip analysis

b) Two-part wedge analysis

c) Log-spiral analysis

Figure 5.21 Limit equilibrium methods for internal stability analysis of reinforced fill slopes (after BS8006:2010).

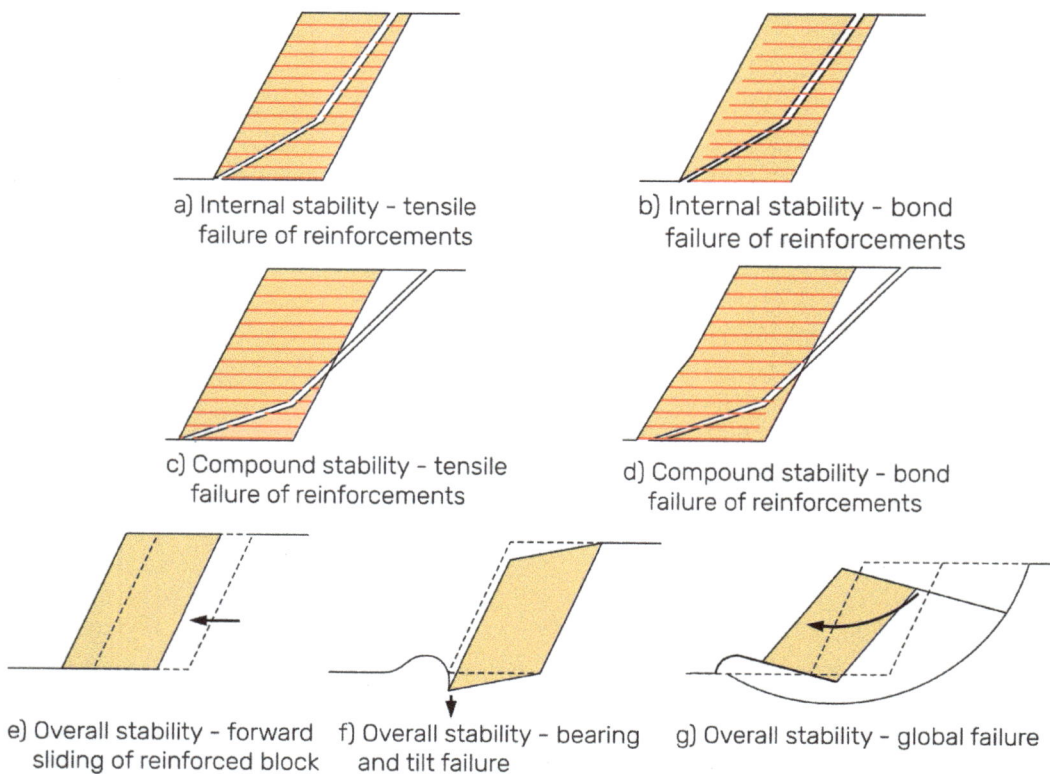

a) Internal stability - tensile failure of reinforcements

b) Internal stability - bond failure of reinforcements

c) Compound stability - tensile failure of reinforcements

d) Compound stability - bond failure of reinforcements

e) Overall stability - forward sliding of reinforced block

f) Overall stability - bearing and tilt failure

g) Overall stability - global failure

Figure 5.22 Potential failure modes for non-standard geometry reinforced fill slopes.

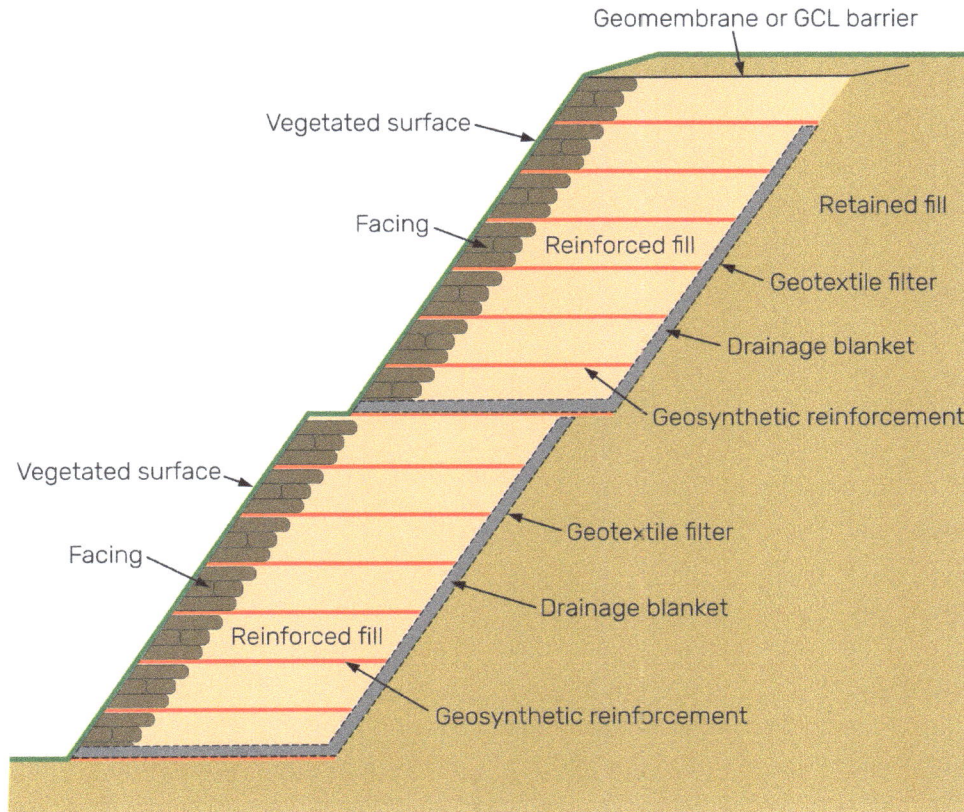

Figure 5.23 Subsurface drainage system layout for a steep reinforced fill slope.

able to deal effectively with this potential mechanism for these more complex geometries.

5.3.7 The importance of good drainage for steep reinforced fill slopes

Many steep reinforced fill slopes are required to have long design lives. As the compacted reinforced fill provides the major shear resistance component in a reinforced slope, it is important that it is protected from external deleterious effects. The major deleterious effect is the action of groundwater and surface water run-off on the reinforced fill zone over time. The presence of water in the reinforced fill and retained fill zones reduces the internal shear resistance and increases deformations and should be prevented by the provision of good quality drainage measures within the reinforced fill slope structure.

Figure 5.23 shows an example of good quality drainage measures provided for a steep reinforced fill slope. To maintain the reinforced fill zone in a stable, dry, condition a quality drainage system layout that encapsulates the reinforced fill zone should be provided. The major structural drainage component is the sloping drainage blanket at the rear of the reinforced fill zone because this acts to reduce the groundwater level before it enters the reinforced fill zone. This ensures a low groundwater head within the base of the reinforced fill zone, thus maintaining its structural integrity. The

drainage layer across the base of the slope tier transfers the groundwater seepage to the face of the slope where it can then be discharged.

Figure 5.23 also shows the provision of a barrier across the crest of the reinforced fill zone to prevent the ingress of surface water run-off into the top of the reinforced fill. This can be provided by a geomembrane or a GCL.

5.4 Reinforced fill retaining walls

5.4.1 Introduction and brief historical overview

Geosynthetic reinforcement can be used to enhance the stability of reinforced fill retaining walls by resisting the forces that would normally cause failure. The technique, shown in Figure 5.24, is the same as that for reinforced fill slopes shown in Figure 5.15, and utilizes horizontal layers of geosynthetic reinforcement placed in the compacted reinforced fill at predetermined vertical spacings. The layers of geosynthetic reinforcement provide additional shear resistance in combination with that of the compacted reinforced fill which results in a stable, constructed retaining wall.

The layers of geosynthetic reinforcement work by intersecting the critical failure plane in the reinforced fill which results in tensile stresses generated in the geosynthetic reinforcement. These tensile stresses are then dissipated back into the surrounding reinforced fill

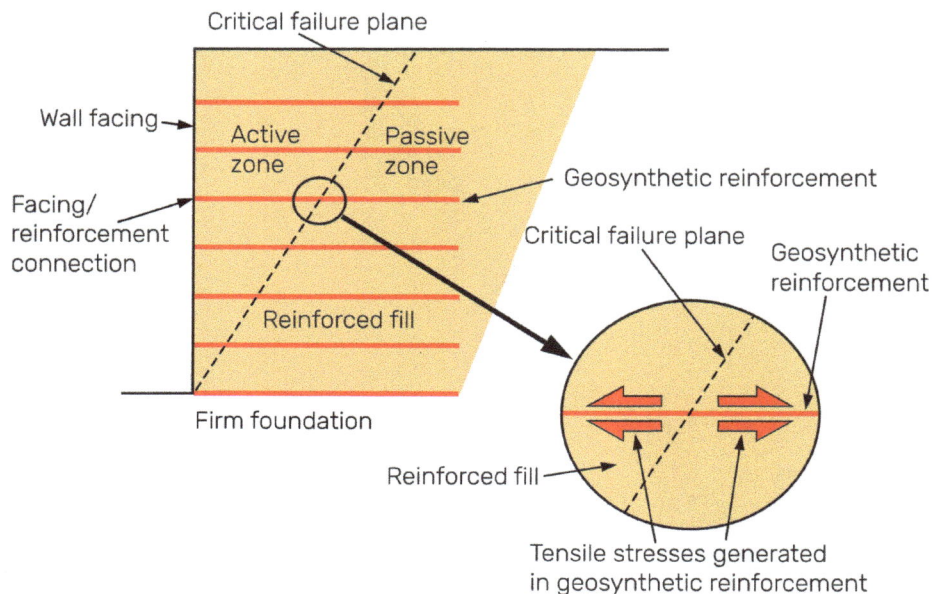

Figure 5.24 Basic features of a reinforced fill retaining wall.

by the action of reinforcement/reinforced fill bond further away from the critical failure plane. This dissipation of reinforcement tensile stresses occurs in both the active zone (towards the wall face) and passive zone (away from the wall face).

As with reinforced fill slopes (Section 5.3), most of the shear resistance of reinforced fill retaining walls is provided by the compacted reinforced fill (this can be as much as 80% of the total) while the minority is provided by the geosynthetic reinforcement (which can be as small as 20% of the total). Thus, for good long-term stability it is important that the behaviour of the reinforced fill remains constant and does not deteriorate over time. This requirement can place constraints on the types of suitable reinforced fill used for long-term reinforced fill retaining walls.

Reinforced fill retaining walls are commonly known as Mechanically Stabilized Earth (MSE) structures in North and South America but are normally referred to as reinforced fill (or soil) retaining walls in the rest of the world.

The first reinforced fill retaining wall using geosynthetic reinforcement was constructed in France in the early 1970's. Here, medium modulus woven polyester yarn strips were cast directly into the wet concrete facing panels. It was observed that over a relatively short period of time the strips showed deterioration in the vicinity of the concrete panels due to alkaline hydrolysis of the polyester yarns. This gave rise to suspicion of the long-term durability of polyester reinforcements.

During the late 1970's reinforced soil retaining walls using geostrip geocomposite reinforcements began to be used in the United Kingdom (UK), (Jones 1996). These geostrip geocomposites consisted of high modulus polyester yarns encased in polyethylene polymer which provided an ideal combination of strength, tensile stiffness and durability. However, because of intellectual property litigation, the use of geostrip reinforcements in reinforced fill retaining walls were subsequently curtailed and did not achieve major growth opportunities until the mid-1980's when the original patents on the Terre Armée metallic strip system had expired. Today, geostrip reinforcements are used in reinforced fill retaining walls throughout the world and have been proven advantageous in different soil environments.

During the early 1980's several generic systems using full-height concrete panels and others using gabion facings were beginning to be used. Some of these were later based on proprietary intellectual property.

Also, during the early 1980's reinforced fill retaining walls began to be constructed using HDPE sheet geogrid reinforcements in the United Kingdom and the United States (e.g. Berg, et al, 1986). By the mid-1980's reinforced fill retaining walls were also being constructed using high modulus polyester geogrids and high modulus polyester woven and knitted geotextile reinforcements. Today, the use of these different geosynthetic reinforcements has become common practice in reinforced fill retaining walls.

Much of the impetus for reinforced fill retaining wall development has been led by the patenting of novel proprietary retaining wall systems. This first started in the late 1960's when the Terre Armée system was originally developed in France and which used concrete segmental facing units with metallic strip reinforcements, Schlosser and Vidal (1969). By the early 1980's several generic systems using full-height concrete panels and gabion facings were beginning to be used. Also, by the mid-1980's the first patented proprietary reinforced segmental block retaining wall systems began to be

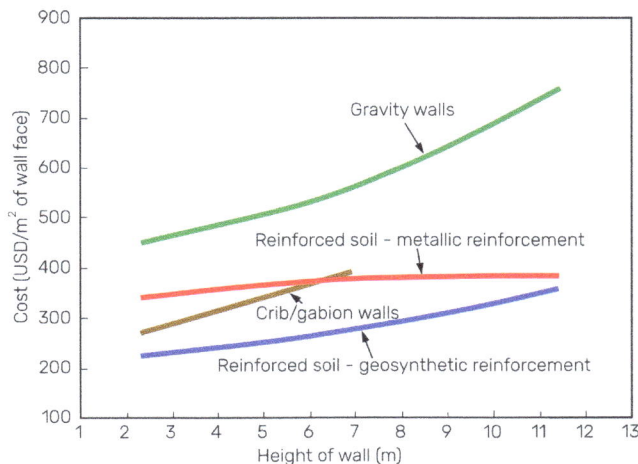

Figure 5.25 Cost of different retaining wall systems (after Koerner, 2012).

used in the United States using geogrid reinforcements, Bathurst and Simac (1994).

Today, the reinforced fill retaining wall technique is ubiquitous the world over and has become standard construction practice for transportation and mining infrastructure. The attractiveness of the reinforced fill retaining wall solution is based around one of relative cost-effectiveness compared with other retaining wall systems. Koerner (2012) has published relative cost data on different retaining wall systems in the USA (shown in Figure 5.25) where geosynthetic reinforced fill retaining walls are shown to provide the lowest cost solution for a wide range of wall heights.

One of the major cost advantages of reinforced fill retaining walls is that their construction is considered to be more in line with an earthworks operation, rather than a more-conventional structural operation (such as for traditional concrete gravity walls). This results in quicker construction using standard earthworks equipment, and hence lower cost.

5.4.2 The components of reinforced fill retaining walls

Reinforced fill retaining walls are composed of components which act together as an overall system. It is this resulting system that provides the retaining wall with its economics, performance, and aesthetics. (This is the same as occurs with steep reinforced fill slopes - see Section 5.3). The individual components of a reinforced fill retaining wall system comprise the wall facing, the geosynthetic reinforcements, the connection between the geosynthetic reinforcements and the wall facing, and the reinforced fill.

As discussed already, many reinforced fill retaining wall systems began as patented proprietary systems based on specific component inventions and use. Over time, and with the expiry of the associated intellectual property, many of these (original) proprietary systems

have become generic (and commonly utilized) systems today.

5.4.2.1 The facing units

Figure 5.26 shows four examples of different reinforced fill retaining wall facings – concrete segmental panels (Figure 5.26a), concrete segmental blocks (Figure 5.26b), wire mesh (Figure 5.26c) and geotextile wrapped (Figure 5.26d). Concrete segmental panels and concrete segmental block facings were developed as part of proprietary systems, while wire mesh facings were developed from a combination of proprietary and generic systems. Geotextile-wrapped facings were developed as part of generic systems.

Concrete segmental panel facings (Figure 5.26a) are precast units made of reinforced concrete. They typically have cruciform, rectangular, T-shaped or hexagonal face shapes depending on the specific proprietary system used. The segmental panels are around 75 mm in thickness and have a face area of around 1 m²/unit. Wall facing aesthetics can be enhanced by the casting of relief patterns onto the segmental units (e.g. see Figure 5.26a).

Concrete segmental block facings (Figure 5.26b) are precast units made of unreinforced concrete blocks. They are more commonly made by dry-casting offsite although they may be made by wet-casting onsite. The segmental blocks come in a variety of sizes (20 kg to 100 kg) and shapes depending on the specific proprietary system involved. They can have smooth or rough "split-stone" surface finishes to enhance aesthetics and can be produced in different colours.

Wire mesh facings (Figure 5.26c) consist of sheets of semi-rigid wire mesh that are bent to form the wall facing at the required face angle (70° to 90°). In this regard they are identical to the wire mesh facings used for steep reinforced fill slopes shown in Figure 5.17b. The height of these units is typically around 0.5 m, but they may be less or greater than this. Inside the wire mesh facing an open geotextile is placed to prevent wash-out of the retained topsoil which acts as the medium for surface vegetation growth. This facing type normally forms part of a vegetated surface finish following completion of construction.

Geotextile-wrapped facings (Figure 5.26d) have the geotextile reinforcement wrapped around at the face of the wall to contain the reinforced fill. This type of retaining wall facing is normally limited to temporary retaining walls (several months to 1 to 2 years design life) as the geotextile is left continually exposed at the wall face.

5.4.2.2 The geosynthetic reinforcements

With the geosynthetic reinforcements supporting a minority of the total shear resistance in a reinforced fill retaining wall (and the reinforced fill supporting the majority), the resulting tensile loads generated

a) Concrete segmental panel facings

b) Concrete segmental block facings

c) Wire mesh facings

d) Geotextile-wrapped facings

Figure 5.26 Four examples of different reinforced fill retaining wall facings.

in the reinforcements are relatively low compared to other reinforced soil applications, e.g. basal reinforced embankments on soft foundations discussed in Section 5.2. Consequently, the types of geosynthetic reinforcements used in reinforced fill retaining walls are either relatively low-strength woven and knitted geotextiles (see Figure 5.27a), geogrid sheets or bonded strips (see Figures 5.27b and 5.27c) or geostrips (see Figure 5.27d). Further, to ensure the geosynthetic reinforcement and reinforced fill behaves as a composite system within the retaining wall structure, the vertical spacings of the reinforcements are commonly limited to between 0.25 m and 0.5 m depending on the vertical location of the facing/reinforcement connection points.

The load carrying polymers used in the geosynthetic reinforcements for reinforced fill retaining walls are either high modulus polyester (PET) yarns, polyvinyl acetate (PVA) yarns, high modulus polypropylene (PP) yarns (for some generic wall types) or uniaxial high-density polyethylene (HDPE) geogrid sheets. These geosynthetic reinforcement types are discussed in Chapter 1 of this handbook and in Section 5.1. These polymer types

provide the ideal combination of geosynthetic reinforcement properties well-suited to reinforced fill retaining walls (in terms of strength, strain and durability).

5.4.2.3 The connections between geosynthetic reinforcements and facing units

The connections between the geosynthetic reinforcements and the facing units transfer any residual tensile stresses in the reinforcements at the wall face to the facing units. This requires the facing units to function as some form of structural component at the wall face.

Depending on the type of connection used at the face, three different types of mechanical behaviour can occur, see Figure 5.28. The difference in mechanical behaviour can be described in terms of the reinforcement connection load T_{con} versus the vertical load acting on the wall connections P_v, and are:

> Full positive connection behaviour, Figure 5.28a. Full positive connection behaviour occurs where the connection capacity, T_{con}, is constant and is independent of the wall facing load

a) Knitted geotextile sheet reinforcement

b) Geogrid sheet reinforcement

c) Geogrid strip reinforcement

d) Geocomposite geostrip reinforcement

Figure 5.27 Four examples of different geosynthetic reinforcements used for reinforced fill retaining walls. Photos courtesy of Solmax.

acting at the wall connection, P_v. An example of facing units with full positive connection are the segmental block units shown in Figure 5.28a where the geogrid reinforcement is constrained within the segmental block facing units to give a full positive connection. It is also common for concrete segmental panel facings having strip reinforcements to have full positive connections (as shown in Figures 5.27c and 5.27d). Likewise, woven and knitted geotextile-wrapped walls also behave as full positive connections at the wall face, as shown in Figure 5.26d.

> Combination connection behaviour, Figure 5.28b. Combination connection behaviour occurs where the connection capacity, T_{con}, has a positive connection component followed by a friction component. A typical example of this is the segmental block units shown in Figure 5.28b where rigid polymer pins provide positive connection resistance to the geogrid reinforcement and where the horizontal surfaces of the

blocks and the stone infill provide the additional frictional resistance. Here, the rigid polymer pins may provide limited positive connection capacity while the frictional connection component provides additional connection capacity depending on the wall facing vertical stress.

> Frictional connection behaviour, Figure 5.28c. Frictional connection behaviour occurs where the connection capacity, T_{con}, increases linearly as the vertical facing load acting at the wall connection P_v, and increases until the full frictional resistance is mobilized, and where the reinforcement can then pull out of the connection. A typical example is the segmental block units shown in Figure 5.28c where the horizontal surfaces of the blocks provide the frictional surfaces for the geogrid reinforcement connection. Several different concrete segmental block units have frictional connections as do wire mesh and gabion facings.

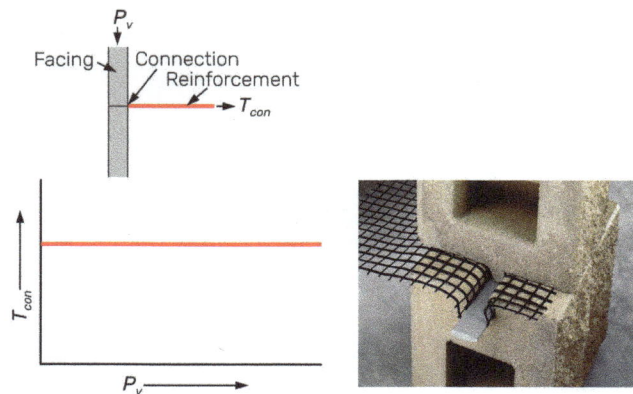

a) Full positive connections (photo courtesy Anchor Wall Systems)

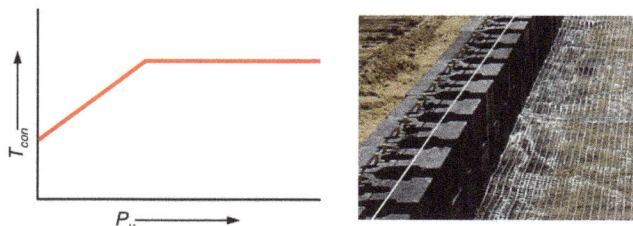

b) Combination connections

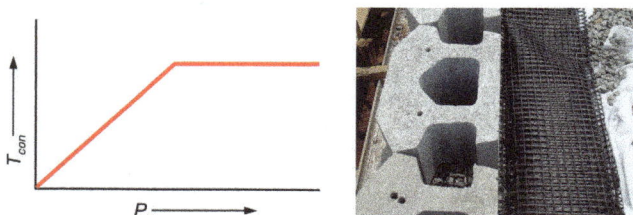

c) Frictional connections

Figure 5.28 Three different types of connection behavior (after Lawson, 2022).

5.4.2.4 The reinforced fill

With the compacted reinforced fill contributing most of the shear resistance in a reinforced fill retaining wall, it is important that its behaviour remains constant over the required design life of the structure. The required design lives for reinforced fill retaining walls can be as long as 120 years (e.g. see Table 5.1), thus for walls designed for these long periods of time it is important that the reinforced fill remains stable over these long time periods.

The reinforced fills used for reinforced fill retaining walls are considered to be purely frictional where the peak shear resistance is provided by the peak frictional resistance f'_p – any cohesive component is neglected.

The reinforced fills used for reinforced fill retaining walls can be divided into two categories – frictional fills and cohesive-frictional fills, with the typical gradation limits for these two types shown in Figure 5.29. Frictional fills have lower fines content than cohesive-frictional fills and are more free-draining and less moisture susceptible. They are more commonly manufactured to meet specific fill performance requirements and are thus generally more costly. Cohesive-frictional fills have

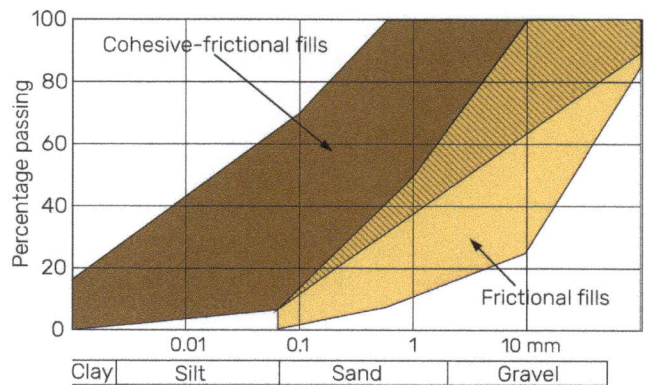

Figure 5.29 Typical grading envelopes for different types of reinforced fills for retaining walls (after Lawson, 2022).

gradations that are typical of many locally available "granular" fills, and thus are not normally manufactured (except for some primary crushing) which gives them their relative cost advantage.

Compacted frictional fills have very good shear resistance with compacted insitu f'_p values ranging from 45° to 55° at low strains of ≤5%. This high insitu shear resistance provides good stability and margin of safety when designing reinforced fill retaining walls, and consequently, these fills are well-suited for good quality reinforced fill retaining structures of long design lives.

Cohesive-frictional fills also have acceptable shear resistance with compacted insitu f'_p values ranging from 30° to 36° but at higher strains of 8% to 15% (c.f. compacted frictional fills). While these fill types are more readily available than frictional fills, it is important to realise that they don't provide the same margin of safety as frictional fills, and consequently, more care should be taken with their design and use, such as close compaction controls and provision of good drainage measures (both surface and subsurface) surrounding the reinforced fill zone. Various National authorities place limits on the amount of fines (clay fractions) allowed for cohesive-frictional fills based on experience. For example, the U.S. Federal Highway Administration (FHWA, 2021) have placed limits on the percentage of the clay fraction allowed in cohesive-frictional fills by limiting the maximum amount to <15% passing the No 200 sieve. This prescribed limit is the result of experience with using strip reinforcements in long-term retaining walls and is a more conservative grading limit than what is shown in Figure 5.29. Conversely, the U.S. National Concrete Masonry Association (NCMA, 2009) has placed an upper limit on fines of <35% passing No 200 sieve (clay) for retaining walls that utilize segmental block facings with geogrid and geotextile reinforcements. When using cohesive-frictional fills care should be taken with regard to the percentage of the clay component as well as close attention to compaction and drainage.

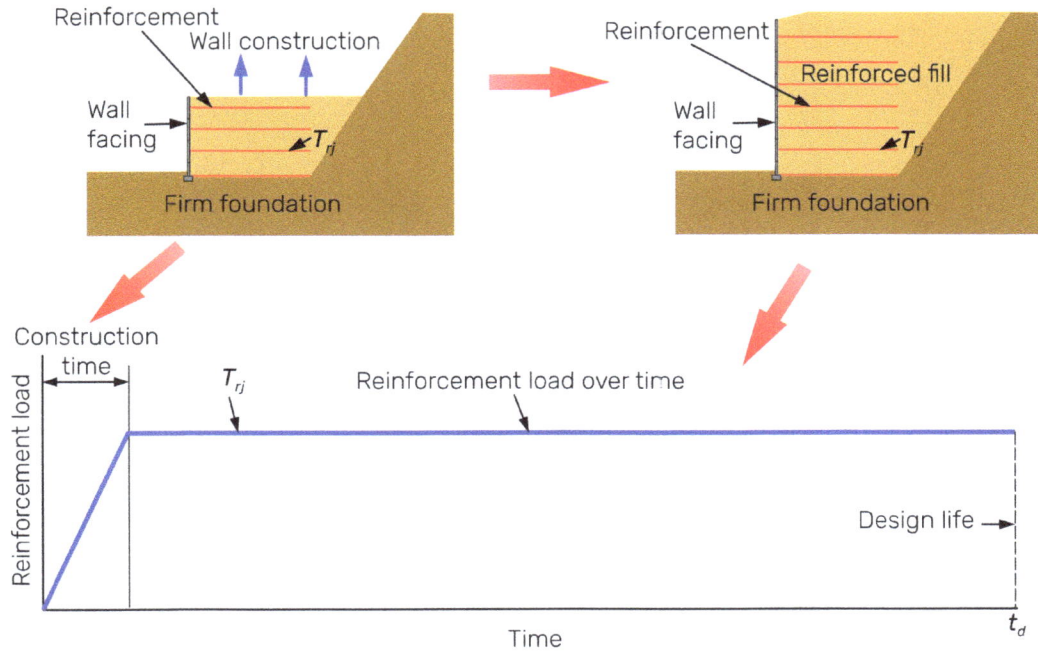

Figure 5.30 Tensile load in geosynthetic reinforcements over time for reinforced fill retaining walls (after Lawson, 2022).

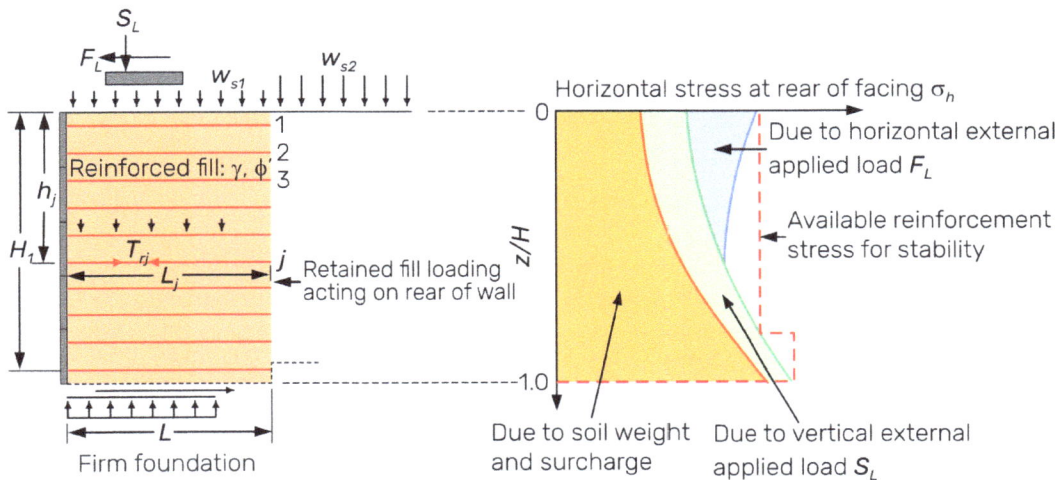

Figure 5.31 "Standard" geometry reinforced fill retaining walls — stability effects to be considered (after Lawson, 2022).

5.4.3 Tensile loads in the geosynthetic reinforcements over time

The tensile load profile in the geosynthetic reinforcements in reinforced fill retaining walls is shown in Figure 5.30 and is identical to that for steep reinforced fill slopes shown in Figure 5.18. During retaining wall construction, the tensile loads increase in the geosynthetic reinforcement layers until such time as the wall construction is complete. Following this, the tensile loads (denoted by T_{rj}) remain constant with time for the remaining design life of the reinforced fill retaining wall (denoted by t_d). The required design lives for reinforced fill retaining walls may range from several years for temporary walls, to many years (75 to 120 years) for long-term structures. Thus, it is important for the geosynthetic reinforcements to carry these design tensile loads over the required design life of the reinforced fill retaining wall.

5.4.4 "Standard" geometry versus "non-standard" geometry reinforced fill retaining walls

Standard geometry reinforced fill retaining walls are the most used structures for infrastructure applications. The typical "standard" geometry is shown in Figure 5.31 where the retaining wall is founded on a firm, flat foundation, and supports vertical and horizontal externally applied loads commonly in the form of parapet loadings for highway structures. Consequently, the loads that generate the horizontal out-of-balance stresses at the rear of the wall facing are due to self-weight of the reinforced fill and surcharge loading, vertical external

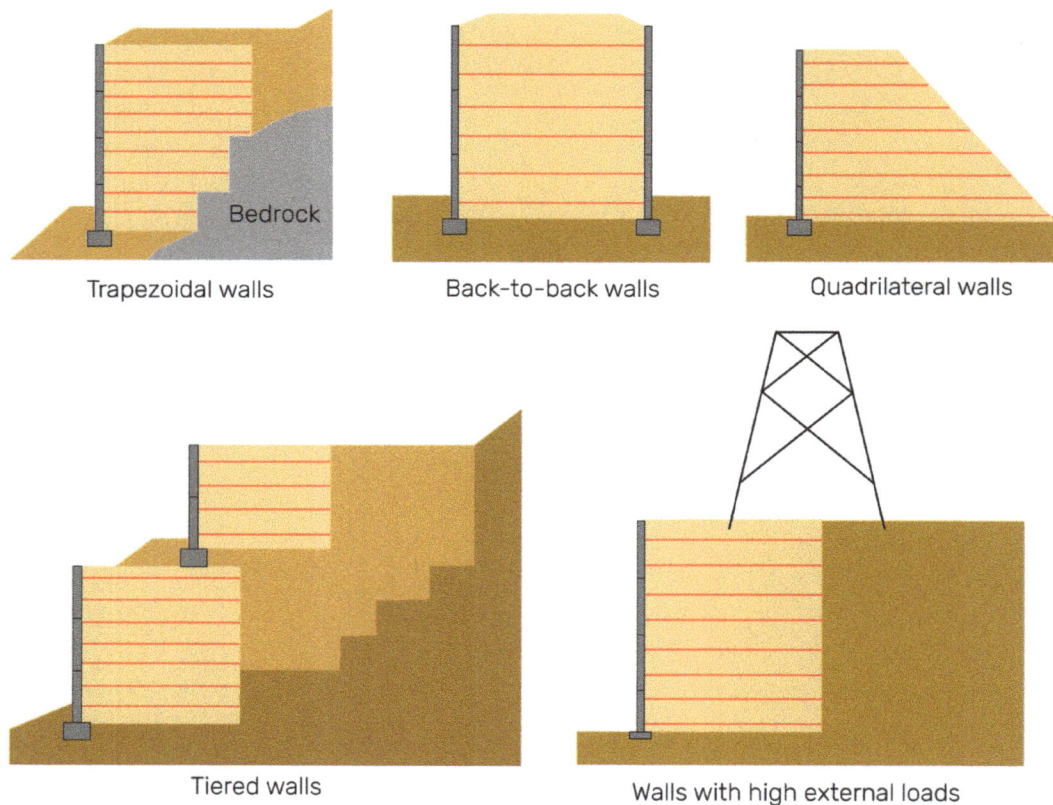

Figure 5.32 Some examples of "non-standard" geometry reinforced fill retaining walls (adapted from BS8006:2010).

applied loads S_L and horizontal external applied loads F_L on the crest of the wall.

The combination of these three load effects provides the total horizontal stress that must be resisted by the layers of geosynthetic reinforcement in the retaining wall. The red dashed line in the right-side diagram of Figure 5.31 is the required reinforcement load profile for stability throughout the wall height that resists the horizontal stresses due to these three out-of-balance load effects.

A common feature of standard geometry reinforced fill retaining walls is a firm, horizontal foundation across the base of the wall ensuring that no critical failure plane can pass through the foundation. In this case, the critical failure surface passes through the toe of the retaining wall, and standard methods of design/analysis exist for this type of reinforced fill retaining wall.

Non-standard geometry reinforced fill retaining walls are normally classed as those walls that have unusual geometries due to either their shape or due to the specific subsurface strata present, or they are subject to unusual external loads. The design of these walls must be handled with care and in a more detailed manner compared to standard geometry walls to ensure they are constructed to the required level of safety. Figure 5.32 provides some examples of non-standard reinforced fill retaining walls.

5.4.5 The importance of good drainage for reinforced fill retaining walls

Many reinforced fill retaining walls are required to have long design lives. To perform effectively over these long periods of time it is important that the compacted reinforced fill environment in which the structure is located remains constant with time. This ensures the reinforced fill system performs in a consistent, predictable, manner over its design life.

The major deleterious effect on retaining wall performance is the occurrence of groundwater and surface water run-off behind and above the reinforced fill zone. The presence of water in the reinforced fill and retained fill zones reduces the internal shear resistance of the fill and increases deformations and should be prevented by the provision of good quality drainage measures.

Figure 5.33 shows the provision of good drainage measures surrounding a reinforced segmental block retaining wall and can be considered typical of reinforced fill retaining walls in general. To maintain the reinforced fill zone at constant, dry, conditions a quality drainage system encapsulates the reinforced fill zone. The major structural drainage component is the sloping drainage blanket at the rear of the reinforced fill zone which intersects and reduces the groundwater level at this location. This ensures a low groundwater head within the reinforced fill zone with groundwater ingress prevented from entering. This maintains the structural

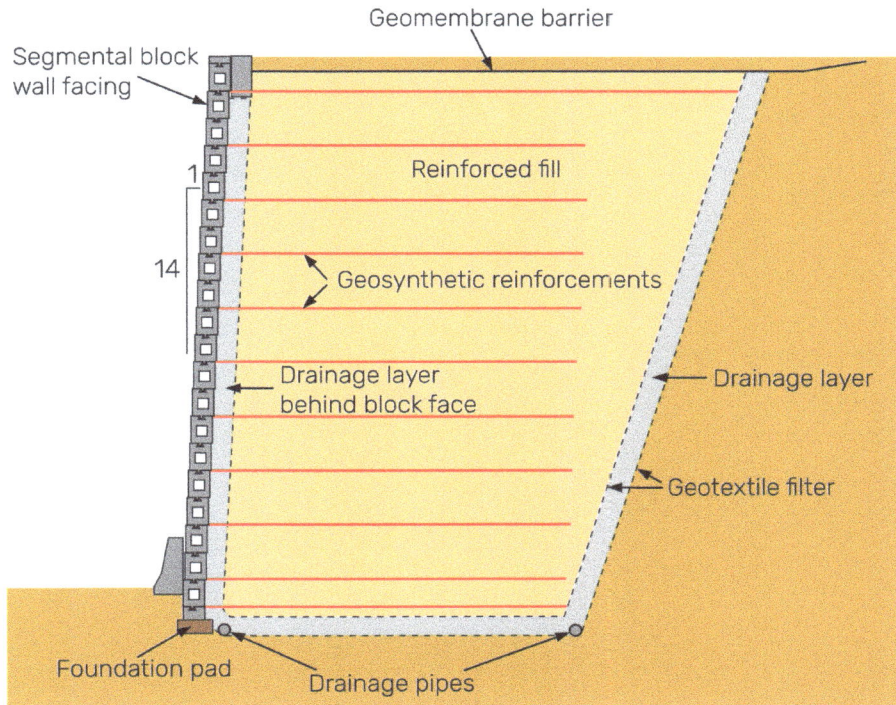

Figure 5.33 Good drainage measures for reinforced fill retaining walls.

integrity of the compacted reinforced fill. The drainage layer across the base of the wall transfers the groundwater seepage to the face of the wall where it can be discharged. Chapter 3 provides details regarding good drainage practice.

Figure 5.33 also shows a vertical drainage layer immediately behind the segmental block facing. It should be noted that this inclusion is good practice although it does not have a structural function for the retaining wall. This vertical drainage layer acts to allow seepage to pass out through the wall facing without erosion of the reinforced fill.

In addition, Figure 5.33 shows the provision of a geomembrane barrier across the crest of the retaining wall to prevent the ingress of surface water run-off into the top of the reinforced fill zone.

5.5 Reinforced soil fill berms

5.5.1 Introduction

Stability berms are constructed of fill materials at the toes of slopes to provide a counterweight to resist rotational and sliding failures. Reinforced fills provide an effective solution as counterweight berms to stabilize slopes because their geometry can be adapted according to land boundaries. One of the most common uses of this technique is to enable the vertical expansion of waste landfill facilities. Figure 5.34 shows two examples, using different stabilizing geometries, where reinforced fill berms have been used to vertically expand the capacity of waste landfill facilities.

5.5.2 Reinforced fill berm details

Figure 5.35 shows the principle of use of a peripheral counterweight berm to enable the vertical expansion of a landfill facility. Here, the stabilizing weight of the peripheral berm W provides an inward horizontal resistance of $W \tan f'$ across the surface of the foundation, and this resists the destabilizing outward thrust of the vertical landfill expansion F_w. The magnitude of the required peripheral berm weight can be determined according to the magnitude of the destabilizing outward thrust of the landfill expansion.

Depending on the berm dimensions, the layers of geosynthetic reinforcement may pass all the way through the containment berm, as shown in Figure 5.36, or may reinforce the sides of the berm slopes only in the same manner as discussed for reinforced fill slopes in Section 5.3.

For effective performance, the facing details of the berm is just as important as the structural behaviour of the reinforced fill. On the internal slope of the berm face an effective landfill barrier system along with landfill drainage should be installed, as shown in Figure 5.36. This ensures a proper landfill barrier system is in place for the landfill expansion. Also, on the external slope of the berm face an effective vegetated surface is to be provided for good, long-term, surface erosion control.

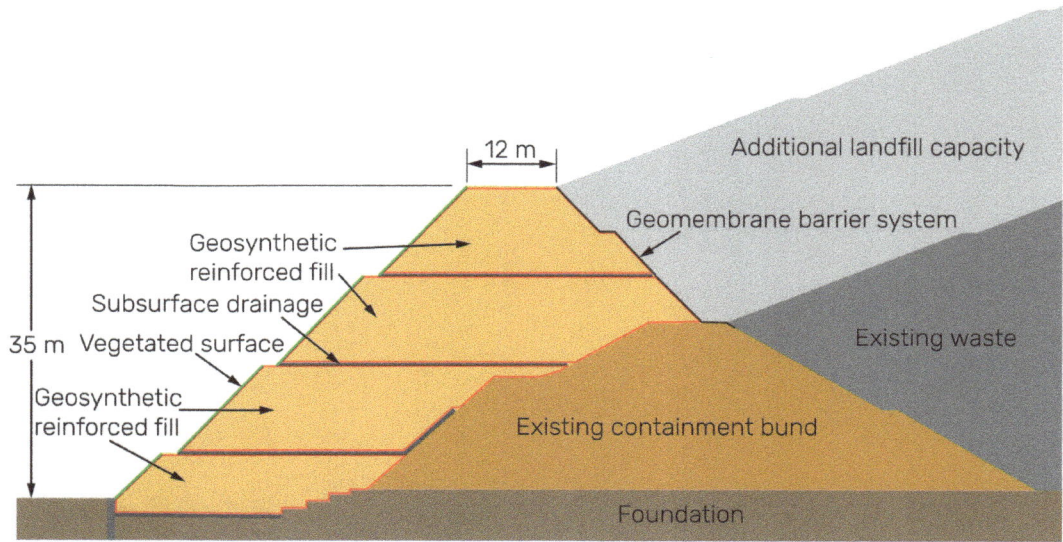

a) Xinfeng landfill expansion, China

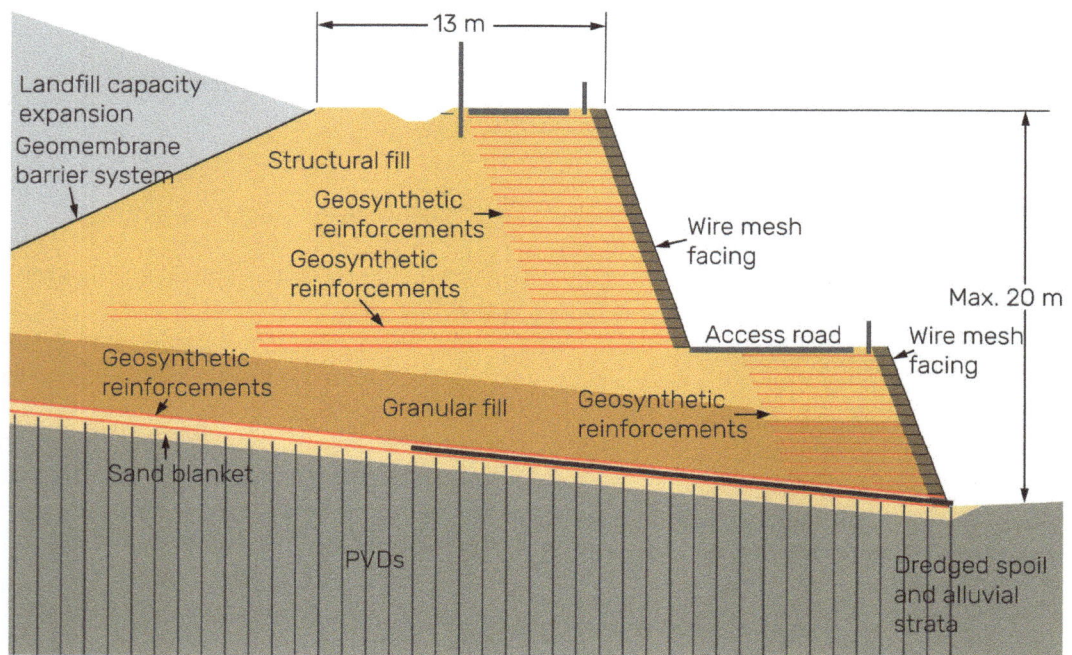

b) Cherry Island landfill expansion, USA

Figure 5.34 Two examples of the use of reinforced fill berms to enable the vertical expansion of waste landfill facilities. Diagrams courtesy of Solmax.

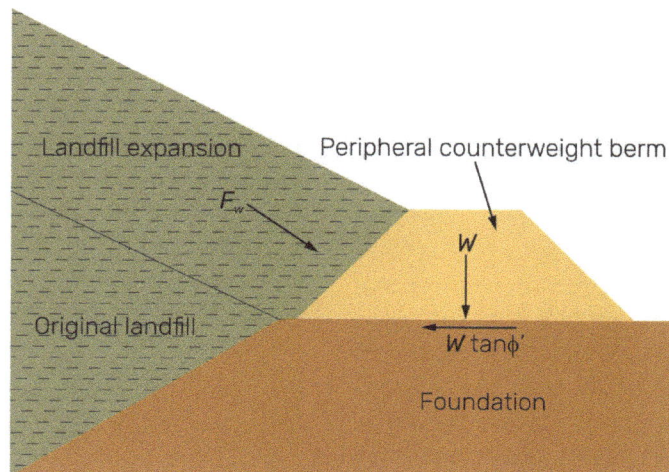

Figure 5.35 The use of a counterweight berm to enable the vertical expansion of landfill capacity.

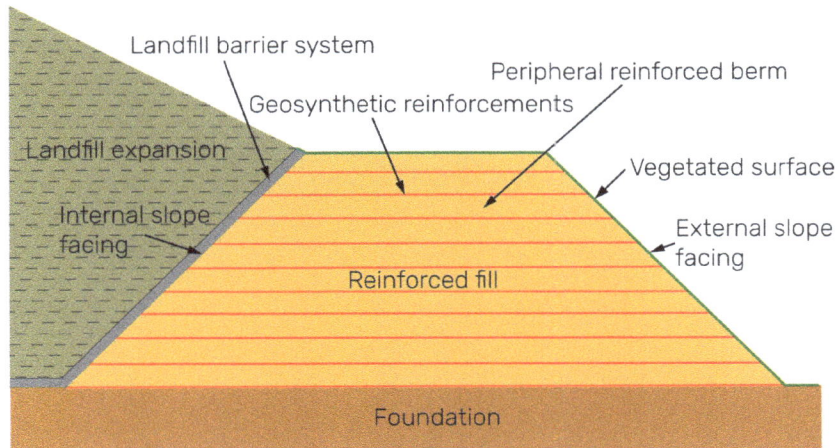

Figure 5.36 Typical reinforced fill containment berm for vertical landfill expansion.

5.6 References

Bathurst, R.J. and Simac, M.R. (1994). Geosynthetic reinforced segmental retaining wall structures in North America, Proceedings Fifth International Conference on Geotextiles, Geomembranes and Related Products, Singapore.

Berg, R.R., Bonaparte, R., Anderson, R.P. and Chourey, V.E. (1986). Design, construction and performance of two geogrid reinforced soil retaining walls, Proceedings Third International Conference on Geosynthetics, Vienna, Austria.

Briançon, L., Faucheux, G. and Andromeda, J. (2008). Full-scale experimental study of an embankment reinforced by geosynthetics and rigid piles over soft soil, Proceedings EuroGeo 4, Edinburgh, UK, 8pp.

BS8006:2010. Strengthened/reinforced soils and other fills, British Standards Institution, UK.

BS8006:1995. Strengthened/reinforced soils and other fills, British Standards Institution, UK.

Collin, J.G., Watson, C.H. and Han, J. (2005). Column-supported embankment solves time constraint for new road construction, American Society of Civil Engineers, Geotechnical Special Publication 131, USA, pp.1-10.

EBGEO (2011). Recommendations for design and analysis of earth structures using geosynthetic reinforcements, Ernst and Sohn, Germany.

Fowler, J. (1982). Theoretical design considerations for fabric-reinforced embankments, Proceedings Second International Conference on Geotextiles, Las Vegas, USA, IFAI, pp.665-670.

Gartung, E. and Verspohl, J. (1996). Geogrid reinforced railway embankment on piles – monitoring, Proceedings IS Kyushu '96, Fukuoka, Japan, Balkema, pp.209-214.

FHWA (2021). Mechanically stabilized earth (MSE) wall fills, Publication no. FHWA-HIN-21-002, Federal Highway Administration, USA.

Holtz, R.D. and Massarsch, K.R. (1976). Improvement of the stability of an embankment by piling and reinforced earth, Proceedings Sixth European Conference or Soil mechanics and Foundation Engineering, Vienna, Austria, Vol. 1, pp.473-478.

Jewell, R.A. (1996). Soil reinforcement with geotextiles, Special Publication 123, CIRIA, UK.

Jones, C.J.F.P. (1996). Earth reinforcement and soil structures, Third Edition, Thomas Telford Ltd, UK.

Koerner, R.M. (2012). Designing with geosynthetics, Sixth Edition, Xlibris Corporation, USA.

Lawson, C.R. (2022). Geosynthetic reinforcements, Lawson Publishing, Malaysia.

Leschinsky. D. and Boedecker, R.H. (1989). Geosynthetic reinforced earth structures, Journal Geotechnical Engineering Division, ASCE, Vol. 115, No. 10, pp1459-1478.

Lui, H.L., Ng, C.W.W. and Fei, K. (2007). Performance of a geogrid-reinforced and pile-supported highway embankment over soft clay: case study, Journal of Geotechnical and Geoenvironmental Engineering, Vol. 133, No. 12, ASCE, USA, pp.1483-1493.

NCMA (2009). Design manual for segmental retaining walls, National Concrete masonry Association, 3rd Edition, USA.

Reid, W.M. and Buchanan, N.W. (1984). Bridge approach support piling, Piling and Ground Treatment, Thomas Telford Ltd, UK, pp. 267-274.

Risseeuw, P. and Voskamp, W. (1993). Geotextile reinforcement for a highway embankment across swamp land, IR Sediyatmo Highway, Indonesia, Geosynthetics case histories, Bitech Publishers Ltd, Canada, pp. 216-217.

Schlosser, F. and Vidal, H. (1969). Le Terre Armée, Bulletin de Liaison Laboratoire Routes, Ponts et Chaussées, No. 41, Paris, France.

Tan, S.B., Tan, S.L., Poh, K.B., Yang, K.S. and Chin, Y.K. (1985). Settlement of structures on soft clays in Southeast Asia, Geotechnical engineering in Southeast Asia, Edited by A.S. Balasubramaniam, D.T.

Bergado and S. Chandra, Balkema, Netherlands, pp.127-145.

Taylor, D.W. (1948). Fundamentals of soil mechanics, Wiley, New York.

van Eekelen, S.J.M. and Brugman, M.H.A. (2016). Design guideline basal reinforced piled embankments, CRC Press Balkema, Netherlands.

van Eekelen, S.J.M., Bezuijen, A. and Alexiew, D. (2010). The Kyoto Road, monitoring a piled embankment, comparing 3 1/2 years of measurements with design calculations, Proceedings Ninth International Conference on Geosynthetics, Brazil, pp.1941-1944.

Volman, W., Krekt, L. and Risseeuw, P. (1977). Tensile reinforcement by textiles – a new method for improving the stability of large fills on soft soil, Proceedings International Conference on the Use of Fabrics in Geotechnics, Paris, France, LCPC, pp.55-59.

Chapter 6

Geosynthetic Barriers in Seepage Control Systems

6.0 Introduction

The chapter on geosynthetics in seepage control systems covers applications such as ponds, reservoirs, canals and levees for the planning, construction and maintenance of geosynthetics, mainly with the use of geosynthetic barriers. The recommendations do not apply to chemical storage facilities, landfills and environmental protection. Other similar seepage control system applications may follow descriptions in this chapter but might need an application related evaluation. Such similar applications could be artificially sealed standing and flowing bodies of water in residential areas and in the open countryside, such as:

> Streams;
> Constructed wetlands;
> Fire-fighting ponds;
> Snowmaking ponds and basins;
> Fishing facilities;
> Sea dykes

6.0.1 General requirements

National applicable standards and other regulations should be specified in the contract specification. Deviations from the state-of-the-art documents should also be noted with reasons. In any case the client should be informed of any known risks. This applies in particular to the use of new materials/components and construction methods that have not yet been sufficiently tested.

6.0.2 Construction regulations or requirements

During the planning process, it must be checked which relevant authority requirements might be required, e.g.:

> Building regulations of the federal states, e.g. on the set-up of a construction site;
> Approval procedures of the federal states and municipalities depending on the size and use of the planned seepage control construction;

> Neighboring rights of the federal states for determining boundary distances;
> Nature and environmental protection law;
> Animal protection law;
> Hygiene and health laws

6.0.3 Proof of suitability

As proof of suitability for supplied geosynthetics the specification should require the type and scope of the proof of su tability as well as the type and number of samples for manufacturing, as well as construction and installation quality and assurance control. In addition, it must be agreed when and under what conditions the results of self-monitoring tests can replace control tests.

6.1 Geosynthetic Barriers in Seepage Control Systems

Seepage control is a critical aspect of engineering, especially in projects involving water retention structures such as ponds, reservoirs, canals and levees. These applications are exclusively covered in this chapter. Geosynthetic barriers play a pivotal role in enhancing the effectiveness of seepage control systems. These geosynthetic materials, which include geomembranes and geosynthetic clay liners, offer several advantages (von Maubeuge et al, 2023), especially if compared to traditional earthen construction methods and include:

> Reliability: high-quality control standards, lifetime verification and multiple proven project applications
> Ecology: significantly lower CO_2 emissions, supporting worldwide climate goals, lower energy consumption, reduction of transport amount or kilometers travelled
> Sustainability: limits the use of all resources, e. g. natural clay resources, energy demand, noise impact
> Cost-effectiveness: reduced building cost compared to traditional methods, even greater

when natural materials have to be quarried off site and brought in, longer service life, less maintenance

> Ease of installation: easy to handle and install on project sites, saving time in the construction process and reduced excavation required, e. g. less fill required, less land disturbed, no need to compact and test
> Resilience: improved structural behavior with the ability to respond, absorb, adapt or recover from extreme load cases caused by natural disasters or extreme weather caused by climate change
> Safety: increased serviceability and protection
> Quality: more homogeneous than soil and clear, established quality controls from pro-duction through installation

This chapter explores the types, applications, and design considerations of geosynthetic barriers in seepage control systems.

6.2 Applications of Geosynthetic Barriers in Seepage Control

6.2.1 Ponds

Ponds are typically small reservoirs with the purpose of storing water and managing water resources especially in areas with seasonal variations in rainfall. Geosynthetic barriers find extensive use in ponds to prevent water seepage out of its storage/collection area. Ponds are versatile structures that contribute to water management Their applications vary depending mainly on environmental conditions and local ecosystems. By installing geosynthetic barriers and other non-barrier geosynthetic materials, engineers can enhance the impermeability of such structures, ensuring the integrity of the water storage system. Ponds can be used to fulfil various purposes. Here are some common applications of ponds:

> Stormwater management: Ponds (Figure 6.1) can function as stormwater management systems, helping to control flooding by collecting and slowly releasing excess rainwater. They act as temporary storage during heavy rainfall events.
> Water storage: Ponds can be constructed to store water for irrigation purposes. This can be particularly important during dry seasons or in regions with unreliable rainfall. In private gardens or public parks ponds contribute to the aesthetic appeal of landscapes but can also be used for recreational activities such as fishing, boating, and bird-watching.
> Water treatment: Ponds can act as natural filters, helping to purify water by trapping sediments and facilitating the breakdown of

Figure 6.1 Installation of a geosynthetic clay liner in a stormwater retention pond. Photo courtesy Naue.

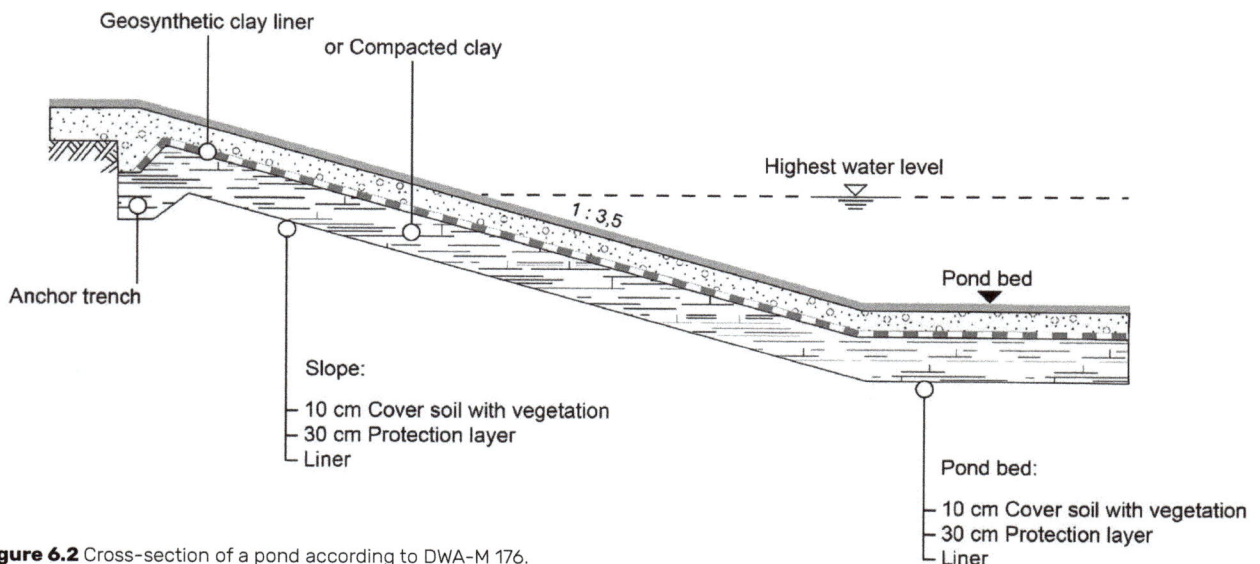

Figure 6.2 Cross-section of a pond according to DWA-M 176.

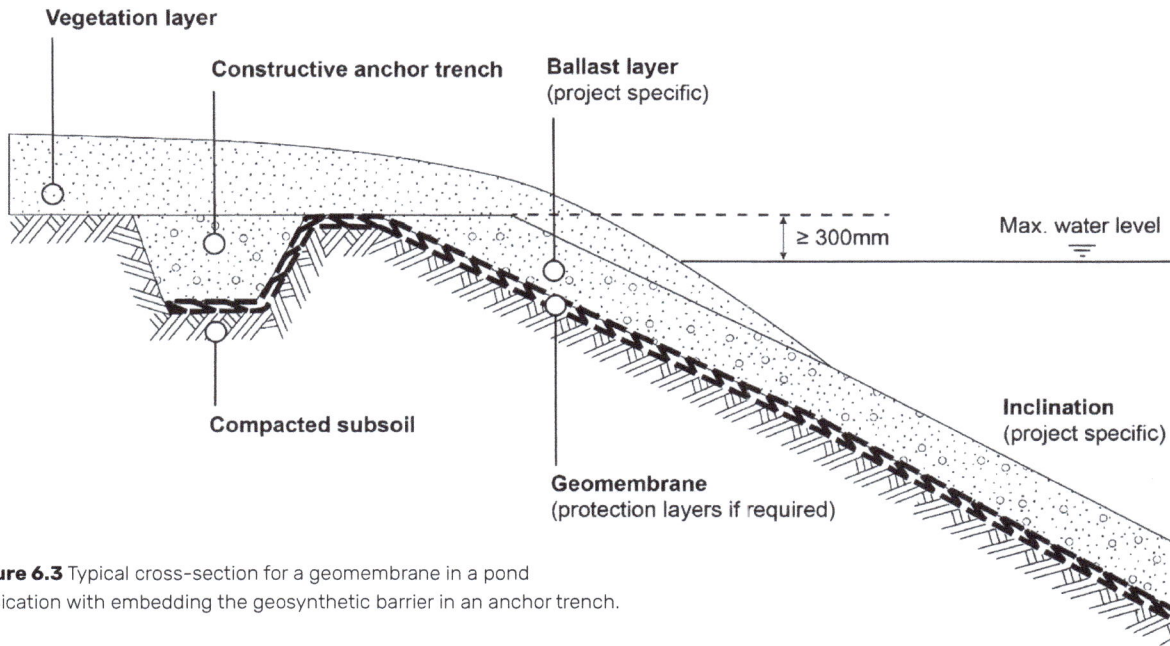

Figure 6.3 Typical cross-section for a geomembrane in a pond application with embedding the geosynthetic barrier in an anchor trench.

Figure 6.4 Geomembrane in a pond application with embedding the geosynthetic barrier in an anchor trench. Photo courtesy IGS 2022 Young Members Photo Contest.

pollutants through biological processes and therefore can improve the water quality.

> Ecosystem support: Ponds can also be used to provide habitats for plants and animals by supporting aquatic life such as fish, amphibians and insects. At the same time ponds can influence their surrounding microclimate.

The purpose of a national design code, e.g. DWA-M 176 (DWA 2013) or FLL (2022) is to provide the designer of stormwater basins with rules that will allow stormwater treatment and retention structures to be designed (Figures 6.2 and 6.3), or existing structures to be upgraded, according to structural, equipment, operational and economic considerations. It also makes clear statements on the use of geosynthetic barriers (GCL and geomembranes).

Further important details for such pond applications are cover soil thickness over the barrier, constant or fluctuating water head, height of water level, slope inclination, etc.

6.2.2 Reservoirs

Reservoirs are large artificial bodies for water collection from rivers or streams. They are critical components of water resource management. Geosynthetic barriers find extensive use in reservoirs to prevent water seepage out of their storage/collection area. Reservoirs are versatile structures that contribute to water management. Reservoirs provide numerous benefits, however their construction as well operation can have significant ecological and environmental impacts. To achieve sustainable water resource management, it is important to balance all relevant factors. By installing geosynthetic barriers and other non-barrier materials, engineers can enhance the impermeability of such structures, ensuring the integrity of the water storage system. Reservoirs can be used to fulfil various purposes. Here are some common applications of reservoirs:

> Water supply: Reservoirs are often used as a primary source of drinking water but during drought periods, reservoirs can provide stored water for various uses, such as for industries and agriculture. Reservoirs can also serve as an emergency water source for firefighting efforts.
> Industrial use: Hydropower reservoirs store water and allow a release through turbines to generate sustainable electricity. Reservoirs may be used by industries for e.g. cooling water for power plants or manufacturing.
> Flood control: During periods of heavy rainfall reservoirs can help to control river flow by storing water. Downstream flooding can be controlled by regulating the release of water.

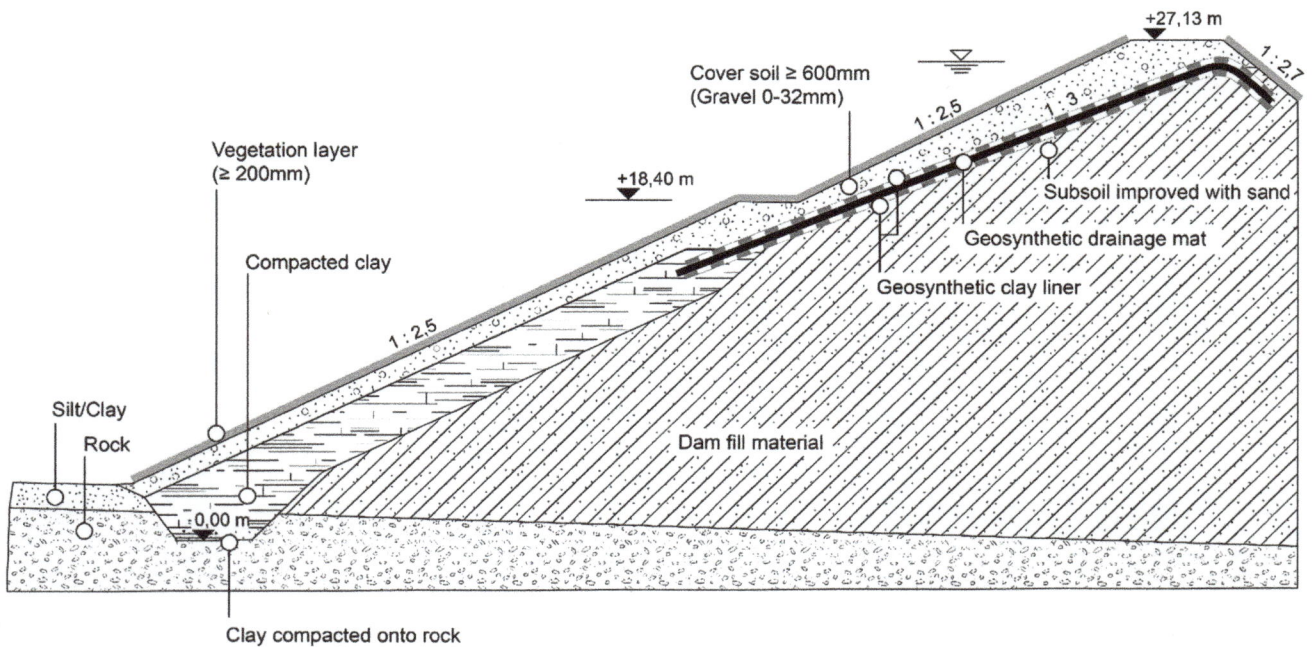

Figure 6.5 Cross-section of hydropower water reservoir with a double-lined geosynthetic clay liner and a leak detection geosynthetic drainage composite.

Figure 6.6 Installation of a geosynthetic clay liner and a leakage collection geosynthetic drainage mat, as well as the cover soil placement in a power water reservoir. Photo Courtesy of Naue. (Note: special considerations are required when planning to install GCLs underwater.)

> Water transportation and recreation: Reservoirs can be used for boats and ships, supporting water transportation. Reservoirs can also provide opportunities for recreational activities such as boating, fishing, and water sports.

6.2.3 Canals

A canal is an artificial waterway constructed to convey water from one location to another for e.g. water supply, industrial uses or to allow boat or ship transportation. Geosynthetic barriers should meet special safety requirements, as the canals depend on their functionality and enormous protective assets and human lives can be affected if the canal systems fail. By integrating geosynthetics into the design, engineers enhance the hydraulic performance of the structure. The low permeability of geosynthetic barriers ensure that water remains within designated areas, preserving the structural integrity of the canal. Canals are an important part of engineering and include:

> Irrigation canals: Transportation of water to agricultural fields for irrigation purposes.
> Navigation canals: Facilitation of boats and ships movement.
> Industrial canals: Transportation of water e.g. to turbines for electricity generation, to water-cooling systems of power plants or for manufacturing purposes.
> Drainage canals: Management of water levels to prevent flooding and drain water from areas.

Government agencies such as the United States Bureau of Reclamation (BuRec) indicate that seepage from unlined irrigation canals and waterways may be substantial and costly; and that geosynthetic barriers offer economically flexible and highly effective performance enhancement for canals (Swihart & Haynes 2002). They are effective alternatives or complements to concrete, asphalt or compacted clay soils.

No matter the construction, the consistent revelation is that geosynthetic liners and systems have outperformed traditional lining methods in longevity and project economics in canal systems.

In underwater installation applications with geosynthetic clay liners (Figure 6.7), special considerations need to be made, such as considering the placement of the overburden confinement in a timely manner. Regular GCLs need to be covered immediately with the cover soil material, whereas as sand mat/GCL composite products might allow a later placement of the cover soil material.

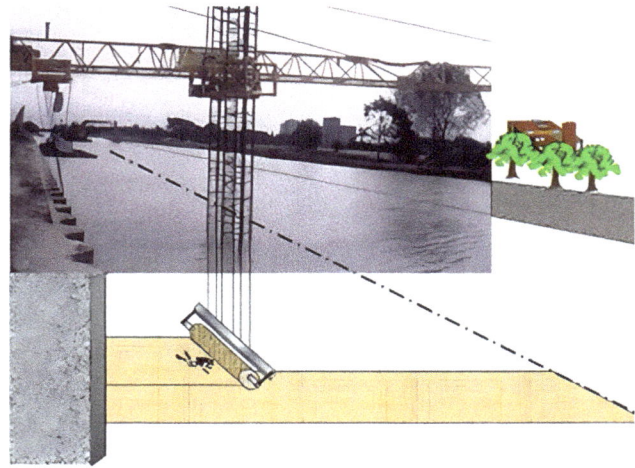

Figure 6.7 Underwater installation of a geosynthetic clay liner in a canal. Photo Courtesy of Naue.

Figure 6.8 Geomembrane installation for a canal providing water for the electricity power plant, Photo Courtesy Naue.

6.2.4 Levees

Levees are embankments - often improved with geosynthetic barriers - constructed to prevent the overflow of water and protect adjacent lands from flooding, such as housing, industrial or agricultural areas. Their main purpose is to confine rivers, streams, lakes, canals and reservoirs within their natural or artificial boundaries. By installing geosynthetic barriers and other non-barrier applications, engineers can enhance the impermeability of such structures, ensuring the integrity of the levee system. The geosynthetics should meet special safety requirements, as the flood protection of entire regions can depend on their functionality and enormous protective assets and human lives can be affected if the flood protection systems fail.

Levees are typically structures made of earth (also called flood defence structures, dykes, dikes or embankments). Their primary objective is temporary flood protection in the event of a flood near by a river or stream. In some cases they can include pumping stations or flood gates.

Many levees worldwide have been constructed centuries or many decades ago with local fill material and were designed for a short-term flood protection and also might not meet current design standards. In levee construction, the use of e.g. GCLs as an alternative to a mineral sealing layer, constructed from compacted, cohesive soils, has produced positive experiences.

"It should be noted that levees may stand for much of their lives without being loaded to their design capacity. This can create a false sense of security in the level of protection they will provide." (CIRIA C731, 2013).

Figure 6.9 Installation of a geosynthetic clay liner on the upstream side of a river levee in a 3-zone levee, Photo Courtesy of Naue.

As flood events around the world increase and last longer than decades ago, current old levee failures seem to occur more often, sometimes resulting in tragic losses of life and the devastation of large areas (CIRIA C731, 2013).

Geosynthetic barriers (Figure 6.9), as well as other geosynthetic materials are often used to improve the levee performance and increase its design lifetime. One concept is the 3-zone levee (Figure 6.10), constructed with an impervious first zone, the permeable core as the second zone and a downstream very permeable zone 3 often also built as a levee defence road (according to DIN 19712, 2013).

One additional option as an example is to incorporate reinforcement layers into the waterside slope or the core of the levee. As Han et al (2008) identify, these layers of reinforcement can increase the factor of safety of the levee against rapid draw-down failure to an acceptable level (Figure 6.11).

6.3 Design Considerations

Designing geosynthetic barriers in seepage control systems requires the consideration of various factors to ensure their effectiveness in the specific applications of ponds, reservoirs, canals and levees. Designing geosynthetic barriers requires a comprehensive understanding of the project conditions, possible regulatory requirements, and most importantly, the performance

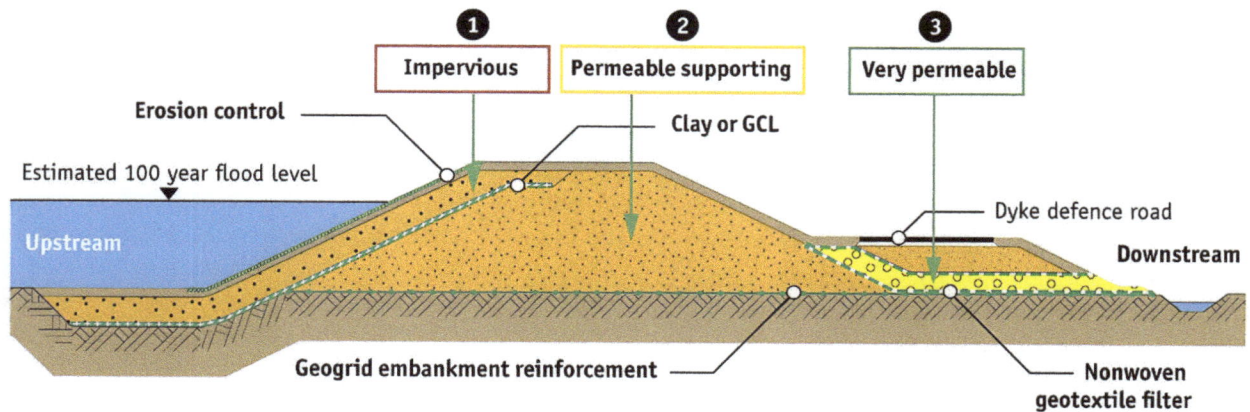

Figure 6.10 Cross section of a 3-zone levee, Photo Courtesy of Naue.

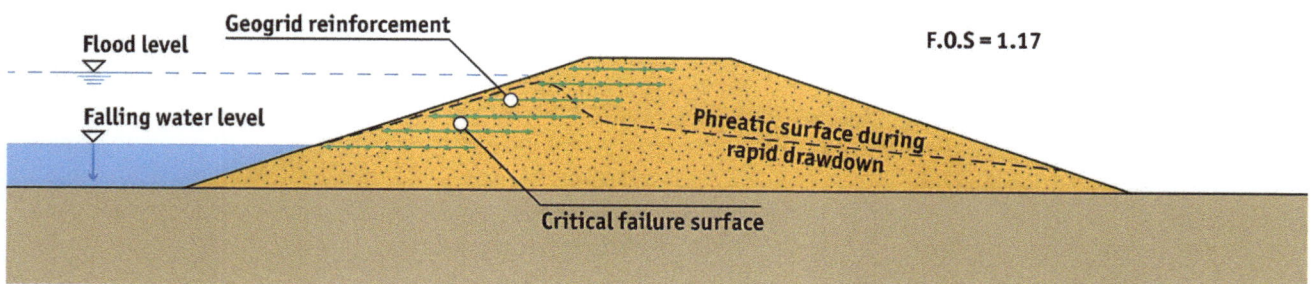

Figure 6.11 The use of geotextile reinforcement to overcome rapid draw-down (after Han et al, 2008)

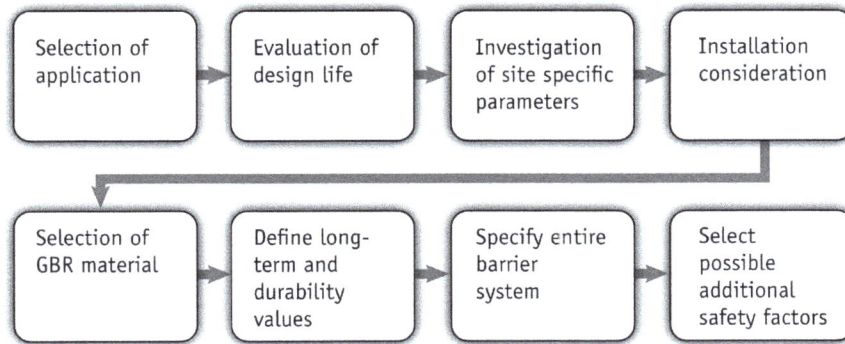

Figure 6.12 Simplified flow-chart for topics to consider during design with geosynthetic barriers (ISO/TR 18228-9:2022 (2022).

characteristics of different geosynthetic barrier materials. The success of the barrier system also involves the exchange between all involved engineers (e.g. geotechnical, environmental, design, etc.), manufacturers and geosynthetic experts. Figure 6.12 illustrates the topics to consider when working on a design with geosynthetic barriers.

Designers benefit from having a wide variety of geosynthetic barriers materials to select from. For example, some geomembranes feature a textured surface to increase friction with adjacent materials. Others feature specific formulations to better accommodate high temperature or long-term exposure applications.

The design criteria for all geosynthetic barriers typically include the following aspects.

6.3.1 Hydraulic resistance

The hydraulic resistance of a geosynthetic barrier in a seepage control system is a crucial parameter that influences the flow of water through the barrier. This resistance (hydraulic conductivity, index flux and permeability coefficient) is essential for preventing or controlling the movement of liquids through the barrier and depends on many key factors.

6.3.1.1 Geomembrane – Hydraulic Resistance

The hydraulic resistance of a geomembrane is influenced by several factors and understanding these factors is essential for the overall geomembrane's performance. Key considerations include:

> Geomembrane Polymer Type and Formulation: product permeability is defined by diffusion through the intact geomembrane in vapor phase.
> Subgrade and Cover Soil Conditions: Coarse soils can cause penetration of the geomembrane during installation and the lifetime and significantly influence the hydraulic resistance. Geotextile protection layers or fine grain

soils (e.g. sand) can prevent geomembrane puncture.
> Quality of Installation: Properly executed seam and weld integrity during installation maintains the geomembrane's hydraulic resistance. Otherwise, inadequate seams can cause leakage for water flow.

6.3.1.2 Geosynthetic Clay Liner – Hydraulic resistance

Geosynthetic clay liners (GCLs) are manufactured hydraulic barriers consisting of clay (typically sodium bentonite) bonded to a layer or layers of geosynthetic materials, e. needle-punched between two geotextiles or bentonite adhesive bonded to a thin geomembrane. The hydraulic resistance of a geosynthetic clay liner is critical for its effective performance in containment applications, such as landfills, lagoons, and other environmental engineering projects. Here are the key considerations for the hydraulic resistance of geosynthetic clay liners:

> Index Flux: The primary component influencing hydraulic resistance is the bentonite clay layer. Sodium bentonite typically has low index flux values. However, the index flux is measured with a confining stress of 27.5 kPa and a water gradient of 1.5m, which might not represent site conditions.
> Bentonite Properties: The quality of the bentonite clay, including its free swell value, fluid loss value, bentonite mass per unit area and particle size, might impact the liner's hydraulic resistance performance.
> Quality of Installation: Properly executed seams and overlaps are crucial to prevent the creation of preferential flow paths within the GCL
> Temperature and Conditions: On-site temperature conditions could influence the bentonite's performance (dry/wet or freeze/thaw cycles) as they could influence the swelling and hydraulic resistance of the bentonite.

6.3.2 Raw Material Properties

The raw material properties of geosynthetic barriers play a crucial role in determining their short-term and long-term performance and effectiveness. Here are key raw material properties that are significant for geomembranes and geosynthetic clay liners

6.3.2 Geomembranes – Raw Material Tests

Different polymers offer varying degrees of chemical resistance, flexibility, and durability. The choice of polymer depends on the specific environmental conditions and the type of containment or seepage control application.

> Polymer density influences the mechanical properties of geomembranes. Example: High-Density PE vs. Low-Density PE: High-density polymers generally provide higher tensile strength and chemical resistance but less flexibility.
> Additives (carbon black, antioxidants, UV stabilizers, etc.): Additives are incorporated to enhance UV resistance, oxidative stability, and other properties. Carbon black, for instance, helps prevent UV degradation.
> Oxidative Induction Time (OIT) and High Pressure OIT (HPOIT) are a measure of a material's resistance to oxidation. The methods have found widespread use in the geosynthetic industry as a quality control and fingerprinting tool and the test is primarily restricted to polyolefin resins such as HDPE. By using the generated data and a modified Arrhenius model a lifetime prediction based on the resistance to oxidative decomposition is often possible.

6.3.2.2 Bentonite in the GCL – Raw Material Properties

Bentonite, the mineral component of a GCL, is the key component for its sealing effect. Bentonite contains a high proportion of highly swellable clay minerals (montmorillonite). This is a three-layer silicate with water absorption capacity. The fine-grained granulometry of the bentonite and the property of montmorillonite to swell when it absorbs water are the reasons for the low water permeability of GCLs.

Montmorillonite can be differentiated into calcium and sodium montmorillonite according to its ionic composition. Bentonites with a high proportion of sodium montmorillonite (Na-bentonites) have a higher capacity for water absorption and swelling of the clay minerals and a lower permeability (index flux value) than bentonites with predominantly calcium montmorillonite (Ca-bentonites).

It is therefore understandable that the bentonite component undergoes a stringent quality control process before the bentonite is used for the production of the GCL. The following test procedures describe the state-of-the-art bentonite testing used for GCLs. It must be understood that these tests are not stand-alone tests and in the GCL industry the result of one single test does not allow an interpretation of the bentonite. However, taking all results together in account they allow a determination of an expected performance of the bentonite.

Sodium bentonite is commonly distinguished by its ability to swell several times its natural volume – when exposed to water. The test method used for quantifying the swelling property for use in GCLs is ASTM D5890 – Standard Test Method for Swell Index of Clay Mineral Component of Geosynthetic Clay Liners. This index test is useful for establishing the relative quality of a clay for use in a GCL. The current industry standard is ≥ 24 ml.

For most environmental applications, sodium bentonite is also evaluated for use based upon its ability to create a seal. This test is ASTM D5891 – Standard Test Method for Fluid Loss of Clay Component of Geosynthetic Clay Liners. Many consider this index test to be a quick qualitative test, suggesting the bentonite's ability to work effectively in a GCL. The current industry standard is ≤18 ml.

6.3.3 Product Specific Considerations

In seepage control systems, geosynthetic barriers have the task of retaining water and preventing flow out of the construction. As a basis for the evaluation of geosynthetic barriers with regard to the recommended requirements, the target issues of performance efficiency, system reliability, allowable leakage rate, economic efficiency, local characteristics and operational aspects should also be considered.

6.3.3.1 Geomembranes – Considerations

When designing with a geomembrane, various property considerations need to be taken into account to ensure the effectiveness, durability, and safety of the containment system. Some key property considerations could be:

> Polymer composition: Different geomembrane materials, such as high-density polyethylene (HDPE), polyvinyl chloride (PVC), ethylene propylene diene monomer (EPDM), and others, have distinct properties that affect their suitability for specific applications.
> Surface color: White geomembranes or white surface geomembranes may have benefits regarding sunlight reflection and surface temperature (e.g. less thermal expansion).

> Surface texture: Textured surfaces have benefits regarding interface friction; differing production processes need to be investigated regarding their short-term and long-term performance.
> Chemical resistance: Depending on the environmental characteristics the resistance of the geomembrane material to present chemicals, considering factors such as pH, temperature, and specific chemical concentration, should be evaluated.
> Thickness: Based on the specific requirements of the project an appropriate geomembrane thickness has to be selected. Thicker geomembranes may provide enhanced durability and better resistance to mechanical damage.
> Hydraulic resistance: Based on the geomembrane's properties and the water head conditions, the hydraulic performance of the geomembrane must be evaluated for its long-term performance, e.g. considering allowable leakage rates.
> Strength and elongation: Assess the strength properties (puncture, tensile) of the geomembrane might be a value to investigate as it might be needed to ensure that the geomembrane can withstand expected loads (e.g. installation stress, settlement, confining stresses, etc.).
> Elongation at yield: This mechanical property describes the amount of deformation or stretching a material can undergo before it experiences a significant change in its mechanical behavior and is an important value for the designer.
> UV resistance: If a geomembrane will be exposed to sunlight, its resistance to ultraviolet (UV) radiation is a value of importance. Some geomembrane materials may require UV stabilizers or protective layers to enhance their resistance to UV exposure.
> Temperature resistance: The expected temperature variations at the site should be considered, when selecting a geomembrane material.
> Flexibility at low temperatures: The geomembrane's flexibility at low temperatures, especially in cold climates should be assessed.

Designing with a geomembrane requires a comprehensive understanding of these material properties and their implications for the specific project. Collaboration with experienced geotechnical and civil engineers, as well as adherence to relevant industry standards, is essential for successful geomembrane designs.

6.3.3.2 Geosynthetic Clay Liners –Considerations

According to national requirements or regulations, the GCL must fulfill specific requirements as listed in GRI-GCL3 or similar. Typically, the basic properties, which influence the GCL performance are:

> Bentonite minimum mass per unit area [g/m²], ASTM D5993 or EN 14196
> Bentonite free swell value [ml/2g], ASTM D5890
> Bentonite fluid loss value [ml], ASTM D5891
> GCL internal peel strength [N/m], ASTM D6496
> short-term index flux value [m³/m²/s], ASTM D5887 or EN 16416
> long-term index flux value, according to GRI-GCL3 or EN 13361 (section A.4.6.2.1)

However, it is also necessary to investigate the performance efficiency of the barrier system with regard to possible effects against e.g. dry-wet changes, frost-thaw cycles and if relevant exposure to salt solutions or other relevant liquids. Therefore, when using any sealing system, e.g. a GCL, the required long-term permittivity should also be achieved against other occurring liquids.

In order to achieve a reliable, permanently efficient barrier system, further requirements are necessary in addition to prove performance efficiency. A minimum 0.8m soil cover thickness over the GCL is considered as sufficient to protect the GCL from desiccation and frost effects.

Geosynthetic clay liners can also feature a number of constructions to assist the design. Some of these include:

> formulated clay additions to assist in performance in chemically aggressive applications (however their long-term performance need to be investigated).
> enhanced needle punching may increase internal stability
> higher bentonite mass per unit area improves the short-term and long-term permeability performance

A multicomponent GCL (GCL with an attached film, coating, or membrane decreasing the hydraulic conductivity or protecting the clay core or both) can significantly increase the performance and reliability of the sealing system and would also allow a reduction of the cover soil thickness, depending on local conditions. Advantages of multi-component GCLs are:

> Prevention of root penetration
> Increasing resistance against desiccation
> Bentonite piping resistance under high water gradients

> Lower permeability
> Barrier against ion exchange

The application of a GCL is considered as uncritical, if the soil pH value is within a range of 5 - 11. In any other cases an investigation of the suitability of the GCL is recommended.

Barrier systems are not designed to handle downhill tensile forces like a geogrid. For that reason, tensile strength properties for a GCL are more of a quality control measure and are only needed to handle installation stresses. The recommended minimum tensile strength (mainly for installation purposes) should be ≥ 10kN/m.

If the GCL is used on slopes, it is necessary to investigate all friction properties (internal and inter-face). This is also valid for multicomponent GCLs.

In applications with sloped embankments, all shear forces parallel to the embankment must be investigated both on the interface as well as in the GCL (internal shear strength) in the hydrated state of the bentonite. Alternatively, if the necessary shear strengths are not achieved, geogrid reinforcement can be used to ensure slope stability.

Further GCL requirements should refer to the selected concept that the designer followed, including an approach to looking over the entire situation and considering the following additional evaluation criteria (further details listed in Figure 6.13), such as

> Effectiveness of the system
> Performance reliability and durability
> Economics
> Control and maintenance
> Local conditions

Geosynthetic clay liners can be incorporated into many seepage control systems in order to reduce seepage and internal erosion. GCLs have been used for many years as liners for these applications and have found recognition in many national guidelines and regulations. The offer has advantages such as:

> Low permeability
> Self-healing behaviour
> Good durability
> Good friction behaviour for embankment slopes
> Good control of quality of a manufactured product.

However, the performance of GCLs can be affected adversely by root penetration and damage caused by

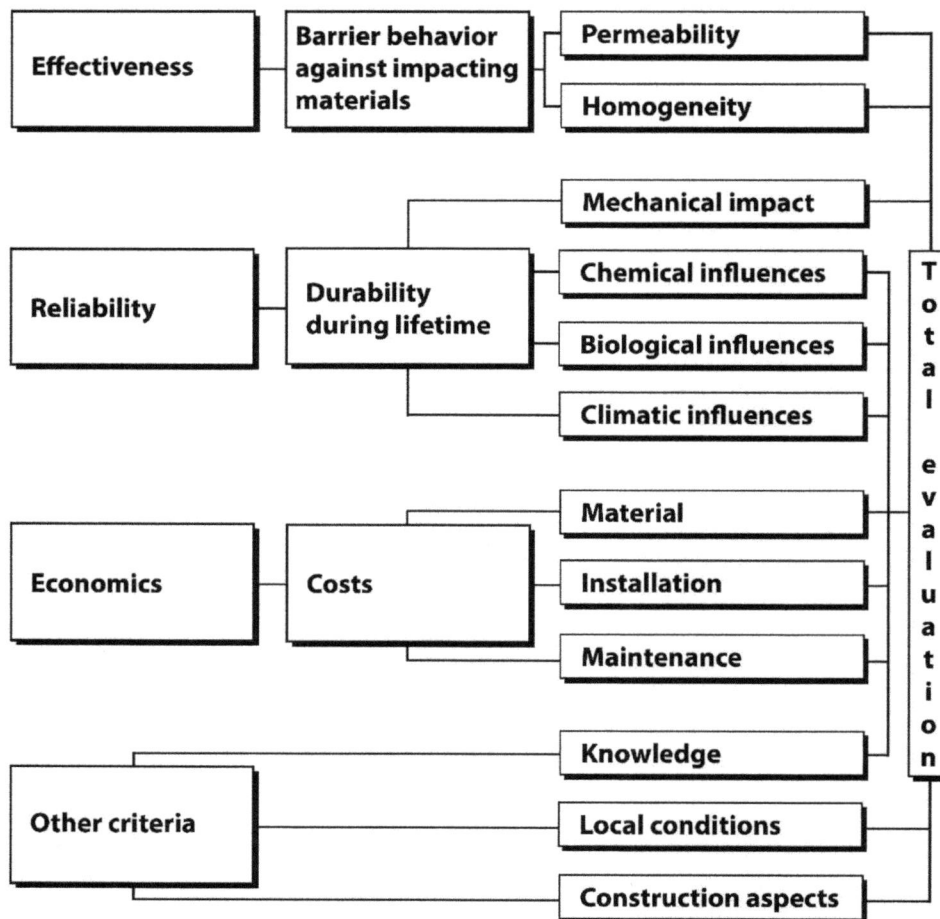

Figure 6.13 Evaluation criteria for seepage control systems (von Maubeuge, 2028).

burrowing animals. These impacts can be counteracted by:

> An appropriate design with sufficient cover soil material.
> A non-cohesive cover layer soil which is unattractive to burrowing animals.
> additional engineering measures.

Typically, an 800 mm cover layer over the GCL should be sufficient, in order to withstand climatic influences.

6.4 Installation

The installation of geosynthetic barriers is crucial for the success and effectiveness of seepage control applications. Proper installation is essential for several reasons:

> Containment integrity: The primary purpose of geomembranes is to provide a barrier against the migration of liquids. Proper installation ensures that the geomembrane forms a continuous, impermeable layer, preventing leaks and maintaining the integrity of the seepage control system.
> Structural stability: Geosynthetic barriers are also used in conjunction with other geosynthetic materials and engineering components. Proper installation ensures that the geomembrane functions as part of a stable and reliable structure.
> Longevity and durability: Correct installation practices, including proper seaming, anchoring, and detailing, contribute to the long-term durability of barrier system.
> Cost-effectiveness: Correctly installed barrier systems are more likely to perform as intended, reducing the need for costly repairs or replacements. Investing in proper installation practices upfront can save significant expenses in the long run.
> Safety considerations: The installation process involves working with large barrier panels and, in some cases, electrical and driving equipment. Following safety protocols during installation helps prevent accidents and ensures the well-being of the installation team.

6.4.1 Geomembrane Installation

The installation of geomembranes involves careful planning, preparation, and precise execution to ensure the effectiveness and longevity of the containment system. A general overview of the installation process can be as follows:

> Transportation: Loading and unloading must be carried out using special lifting equipment so that point or linear loading and thus damage to the rolls is ruled out. The rolls must be transported to the construction site in such a way that they cannot be damaged.
> Storage: The rolls must be stored on site on surfaces that are level and free of stones. If the rolls are stacked, they must be stacked parallel to each other. The maximum stacking height is limited according to manufacturer's guidelines.
> Site preparation: Clearing the site of any debris, rocks, or sharp objects that could potentially damage the geomembrane during installation. Grading and compacting the subgrade create a smooth and stable foundation. Inspection of the subgrade for any irregularities, sharp objects, or potential sources of puncture that could compromise the integrity of the geomembrane.
> Geotextile placement (if applicable): In some installations, a geotextile laying between the geomembrane and the subsoil may be used to provide protection to the geomembrane and prevent short-term or long-term damage from subsoil objects.
> Geomembrane unrolling: Damaged geomembranes may not be installed. No detrimental ruts may be created during the installation process. After unrolling, especially prior to cover soil placement, the geomembrane should have no wrinkles, folds, or any other irregularities. Laying is generally carried out using special laying vehicles. No detrimental ruts may be created during mechanical laying.
> Anchor trench: Trenches for geosynthetic components of the sealing system (e.g. membranes, geogrids or protection geotextile) are usually arranged at the top of the embankment to prevent possible displacement of the geomembranes during the installation phase. It should be noted that this trench is only to be regarded as a temporary support during construction, but by no means as an anchorage for a permanent tensile force in the geomembrane. The sealing system must be designed in such a way that in slope areas the downslope forces from the overburden on the sealing system can be transferred directly via a sufficient shear bond between the components of the sealing system. Appropriate verifications of stability against sliding must be carried out for this purpose.
> Seaming: By using approved welding crews, suitable seaming techniques (e.g. extrusion welding, hot wedge welding, etc.) should join adjacent geomembrane panels together.

> Quality control: Quality control checks during and after seaming should be carried out to ensure that the seams are secure, free of defects, and meet specified standards.
> Ballast Placement: Securing the edges of the geomembrane by burying them in an anchor trench or using ballast materials such as soil, sand, or sandbags. This also helps prevent wind uplift.
> Penetrations and details: For penetrations, seams around structures, and other similar project-specific requirements manufacturers guidelines should be followed, if not described in the specification. The waterproof integrity of these details must also be ensured.
> Final inspection: Prior to cover soil placement a final inspection should be conducted to verify that the geomembrane is properly installed, free from damage, and meets the project specifications.
> Cover material placement (if applicable): In the most seepage control systems cover soil material may be placed over the geomembrane to protect it from UV exposure, mechanical damage, and weathering.
> On-site quality assurance with leak detection in geomembranes involves using various methods, such as electrical leak location (ELL) testing, to identify and address breaches or defects in the geomembrane, ensuring their integrity and effectiveness. This proactive approach helps prevent structural failures.
> Documentation: Maintain detailed records of the installation process, including quality control checks, seam testing, and any modifications made during installation.

It's important to note that the specific installation procedures may vary based on the type of geomembrane material, site conditions, and project requirements. Always follow manufacturer guidelines and industry standards during the installation process. Consulting with experienced professionals in geotechnical engineering or environmental engineering is also recommended for complex projects.

6.4.2 Geosynthetic Clay Liner Installation

GCL products are thin barriers and require a focus during the installation phase to ensure a damage-free installation, ensuring a long-term performance.

Basically, the installation instructions of the manufacturer or the project-specific installation instructions must be observed, as these describe suitable measures to ensure that the GCL can be installed properly.

An execution drawing (laying plan) should also be drawn up prior to installation, which not only shows the GCL positioning, but also specifies the time schedule for the laying.

Unloading, storage and further transportation of the GCL must be carried out in such a way that damage caused by point or linear loads is excluded (e.g. use of special lifting equipment). Damaged roll packaging must be sealed watertight. The roll packaging may only be removed shortly before the rolls are laid. Damaged rolls (mechanical damage, pre-hydrated bentonite) must not be installed.

The flooring must be laid on a level and stable substrate. A widely graded mixture of gravel and sand or finer soils should be preferred as the subsoil.

The subgrade must be leveled, rolled and compacted if necessary. It must also be free of standing surface water. Directly before laying the GCL, the soil surface must be leveled and any large, open pore spaces must be filled.

Overlapping areas (minimum 30cm) must be sealed with the same bentonite as contained in the bentonite mat. All overlaps must be wrinkle-free and free of foreign bodies. If there are pieces of foreign matter on the exposed overlap area, these must be carefully removed. The overlap area must not be walked or driven on. All overlaps must be laid like roof tiles in the direction of the slope. Downhill cross joints (adjacent transverse overlaps) are not permitted; T-joints must be reduced to a minimum. Driving over the installation area must be avoided.

Connections, terminations and penetrations must be carried out in accordance with the installation instructions or manufacturer's specifications.

Mechanical stresses that occur during the installation phase and after completion can influence the effectiveness of any sealing system. The highest static and dynamic loads for the GCL usually occur during floor application. When applying the bulk material, care must be taken to ensure that the floor material is not poured directly onto the GCL, but in front of the head and onto previously distributed material, and then distributed with a suitable device.

A widely graded material with the smallest possible maximum grain size is recommended and round grain is preferable to crushed material.

In principle, a GCL must not be built over in a hydrated state. As a guideline, the water content of bentonite must not exceed 50% by mass.

In the overlap area, care must be taken to ensure that the soil is distributed in the overlap direction, preventing that soil material penetrates into the overlap area.

In order to ensure GCL protection against damage by vehicles, a minimum soil cover of 30cm is required, if the covered area is only crossed once. In the case of multiple traffic, it is advisable to increase the thickness of the cover to at least 60cm, depending on the load and cover soil type.

A cover soil thickness of a minimum of 30cm over the GCL must be installed every working day or prior to heavy rain fall. Overlaps required for the following day must be wrapped overnight with plastic sheet to prevent bentonite swelling without confining stress.

6.5 Quality Control and Assurance of Geosynthetic Barriers

Quality control is a critical aspect of geosynthetic barrier installations to ensure their effectiveness in various seepage control applications. The successful performance of geosynthetics depends on rigorous quality control measures throughout the design, manufacturing, transportation, and installation processes. This chapter explores the key elements of quality control for geosynthetic barriers, emphasizing the importance of adherence to standards and the application of robust inspection procedures.

6.5.1 Quality Control in Manufacturing

The foundation of quality control begins with the selection of high-quality raw materials for manufacturing geosynthetics. Manufacturers should conduct thorough material testing to ensure that the selected polymers meet specified standards for strength, flexibility, biological and chemical resistance, hydraulic resistance and compatibility. Quality control protocols at this stage help eliminate potential material defects and ensure the overall reliability of the geosynthetic barrier.

Stringent control over the manufacturing processes is essential to produce geosynthetic barriers with consistent properties. Continuous monitoring of the manufacturing techniques helps identify variations in e.g. material thickness and distribution, tensile strength, and other crucial characteristics. This ensures that the final products meet the specified engineering requirements.

Description of the processes should be available from the manufacturers and are typically certified according to ISO 9001 or similar standards.

6.5.2 Quality Assurance Testing

Manufacturers should conduct comprehensive third-party quality assurance testing on finished geosynthetics, typically once or twice a year. This may involve tests for mechanical, physical or hydraulic properties. This quality assurance testing is done on raw materials as well as on the finished product and documents production processes, the type and extent of the manufacturing quality control and any other relevant observations. Independent quality control measures at this stage serve as a final check to confirm that the geosynthetic barriers conform to industry standards and project specifications.

6.5.3 Quality Control in Transportation and Storage

Proper packaging and transportation practices are crucial to prevent damage to geosynthetic barriers during transit. Quality control measures include inspecting packaging integrity, securing rolls or panels to prevent shifting, and protecting materials from adverse weather conditions. Any signs of damage or compromised packaging should be addressed before installation.

Geosynthetic barriers must be stored in a controlled environment to prevent exposure to UV radiation, extreme temperatures, and moisture, which can degrade the materials over time. Quality control in storage involves periodic inspections to ensure that storage conditions meet manufacturer recommendations and industry standards.

6.5.4 On-site Verification Tests

Geosynthetics are predominantly standardized and factory quality-assured products, so that special control tests on the construction site are not necessary. Manufacturers are obliged to comply with the prescribed properties as part of product liability. The contractor is responsible for checking the delivered products and monitoring the execution.

However, on-site verification testing for geosynthetics is often done and recommended. It involves assessing the installation and performance of geosynthetic materials at the construction site. Third-party on-site testing is essential to ensure that these materials are properly installed and meet the specified design and performance criteria.

6.5.5 Quality Control during Installation

Before installation, a comprehensive pre-installation inspection is essential. This includes verifying that the received materials match the specifications, checking for any visible defects or damage, and ensuring that the installation site meets the design requirements. Any discrepancies or issues identified during this phase should be addressed before proceeding.

Adherence to proper installation procedures is paramount for the success of geosynthetic barriers. Quality control measures during installation involve verifying that seam welding, adhesive bonding, or other joining techniques meet specified standards. Additionally, technicians should check for wrinkles, folds, or irregularities in the barrier, as these can compromise its performance.

6.5.6 Field Testing

Field testing is a crucial quality control step to confirm the integrity of the installed geosynthetic barriers. This may involve non-destructive testing methods such as electrical leak location surveys for geomembranes or puncture testing for geotextiles. Field testing provides real-time feedback on the barrier's performance and helps identify any areas that may require attention or repair.

6.5.7 Documentation and Record-Keeping

Maintaining detailed documentation throughout the entire process is a fundamental aspect of quality control. This includes records of material testing results, manufacturing quality assurance reports, transportation and storage conditions, pre-installation inspections, installation procedures, and field-testing outcomes. A comprehensive record-keeping system ensures traceability and accountability at every stage of the geosynthetic barrier's lifecycle.

6.5.8 Site-Specific Conditions

The design of a geosynthetic barrier system must consider site-specific conditions, including soil properties, groundwater levels, and climatic factors. Conducting a thorough site investigation is crucial to understanding the geotechnical and hydrogeological aspects that influence the selection and design of geosynthetic barriers.

6.5.9 Compatibility

In most applications described in this section "seepage control systems", ponds, reservoirs, canals and levees are used for the capture, transportation, storage and distribution (Figure 6.14) of all types of liquids but freshwater is the most common (Koerner et al, 2008).

Geosynthetic compatibility refers to the ability of different geosynthetic materials to perform together effectively in a given application without adverse interactions that could compromise their intended functions. Ensuring compatibility between different geosynthetic materials is essential for the long-term performance of the seepage control system.

Key aspects of geosynthetic compatibility include:

> Chemical Compatibility: Geosynthetics may be exposed to a variety of environmental conditions, including different types of soil, water, and chemical substances. Due to the fact that the geosynthetics used for the applications described in this chapter, are mainly exposed to fresh water, chemical compatibility is not the main issue of concern. However, chemical compatibility ensures that the geosynthetic materials do not degrade or experience adverse reactions when in contact with specific chemicals or substances present in the environment.

> Physical Compatibility: Geosynthetic materials often work together in a composite manner to achieve specific engineering objectives.

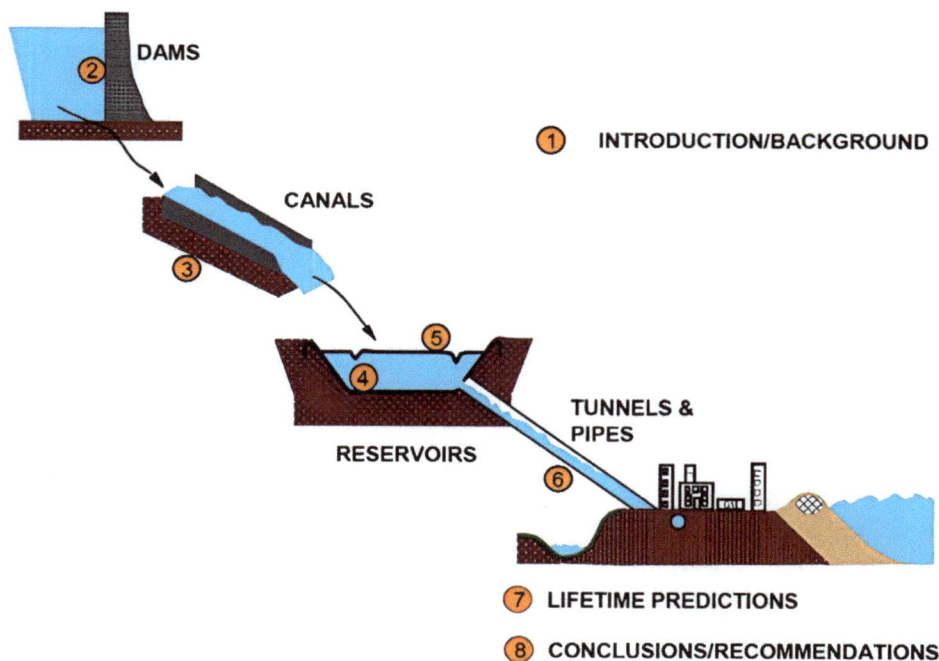

Figure 6.14 The suggested "geosynthetics freshwater cycle" for capture, transportation, storage and distribution of freshwater. Photo courtesy Koerner

Physical compatibility ensures that different geosynthetic components, such as geotextiles and geomembranes, can be effectively combined without causing issues like sliding, puncture or any other reduced performance.

> Temperature Compatibility: Geosynthetics may be exposed to a wide range of temperatures depending on the geographic location. Compatibility with temperature variations ensures that the materials maintain their performance characteristics.

> Durability and Longevity: Compatibility is linked to the ability of geosynthetic materials to withstand environmental factors, weathering, and aging without significant degradation.

Achieving geosynthetic compatibility is essential for the successful design and construction of seepage control applications.

6.5.10 Construction Quality Control

Proper installation and construction quality control are paramount to the success of a seepage control system. Strict adherence to installation guidelines and inspection protocols is necessary to guarantee the integrity of the geosynthetic barriers. Adherence to installation guidelines, meticulous inspection protocols, and the use of advanced construction techniques are essential for ensuring the integrity of the geosynthetic barriers. Continuous monitoring during construction activities

helps identify and rectify potential issues promptly, minimizing the risk of system failure.

6.6 Some Basics with Geosynthetic Barriers

Geosynthetic barriers are relatively thin sealing products (geomembrane from approx. 1mm – 2.5mm; geosynthetic clay liner from approx. 8mm – 20mm), that can be punctured, if proper installation processes are not followed. They can also fail if wrong design measurements were considered or if wrong confining stresses or even inadequate cover soil material were placed on the geosynthetic barrier. Figure 6.15 gives a brief overview of what phases issues should be investigated, considered or dealt with.

During the design phase, the designer or any other responsible person should have a full view of the project and have knowledge of all details as in this stage the product specification has to be prepared and cover all aspects of the design. This would also involve external effects, durability issues and the installation.

Having a proper specification is crucial for various reasons:

> Specifications provide a clear and detailed description of the requirements and expectations for a project and the involved products. They outline the standards, materials, dimensions, and other essential details that

Important issues to consider

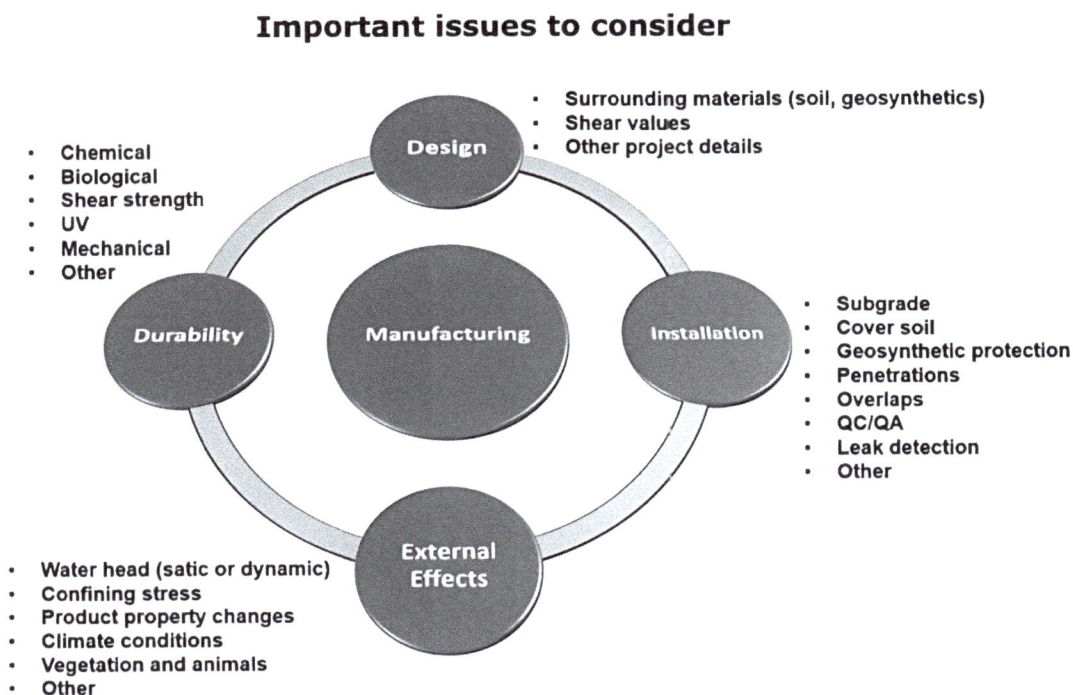

- Chemical
- Biological
- Shear strength
- UV
- Mechanical
- Other

Design
- Surrounding materials (soil, geosynthetics)
- Shear values
- Other project details

Durability

Manufacturing

Installation
- Subgrade
- Cover soil
- Geosynthetic protection
- Penetrations
- Overlaps
- QC/QA
- Leak detection
- Other

External Effects
- Water head (satic or dynamic)
- Confining stress
- Product property changes
- Climate conditions
- Vegetation and animals
- Other

Figure 6.15 Simplified diagram for topics to consider from the design to the time the geosynthetic barrier is in service.

stakeholders need to understand to meet the desired outcome.

> Specifications serve as a communication tool between different parties involved in a project. Clear specifications help ensure that everyone has a good understanding of the needs to be accomplished.
> Detailed specifications help ensure that the final product meets the required quality standards by providing specific criteria.
> Specifications often include regulatory requirements, ensuring that a project or product complies with these specifications, avoiding legal issues.
> Specifications provide the basis for cost estimations.
> Specifications ensure that at the end the project meets the expectations of the client or end-user.
> Specifications serve as a documented record of the project requirements at the beginning and the end of the project as well during its service life.

As an example, there are various methods of inspections that can be carried out to ensure a proper documentation, also for a record keeping:

> Verification that the technical details of proposed product and the accompanying method statements for GCL conform with the specification.
> Initial site visits and meetings with the Contractor to discuss proposals and method statements to ensure the system is placed in accordance with CQA procedures.
> Checking of MQC Certificates and cross-checking with the placed materials on site.
> Regular checking of geosynthetic barrier storage arrangements on site. Inspection of subgrade prior to placement of geosynthetic barrier.
> Inspection of the placement of liners and jointing of seams on site in accordance with manufacturer's recommendations or specification requirements.
> Inspection of subgrade and cover material to ensure it conformed with the specification.
> Daily recording and documenting of weather conditions, plant used, site personnel, deliveries, site meetings.
> Regular photographing of works.
> Daily preparation of lining layout plan showing where panels were placed.

In summary, proper specifications, inspections record keeping are important to successful project management, product development, and construction activities.

6.6.1 Geomembrane – Basic things to follow or do wrong

Figure 6.16 Use overlap protection strip to keep weld areas clean but the welding area might be needed to be cleaned anyway. Photo courtesy Naue

Figure 6.17 This unprotected overlap qualifies for some extra cleaning before start of welding works. Photo courtesy Naue

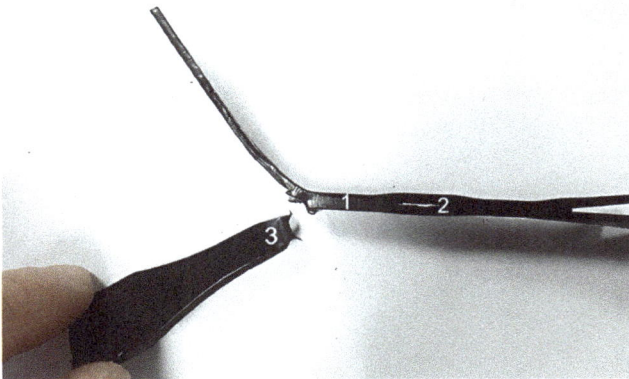

Figure 6.18 Proper welding parameters in alignment with good welding properties of the geomembrane leads to desired ductile failure outside the weld in peel tests. (1 – No separation in the weld; 2 – Air channel intact; 3 – Ductile failure outside of the weld) Photo courtesy Naue

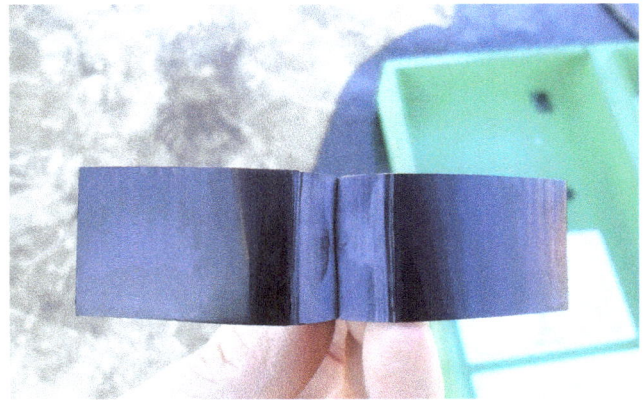

Figure 6.19 Incorrect welding parameters – here presumably too fast welding, especially in combination with poor formulation quality - leads to peel failure in destructive weld testing and ultimately to insufficient weld strength. Photo courtesy Naue

Figure 6.20 Proper placement of sandbags against wind uplift. Photo courtesy GSI

Figure 6.21 Result of neglecting wind uplift on geomembranes. Photo courtesy TRI

Figure 6.22 Protection geotextile under the geomembrane and on top of coarse gravel subgrade. Photo courtesy Naue

Figure 6.23 Insufficient grading and preparation of subgrade with standing water table in installation area – not qualified for installing geomembrane. Photo courtesy Sytec

6.6.2 Geosynthetic clay liners – basic things to follow or do wrong

Figure 6.24 Proper unloading and transportation equipment and following construction safety rules. Photo courtesy Terrafix

Figure 6.25 Improper lifting of a geosynthetic clay liner on site. Photo courtesy Kent von Maubeuge

Figure 6.26 Prior to installation, inspection of delivered material and replacement of damaged material. Photo courtesy of Hallaton Environmental Linings.

Figure 6.27 Prior to installation, if GCL rolls need to be stored for a longer period on site, it is recommended to cover them with a weather resistant and waterproof tarpaulin or plastic sheet. Photo courtesy of Hallaton Environmental Linings.

Figure 6.28 The subgrade should be in accordance with the specification and the manufacturers recommendations. This would be an inadequate subgrade. Photo courtesy Terrafix

Figure 6.29 The cover soil placed over the GCL should be in accordance with the specification and the manufacturers recommendations. The coarse rocks in this case seem inadequate. To allow this type of cover material to be used, two GCL layers were installed. Photo courtesy Sytec

Figure 6.30 The GCL was not anchored in an anchor trench but rather pinned with steel sticks. Photo courtesy Naue

Figure 6.31 Pre-covering the GCLs with soil material prior to cover soil placement is a good recommendation as well as placing the cover soil on top of the GCL rather than pushing. Photo courtesy Naue

Figure 6.32 Cover soil placement typically uphill or angular to the slope but preferrable not downhill and never against the overlap direction. Photo courtesy TRI

6.7 Sustainability

The sustainability of geosynthetic barriers refers e.g. to the environmental, economic, and social considerations associated with the production, use, and end-of-life management of these materials. Geosynthetic barriers, such as geomembranes and geosynthetic clay liners, are widely used in seepage control applications such as ponds, reservoirs canals and levees. Evaluating the sustainability of geosynthetic barriers involves examining several key aspects, especially in comparison to alternative solutions, such as traditional engineering methods or other geotechnical materials, such as compacted clay liners, concrete or asphalt lining systems.

> Environmental impact: Consideration of the environmental impact of extracting and processing raw materials used in geosynthetic barriers.

> Manufacturing processes: Assessing the energy consumption, emissions, and waste generated during the manufacturing of geosynthetic barriers.

> Durability and longevity: Comparing the service life of lining systems, as a longer service life reduces the need for replacements, leading to less material consumption and waste.

> End-of-life considerations: Assessment of the recyclability of geosynthetic barriers.

> Disposal methods: Evaluation of the environmental impact of disposal methods (e.g. incineration or landfilling).

> Installation: Consideration of resources used during installation (e.g. energy consumption, water consumption, amount of waste material).

> Social considerations: Evaluation of the social impact of the geosynthetic project on local communities, including potential job creation or disruption.

> Health and safety: Evaluate if the use of geosynthetic barriers complies with health and safety regulations to protect workers and nearby residents.

> Life Cycle Assessment (LCA): Conduction of a life cycle assessment to evaluate the environmental impacts of geosynthetic barriers from raw material extraction to end-of-life.

It's important to note that the sustainability of geosynthetic barriers can vary based on many factors such as the specific application, material type, and project conditions. Manufacturers, engineers, and project managers play crucial roles in ensuring that geosynthetic projects are designed, implemented, and managed with sustainability considerations in mind.

6.8 Summary

There is every reason to believe that geosynthetic barriers will continue to be adopted into regulations around the world. This has much to do with the innovation and quality control measures in manufacturing and installation in the field. It also has much to do with geosynthetics being used in situations to perform better and/or more economically than traditional geotechnical de-signs. With a large record of data in support of cost and performance measures, and with secondary benefits such as decreased project carbon footprints, the field's growth is assured.

These geosynthetics offer a wide range of physical, mechanical and durability resistance properties. Geomembranes can be compounded for greater resistance to ultraviolet light exposure, ozone and micro-organisms in the soil while GCLs can be produced with various geotextiles for enhanced frictional properties. Different combinations of these properties exist in various geomembranes as well as GCL materials to address a wide spectrum of geotechnical applications and designs. Several methods are used to join or seam large panels of geomembranes and GCLs, in both factory-controlled and field environments. Each material has highly developed quality control techniques and unique characteristics that govern their manufacture and installation.

As advanced products and manufacturing and installation techniques evolve, project economy and performance will continue to improve, both with and in advance of regulatory recognition.

These geosynthetic materials, which include geomembranes and geosynthetic clay liners, offer reliable, ecological, sustainable, cost effective, easy-to-install, safe, durable and resilient solutions for mitigating seepage issues.:

6.9 References

ASTM D5887-22 (2022). *Standard Test Method for Measurement of Index Flux Through Saturated Geo-synthetic Clay Liner Specimens Using a Flexible Wall Permeameter*, ASTM International, United States.

ASTM D5890-19 (2019). *Standard Test Method for Swell Index of Clay Mineral Component of Geosyn-thetic Clay Liners*, ASTM International, West Conshohocken, PA, United States.

ASTM D5891-19 (2019). *Standard Test Method for Fluid Loss of Clay Component of Geosynthetic Clay Liners*, ASTM International, West Conshohocken, PA, United States.

ASTM D5993-18 (2022). *Standard Test Method for Measuring Mass per Unit Area of Geosynthetic Clay Liners*, ASTM International, West Conshohocken, PA, United States.

ASTM D6496-20 (2020). *Standard Test Method for Determining Average Bonding Peel Strength Be-tween Top and Bottom Layers of Needle-Punched Geosynthetic Clay Liners*, ASTM International, USA.

DIN 19712:2013-01 (2013). *Flood Protection works on Rivers*, Deutsches Institut für Normung e. V. (Hrsg.), Beuth Verlag GmbH, Berlin

EN 13361:2018 (2018). *Geosynthetic barriers – Characteristics required for use in the construction of reservoirs and Dams*, Deutsches Institut für Normung e. V. (Hrsg.), Beuth Verlag GmbH, Berlin

EN 14196 (2004). *Geosynthetics - Test methods for measuring mass per unit area of clay geosynthetic barriers;* Deutsches Institut für Normung e. V. (Hrsg.), Beuth Verlag GmbH, Berlin

EN 16416 (2013): *Geosynthetische Tondichtungsbahnen – Bestimmung der Durchfluss-rate – Triaxialzellen-Methode mit konstanter Druckhöhe*. Deutsches Institut für Normung e. V. (Hrsg.), Beuth Verlag GmbH, Berlin

ISO/TR 18228-9: *Design using geosynthetics - Part 9: Barriers*, https://www.iso.org/standard/65550.html

CIRIA C731 (2013). The International Levee Handbook. *US Arme Corps of Engineer*, British Library Cataloguing in Publication Data, London, UK, (ISBN: 978-0-86017-734-0)

DWA. Merkblatt DWA-M 176 - Hinweise zur konstruktiven Gestaltung und Ausrüstung von Bauwerken der zentralen Regenwasserbehandlung und -rückhaltung, ISBN 978-3-942964-99-9, Deutsche Verei-nigung für Wasserwirtschaft, Abwasser und Abfall e. V. (DWA), Hennef, Germany. 2013.

FLL 2022 (Draft). Gewässerabdichtungsrichtlinien - Richtlinien für Planung, Bau und Instandhaltung von Gewässerabdichtungen, Forschungsgesellschaft Landschaftsentwicklung Landschaftsbau e. V. (FLL), Bonn, Germany.

GRI-GCL3. 2016. Standard Specification for Test Methods, Required Properties, and Testing Frequen-cies of Geosynthetic Clay Liners (GCLs). Folsom, PA, USA: Geosynthetic Institute.

Han, J, Chen, J and Hong, Z (2008). Geosynthetic rein-forcement for riverside slope stability of levees due to rapid drawdown". In: H-L Liu, A Deng, J Chu (eds) Geotechnical engineering for disaster mitigation and rehabilitation, Science Press Beijing & Springer-Verlag GmbH, Berlin, Heidelberg (ISBN: 978-3-540-79846-0)

Koerner, R.M., Hsuan, Y.G., Koerner, G.R. (2008). Freshwater and Geosynthetics; A Perfect Marriage, The First Pan American Geosynthetics Conference & Exhibition, 2-5 March 2008, Cancun, Mexico

Swihart, J.J. & Haynes, J.A. (2002). Deschutes - Canal-Lining Demonstration Project, Year 10 Final Report, R-02-03, US Bureau of Reclamation.

von Maubeuge, K. P. (2018). Geosynthetic barriers in regulations and recommendations in line with the ISO design guide? Innovative Infrastructure Solutions, Springer Nature Switzerland AG

Von Maubeuge, K.P, Shahkolahi, A., Shamrock, J. (2023). A brief summary of worldwide regulations and recommendations requiring geosynthetic barriers, 12th International Conference on Geosynthetics (12 ICG), Roma, Italy, 17–21 September 2023, DOI: 10.1201/9781003386889-274

Chapter 7

Geosynthetics in Environmental Protection

7.0 Overview

Providing environmental protection involves keeping solids, liquids, and gases from coming into contact with either the natural and/or human environment when they contain elements or chemicals in a form that could have a significant negative impact on the natural and/or human environment if released. We seek to prevent contaminants from escaping storage or containment by constructing barrier systems. As will be elaborated below, barrier systems have two basic components: a liner system that provides resistance to contaminant escape and a collection system that reduces the motivation (driving force) encouraging the escape.

The geosynthetic materials that we mostly use in a liner system (geotextiles [GTX], geomembranes [GMB], geosynthetic clay liners [GCL]) and the collection system (geotextiles, geopipes, geocomposite drains, geonets, geocells) have been described in general terms in Chapter 1. The geosynthetics are often used together with the sands, gravel, and clay in barrier systems as will be described in this chapter. Often two or more different geosynthetics are used to develop a composite liner or drainage system. A composite liner is typically comprised of a geomembrane in contact with a geosynthetic clay liner or a compacted clay liner. Sometimes the composite liner may involve the geomembrane, a geosynthetic clay liner, and a compacted clay liner [CCL].

7.0.1 Leachate

Solids are relatively easy to keep in place provided they remain in the solid-state. However, on contact with water, contaminants can leach from the solid phase into the water generating what is known as leachate. More generally, leachate is a fluid that could impact ground or surface water quality if it escapes.

There are two primary mechanisms for contaminants in the leachate to escape from a containment facility: advection (leakage) and diffusion (see Rowe et al 2004 for a detailed discussion).

7.0.2 Advection (leakage) of leachate

Advection (leakage) is the movement of contaminants in flowing water. The driving force is the difference in the leachate level in the facility and the water level in the receptor outside. This can be reduced by collecting leachate as discussed below. If the resistance is provided by a clay liner (either GCL or CCL) then leakage is controlled by Darcy's law (flow rate = hydraulic conductivity x hydraulic gradient). A geomembrane is an essentially perfect barrier to advection except where it has holes. Leakage through holes is controlled by head difference, the size of the holes, and what is in contact with the GMB.

7.0.3 Diffusion of contaminants in leachate

Diffusion is the movement of elements or compounds from areas of high concentration (leachate) to low concentration (e.g. groundwater). The diffusive flux through a clay liner (GCL or CCL) = mass of contaminant per unit area per unit time. It is controlled by the diffusion coefficient x the concentration gradient. An HDPE GMB is an almost perfect diffusion barrier to most elements and many compounds. However, certain organic compounds can diffuse very quickly through many GMBs. The diffusive flux through an intact GMB = mass of contaminant per unit area per unit time is controlled by the partitioning coefficient x diffusion coefficient x the concentration gradient.

7.0.4 Contaminants in gaseous phase

Landfill produces landfill gas (mostly methane and carbon dioxide) plus other "trace" gases. These trace gases may include some volatile organic compounds that volatilize from the pure phase (e.g. in contaminated soil) or from the compound being dissolved in leachate into the air or surrounding gas (often mostly carbon dioxide and methane when in a landfill).

Gases can also escape by advection and diffusion. Advection for gases is the movement of the gas from high pressure (inside the landfill) to low pressure (outside). Diffusion is the movement from a high concentration in a landfill to a low concentration outside. For soil, the resistance to both advection and diffusion is related to the degree of saturation. At low degrees of saturation, the rates of advection and diffusion can be very high.

7.0.5 What are we protecting?

The opening sentence of this chapter stated that *"Providing environmental protection involves keeping solids, liquids, and gases from coming into contact with either the natural and/or human environment when they contain elements or chemicals in a form that could have a significant negative impact on the natural and human environment if released"*. In that statement the term *"natural and/or human environment"* is intentionally very broad and nonspecific. But when it comes to actual projects we need to be quite specific.

Most commonly we seek to protect surface and groundwater. Surface water includes streams, wetlands, ponds and lakes. Groundwater is precipitation that enters the soil and migrates through the soil which can be used as drinking water, for commercial use, and plays a critical role in providing base flow to streams and rivers which is especially critical during periods of low precipitation. Thus, groundwater that is contaminated can eventually impact humans and the natural environment. Thus, an objective of the barrier system is to prevent direct contamination of surface water and groundwater in aquifers from leachate or other nonpotable liquid (e.g. pregnant liquor in heap leaching, pore water in tailings, acid rock drainage from the waste rock).

Commonly we seek to protect the air we breathe from gases that could affect us or the ecosystem either directly or indirectly. Gases like hydrogen sulphide (rotten egg gas) and other odorous emissions from a landfill can be detected by the nose at very low concentrations and can be extremely irritating to residents around the landfill. Carbon dioxide and methane in sufficient concentrations can also be deadly but even at much lower concentrations they have an indirect impact on humans and the ecosystem generally by contributing to global warming. One of the many purposes of a cover system is to minimize the escape of gases and allow their collection and management. The liner server system can also serve as a barrier to gases in a gaseous form but also in a dissolved form which upon migrating through the liner system may dissolve and then migrate in a gaseous form. Aquifers have been contaminated by the migration of volatile organic compounds (like benzene, toluene, ethylbenzene, xylenes, vinyl chloride, dichloromethane (DCM), trichloroethylene (TCE), tetrachloroethylene (PCE)) from a landfill through unsaturated soil to the water at depth.

7.0.6 Aquifer

An aquifer is a saturated geologic unit, commonly comprised of either granular material (e.g. cobbles, gravel, sand) or fractured rock from which water can be extracted readily by the excavation of a hole (well). Some definitions of an aquifer define the lower limit as when there's just enough water to run a tap. Practically speaking, an aquifer will have a hydraulic conductivity of 10^{-5} m/s or higher, however granular units with silt may yield sufficient water for minor activities and hydraulic conductivity down to about 10^{-6} to 10^{-7} m/s.

7.0.7 Aquitard

An aquitard is a geologic unit that provides significant resistance to the flow of water. Typically, it involves clays, silty clays, clay silts, and tills (tills being an assorted mixture ranging from boulders to clay size material deposited by a glacier common in parts of the world covered by ice during the last Ice Age around 10,000 to 15,000 years ago). An aquitard typically has a hydraulic conductivity of less than 10^{-9} m/s. An aquitard is a natural barrier and when present can be an ideal component of an engineered barrier system. While typically an aquitard will contain at least 10% clay-size material, not all clays are aquitards. In particular, clays can contain fractures and while the clay itself may have a low permeability in small individual specimens, the unit as a whole does not classify as an aquitard because these fluids can migrate through the fractures. Thus, a fractured natural clay is not necessarily a good barrier in that state but can be made a good barrier by excavation of some of the material and re-compaction as a compacted clay liner.

7.1 Some Lessons to Remember

Society's awakening to the damage being self-inflicted on the environment in which we live began in the 1970s with incidents of people becoming sick due to the chemicals being released into the environment and either coming into direct contact with humankind in liquid (e.g. nonaqueous phase liquids) or gaseous form (e.g. methane or volatile organic compounds) or by the human consumption of food or water which was contaminated. Although there are examples of deliberate release of contaminants into streams, wetlands, other water bodies, or permeable soil with the intent of inexpensive disposal, in large part contamination occurred because of an unintended release.

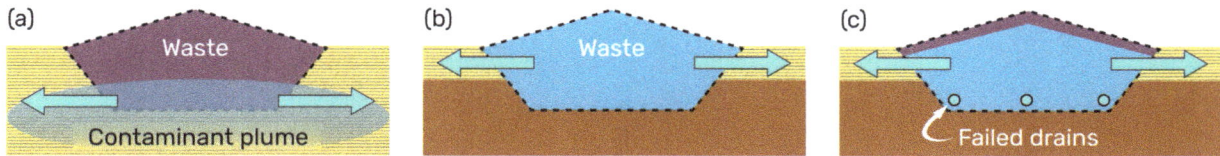

Figure 7.1 Schematics (a) Classic dump, (b) good natural containment below weathered zone, bathtub induced plume in the fractured more permeable upper weathered region, (c) Geotextile wrapped drains control leachate for a while but then clog, cease to control the leachate, and then a contaminant plume develops. Not to scale.

7.1.1 Consequences of a failure to contain- Lesson 1

Examples of unintended contamination include workshops where heavy metals (e.g. lead, mercury) and volatile organic compounds (e.g. dichloromethane) were being used, from storage areas where they were being contained until a valve was left open or a storage tank failed or overflowed releasing contaminants into the surface and groundwater, or because they were disposed of in a waste dump that was often a worked-out sand or gravel pit (Fig. 7.1a). Incidents, such as the infamous Love Canal in Niagara Falls New York (e.g. Rowe 2012a) brought awareness of the dangers and consequences of inappropriate management of waste and chemicals. **Lesson 1:** It may take decades, but waste not properly contained will have a significant impact on the environment and human health.

7.1.2 Leachate collection and nature's revenge - Lesson 2

With the recognition of the need to contain contaminants and waste came restrictions on the waste that could be disposed of in different classes of landfills and the development of the concept of an engineered landfill with some form of containment system. The earliest containment systems would largely be fortuitous situations where a waste disposal site was located in low permeability clay (e.g. Goodall and Quigley 1977). Unfortunately, it was soon recognized that even this fortuitous situation was problematic as the hole became a bathtub and contaminated water was again released into the environment (Fig. 7.1b). This fluid, known as leachate, was essentially water arising either from the degradation of organic waste or precipitation that permeates through the waste. With the recognition that this fluid needed to be collected and treated came the development of leachate collection systems and one of the first uses of geosynthetics in addressing environmental issues.

The earliest collection systems comprised pipes or piles of gravel, often wrapped in a geotextile (Rowe and Yu 2012) and placed at the bottom of the landfill to allow the collection of the contaminated fluid. However, it was not recognized that the bacteria in the leachate would attach themselves to the granular material and geotextile fibres, building up biomass and facilitating the precipitation of calcium carbonate that closed pores in both the granular material and geotextile substantially reducing hydraulic conductivity until it was no longer sufficient to control the level of leachate in the landfill. At this point, the leachate levels build up again in the bathtub, and once more overflow impacts the environment (Fig. 7.1c). **Lesson 2**: When filtering or draining fluids other than clean water, consider the potential biological and geochemical changes that could occur to substantially reduce the effectiveness of your filter or drainage material. This applies to granular materials, geotextiles, geopipes, geocomposite drains, geonets, etc.

7.1.3 The need to understand the system and work with nature - Lesson 3

Advances in the understanding of the mechanisms of leachate collection systems and other drainage systems (e.g. Bass 1986, Brune et al 1991, Koerner and Koerner 1995; Fleming et al 1999; Fleming and Rowe 2004; McIsaac and Rowe 2005, 2006, 2007, 2008; Rowe and Yu 2010, 2012; Yu and Rowe 2021) techniques were developed for predicting their service life (Rowe and Yu 2013) and more robust designs were developed where a geotextile is use sacrificially to encourage biological activity at a noncritical location and protect the more critical location directly above the liner (Rowe et al 2000; Fig. 7.2). **Lesson 3:** Optimized long-term performance of the system.

7.1.4 Diffusion can matter - Lesson 4

Advection (leakage) can be reduced by collecting leachate as discussed above. In some hydrogeologic environments, it is possible to excavate below water level in the surrounding soil and underlying aquifer. In these cases, if the water level in the aquifer is above the leachate level then water flows into the landfill and leachate cannot escape by advection (Fig. 7.3a); this is called hydraulic containment or a hydraulic trap. However, diffusion is the movement of compounds or elements from a high concentration (in the landfill) to a low concentration (in the aquifer). Outward diffusion occurs even though there is inward flow. Thus, diffusion can occur through saturated or unsaturated clay for virtually all contaminants (some may be retarded by

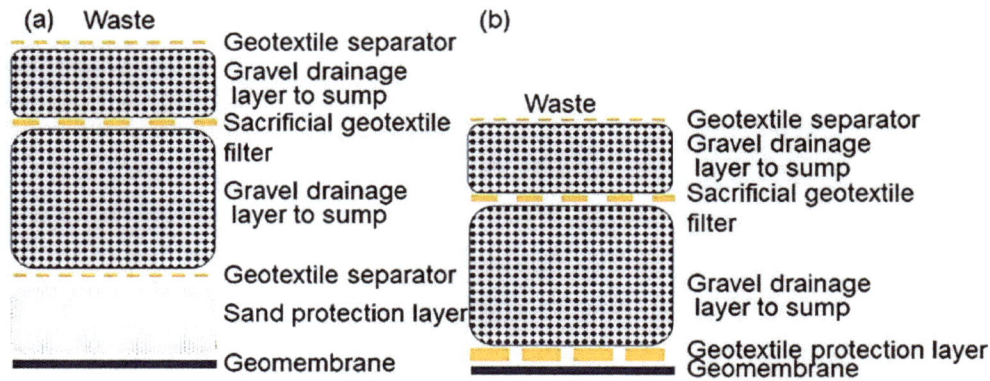

Figure 7.2 Leachate collection system cross-sections for putrescible waste: (a) Long design-life and/or high loads, (b) Short design-life and/or low loads. Not to scale.

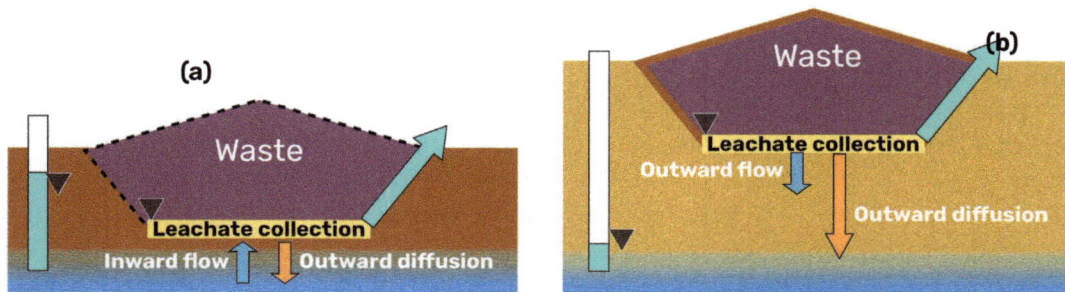

Figure 7.3 Leachate collection system is operating but outward diffusion in both cases: (a) Hydraulic containment (b) Lined landfill in unsaturated sand. Not to scale.

the soil). Thus, one needs to do calculations to establish the potential impact of the design on the aquifer even though there can be no leachate escape while the leachate collection system is operating. The inward flow is controlled by the difference in head and the hydraulic conductivity of the soil between the aquifer in the leachate collection system. Low permeability reduces the flow and hence the volume of groundwater collected to a manageable level. It also reduces inward advection resisting outward diffusion. This system needs either a natural or engineered low permeability layer between the waste in the aquifer. It will likely involve several geotextiles in the leachate collection system (e.g. see Fig. 7.2) but no geomembrane is required provided the level of diffusion against the flow is sufficiently small to meet regulatory requirements.

Figure 7.3b shows the opposite situation in terms of the gradient with a clay or composite-lined landfill over unsaturated sand with the water level in the aquifer well below the landfill. The outward leakage is partially controlled by the clay or composite liner but also because of the very low hydraulic conductivity of unsaturated sand at low water content, it is often assumed that this system will have a negligible impact on the aquifer. However, this assumption is often incorrect irrespective of whether it is a clay liner alone or a composite liner with an HDPE geomembrane. This is because volatile organic compounds (VOCs, e.g. benzene, dichloromethane, vinyl

chloride, trichloroethylene, tetrachloroethylene etc.) in the waste can readily defuse through the geomembrane and then through the near-saturated compacted clay or near dry GCL except where there are holes in the geomembrane. Once they reach the unsaturated sand, the VOCs dissolved in the porewater of the clay will mostly volatilize into the airspace of the unsaturated sand. Diffusion through unsaturated sand can be 4 to 5 orders of magnitude faster than through saturated sand or clay. Consequently, the aquifer can be quickly contaminated by volatile organic compounds even though other contaminants being transported by advection may take centuries or even millennia, if suitable geomembranes are used, to reach the aquifer. **Lesson 4:** even with hydraulic containment (Fig. 7.3b) or a near-perfect composite liner aquifers can be contaminated by diffusion of volatile organic compounds or other compounds and elements in the fluid to be contained.

7.1.5 A geomembrane may be critical even with no leakage - Lesson 5

With hydraulic containment (Fig. 7.3a) diffusion of PFAS compounds like perfluorooctanoic acid (PFOA) may still be significant even with no leachate escape in the absence of a GMB. In cases like this the addition of a GMB (a) reduces the influx of water (i.e. with hydraulic containment) except at holes in the GMB, and (b) is an excellent diffusion barrier to PFOA (Rowe et al 2023; Barakat et al 2024). Thus, although there are inward gradients, the GMB reduces the impact on the aquifer for PFOA. **Lesson 5**: A geomembrane can play a critical role as a diffusion barrier even if there would be no leakage without it.

7.2 Overarching Design Questions

Before embarking on the design of containment structures for solids, liquids and/or gases, six questions must be answered. Without the answers to the following questions, one cannot develop an appropriate design.

1. What are the geotechnical and hydrogeologic conditions of the proposed site?
2. What is being contained?
3. How much leakage and/or fluid escape is acceptable?
4. How long must the containment system last (design life)?
5. What is the environment in which it must function and how can that change during its design life?
6. What can go wrong and what are the unintended consequences of the design?

The following subsections elaborate on these six questions and the associated design considerations.

7.2.1 What are the geotechnical and hydrogeologic conditions of the proposed site?

All containment structures must be stable and fit for purpose. Common considerations include:

a. bearing capacity,
b. stability against sliding,
c. stability of any slopes in the containment structure,
d. the ability to rest and resist water pressures,
e. prevention of piping,
f. liquefaction,
g. blow-out,
h. loss of strength due to gas evolution with unloading,
i. evaluation of the suitability of the site geography and its hydrogeology for monitoring and contingency measures,
j. seismic resistance,
k. erosion, and
l. settlement and differential settlement foundation.

The foregoing considerations need to be evaluated, as appropriate for the given consideration, during the construction and filling of the facility, at closure, and during the post-closure period for the design life of the facility. Particular attention should be paid to extreme events in the context of climate change and the fact that traditional return periods are frequently exceeded by storm events that lead to an ever-increasing frequency of high rainfall, wind loadings, and extreme temperatures (both high and low).

7.2.2 What is being contained?

The answer to this question is not always as obvious as it may appear. The answer to this question is critical with respect to two considerations. The first is assessing the critical contaminant(s) and their pathways for escape as well as the regulatory or best practice limits on what will have a negligible effect on the environment. The second is the effect it may have on the service life of the engineered components of the system.

7.2.2.1 Landfill leachate:

The predominant constituents of landfill leachate of volatile fatty acids, common cations such as sodium, calcium and magnesium and anions such as chloride and sulfate. However, it will also contain small amounts of heavy metals, volatile organic compounds, ammonia/ammonium, and of particular concern in recent times will be the concentrations of Per-and Polyfluoroalkyl substances (PFAS) in the leachate (Rowe and Barakat 2021; Rowe and Zhou 2023).

7.2.2.2 Mining solutions:

Minerals are usually extracted with the use of other chemicals. The predominant mode of mobilization is to substantially increase or decrease pH (which depends on the particular mineral to be extracted). In addition, other chemicals may be added (e.g. Cyanide, a surfactant, or kerosene etc) again depending on the mineral to be extracted. The mineral will often mobilize other minerals which may not be in sufficient concentration to warrant recovery but may be in sufficient concentrations to impact the environment if released prematurely.

For example, ore containing valuable minerals may also contain arsenic and depending on concentration this could be a critical contaminant. Some tailings contain pyrite and even if this is stabilized at the time of disposal there is a potential for generating acid drainage if it comes into contact with oxygen. In addition, while these items just mentioned will show up in a routine chemical analysis there may also be smaller quantities of chemicals added in the processing that could be critical, particularly concerning the service life of the engineered components. Compounds such as wetting agents (surfactants), frothing and anti-frothing agents, etc. may be in relatively small amounts in the porewater compared to the concentration of metals but the main fact is the most problematic component with respect to service life. It may take considerable investigation to identify what exactly has been added and what will be the concentrations in contact with the lining system when placed in storage.

7.2.3 How much leakage and/or fluid escape is acceptable?

As so well stated by Giroud (2016):

> "All liners leak": this was stated by Giroud & Bonaparte(1989a) at the beginning of their paper. This should not be construed as meaning that there is no way to safely store liquids. In fact, recognizing that all liners may leak is the first step to the safe design of liquid containment systems.
> The design of a containment structure cannot be safe if the possibility of leakage is not recognized in the first design step."

There is now good data available on leakage probabilities with and without the leak location survey for primary composite liners in a lined system (Beck 2015; Gilson-Beck 2019; Rowe and Zhao 2023). These probabilities are very useful for consideration in design. Combined with recognizing that liners leak and we need to be able to design based on a reasonable estimate of leakage, we also need to know how much leakage is acceptable to design a barrier system that will keep the leakage below the allowable limit. The amount of acceptable leakage for a dam or portable water reservoir or canal can be orders of magnitude greater than that acceptable for contaminated fluid (e.g. leachate). Potable water storage reservoirs can experience a drop in water level of 5 mm/day would be difficult to measure and even if measured would likely be attributed to evaporation and be of little concern. However, 5 mm/day is 50,000 L per hectare per day (lphd). In contrast, a single composite landfill in Ontario might be expected to exceed allowable levels of per-fluorooctanesulfonic acid (PFOS) in groundwater with a leakage of 40 lphd (Rowe and Zhao 2023).

Thus, while the volume of water that's leaking is important is not sufficient to assess what is an allowable leakage. When dealing with contaminated fluid, be it a landfill leachate, a brine, pregnant liquor, tailings porewater, mineral processing solutions, etc. it is essential to consider what represents the most critical contaminants (i.e. what chemicals in the fluid are likely to be most problematic from an environmental impact perspective and hence control the design because of regulatory restrictions). In selecting the critical contaminants one typically considers three factors:

a. the source concentration relative to the regulatory limit (typically the lowest of that for drinking water, groundwater, and surface water). Some contaminants may be more critical for groundwater and others more critical for surface water. For some contaminants (e.g. ammonia/ammonium, the allowable levels may depend on another parameter (e.g. pH). Other things being equal, the greater the ratio of source concentration to regulatory limit the more critical is the contaminant.

b. Mobility – how easily is it transported by advection (i.e. with flowing water all leachate) and diffusion (under concentration gradient) in air and water? This includes consideration of its potential for sorption either to the soil, organic matter in the soil, or geosynthetics.

c. Persistence – does the contaminant experience radioactive or biologically induced decay and if so what is its half-life? Many pure hydrocarbon compounds can experience biologically induced breakdown(e.g. benzene, toluene, ethylbenzene, xylene) as well as being lost from a dissolved state to a gaseous state in a landfill thereby reducing the leachate concentrations much faster than by dilution. Also, some chlorinated compounds (e.g. perchloroethene (PCE) and trichloroethene (TCE)) breakdown under methanogenic conditions in a landfill daughter products (e.g. cis-dichloroethene [cis-DCE]) but the daughter products can further breakdown to give a more problematic accumulation of vinyl chloride). All of these volatile organic compounds mentioned in this paragraph are of particular concern because of their ability to rapidly diffuse through a traditional PE geomembrane.

7.2.4 How long must the containment system last (design life)?

The design life of a containment facility will significantly affect the design of the containment system and the choice of materials. There may be several stages to the design life.

a. Construction (maybe months to years and in some cases decades)
b. Operations
c. Closure
d. Post-closure maintenance and monitoring.
e. Long-term maintenance and monitoring.

Construction may take months to years and in some cases decades. If the barrier system will be exposed to the elements for any significant time, it must be designed for those conditions and this may notably affect the design and cost.

The period of operations includes all the time the facility is in active use after construction. Most commonly, it may range from a few decades to half a century. However, care is needed to select this conservatively since once the liner is placed (e.g. in heap-leach and landfill applications) it is relatively easy to accommodate a longer operation in the design before the barrier system is constructed but it may be very difficult if not impossible to extend the service life of the drainage layer or liner system after it is covered. The operation period is frequently when the liner is subjected to its highest temperatures and the volumes of fluid collected in their greatest (e.g. the landfill before the final cover is placed), and the greatest chemical exposure (since concentrations often decrease with time).

The closure period may be years to decades depending on the type of application. Pond liners may be removed and disposed of in a relatively short period. A landfill being developed in cells may have final covers being placed over half a century or even more.

Except for those applications where the facility is removed (e.g. ponds) there will be a need for a post-closure monitoring and maintenance period when the most aggressive monitoring is conducted. Any post-closure monitoring should be combined with modelling to predict the likely time to potential impact. Experience has shown that even in old dumps with no engineering it may take several decades before the impacts are detected in borehole monitors. This will include heap leach pads, landfills, and tailings storage facilities. The length of the post-closure monitoring period should be established based on the length of time of greatest risk.

The post-closure maintenance and monitoring will be followed by a period of long-term maintenance and monitoring with less frequent checks and maintenance until the contaminating lifespan is reached, where the contaminating lifespan is, by definition, the time period during which the containment facility must control the escape of contaminants to protect the environment. Thus, the contaminating lifespan is reached when concentrations of the critical contaminants are low enough that they can be released to the environment by natural processes without the presence of the barrier system.

The length of the post-closure and long-term monitoring of each of them will depend on the application, the nature of the waste, and especially on the critical contaminants. For example, in the low-level (radioactive) waste (LLW) facility where the bulk of the waste will have a half-life of less than 30 years, after 60 years level of radioactivity will have dropped to about 25% of its initial value and after 300 years to about 0.1% and after 500 years to about 0.001% of its initial values. A recently approved LLW facility in Canada has an estimated initial construction period (Phase I) of 3 years with a second Phase II construction during the 50-year operating period. After closure (about 1 year) the post-close period is 300 years followed by 200 years of long-term maintenance and monitoring. This gives a total design life of about 550 years and any geosynthetics critical to the performance of the system (the primary, secondary, and tertiary geomembranes and bentonite in the GCL) must have a 550-year service life. Based on test data, the Canadian Nuclear Safety Commission approved the facility on this basis.

Biological decay, although less predictable, as well as the effects of simple dilution of contaminants that do not degrade can all be modelled similarly to radioactivity (e.g. see Rowe 1991, Rowe et al 2004)

The design and choice of materials must be selected to provide a system that meets the required design life whether it is decades, centuries, or millennia.

7.2.5 How can the environment in which it must function change during its design life?

The environment in which the facility must function during its design life includes local topography and the potential for flooding, tsunami, and earthquake that must be considered in the design. It includes the water table, the potentiometric surface in an underlying receptive aquifer, and the native soils that will separate the facility from potential receptors in consideration of the relative performance of those soils and their existing and or more or less saturated state.

A design depending on hydraulic containment (Figure 7.3a) is dependent on the water level (potentiometric surface) in the aquifer remaining within a tolerable range. A design must not only consider what is there today but also how the construction of the facility will affect water levels (the shadow effect, see Rowe et al 2020), and how climate change over the design life will affect the operation of the system. These factors need to be considered to the maximum extent practicable. However, even with careful design, there is still a need for monitoring and contingency measures if either the leachate collection system fails or the water levels in the aquifer drop causing a loss of hydraulic containment.

Water levels and degree of saturation are very important for granular layers in the natural geologic and hydrogeologic systems outside the facility. In a saturated state, they can be a passage for liquids (e.g. leachate) from the facility. In an unsaturated state, the same granular layers can be a ready conduit for gas in both advection and diffusion. A liner system comprised of a geomembrane and clay liner is particularly vulnerable in a zone subjected to wet-dry cycles and changes in water table level, particularly if that level drops for a prolonged period. If there is a hole in the geomembrane, reliance is placed on the clay liner to prevent the migration of gases and contaminants in those gases out of the municipal solid waste landfill. Both compacted clay and GCLs are good diffusion and advection barriers in a saturated state. When the degree of saturation drops below 50% the compacted clay and GCLs can desiccate. Even, even if they don't desiccate, experience indicates that the increase in gas permeability and the diffusion coefficient can be several orders of magnitude (Bouazza et al 2017; Rowe 2020a). Thus, the potential for this to occur needs to be considered and mitigated in design.

The temperatures and conditions to which the barrier system will be exposed during construction, operations, closure, and post-closure can equally affect the long-term performance of a barrier system. Exposure of the system to the sun, freeze-thaw, and wet-dry cycles during construction, operations, or post-closure (especially for covers) can all impact the long-term performance and must be considered and mitigated (Rowe 2020a)

The service life of all geosynthetics is highly sensitive to the temperatures to which they are exposed and especially to the sustained highest operating temperatures (Rowe 2020a,b).

It is often assumed that a GCL will be hydrated. However, except in pond applications, when the GCL is fully hydrated below a low water level, the GCL degree of hydration (saturation) will be highly dependent on the grain size, water content, and mineralogy of the subsoil with which it has intimate contact (Rowe 2020a). Realistic consideration of the actual degree of saturation is required in the design. This is particularly critical when dealing with gas migration as noted above (Bouazza et al 2017b; Rowe 2020a).

7.2.6 What can go wrong and what are the unintended consequences of the design?

It is natural to be enthusiastic about a design that cost-effectively meets the design requirement. However, it is a mistake not to ask the questions "*What can go wrong?*" and "*What are the unintended consequences of the design*". A failure modes and effects analysis is a routine part of the dam design. It should become a routine part of containment facility design.

Things can go wrong because of poor construction (hence the need for construction quality assurance). Things can go wrong because the conditions assumed in the design differ from those present (hence the need for the designer to inspect conditions and ensure their consistent with design assumptions before materials are placed). As previously noted, things can go wrong due to unanticipated exposure of the system to the elements before the liner comes into service.

Careful consideration needs to be given to all potential factors that could affect long-term performance in

the should, as needed, be mitigated by design, and/or monitoring and contingency plans. Things can go wrong because a component of the system fails to perform as expected. Things can go wrong because of climate change.

Sustainable design and environmental containment require design, construction, operations, closure, and post-closure monitoring and maintenance that protect the environment for the contaminating lifespan of the facility.

7.3 Materials Selection

When considering the materials used in environmental protection facilities, one must consider the purpose of the material and the design life required for that material within the overall barrier system. Typically, geosynthetics fall into one of three categories where the geosynthetics:

1. are needed for construction and operations but not needed long-term.
2. must survive for a prescribed period but are not needed for the entire design life of the facility.
3. are essential to meeting fitness for purpose for the entire design life.

These three categories will be discussed in the following sections. In each case, de minimis specifications such as GRI specification represent a good starting point. Material meeting these requirements may be perfectly suitable for certain general applications where the exposure conditions are not particularly challenging and the design life for this component is relatively short. However, these specifications also often contain a caution such as:

"*This standard specification is intended to ensure good quality and performance of ... in general applications but is possibly not adequate for the complete specification in a specific situation. Additional tests, or more restrictive values for test indicated, may be necessary under conditions of a particular application.*"

The caution is particularly appropriate in geoenvironmental applications when one is often dealing with elevated temperatures, and/or aggressive chemicals, and/or higher loads, and/or high heads, and/or extreme exposure etc. Under the circumstances meeting the requirements of the standard specification may be necessary but not sufficient to ensure good long-term performance. Some of those circumstances will be highlighted in the remainder of this document. However, it is cautioned here, that this too is a general guide and cannot anticipate all possible situations. It should not be regarded as complete, and the reader should seek appropriate expertise for projects that are challenging in terms of either exposure conditions and/or required service life.

The service life of the geomembrane is related to changes in the material with time to nominal failure for the material typically being defined as the time when important physical property such as stress crack resistance reduces to 50% of its reference or specified value. The service life is related to the material in terms of the time to nominal failure but also depends on the strains developed in the geomembrane. Strains above about 3-6% may result in a service life shorter than the time to nominal failure while smaller strains may result in a longer time to nominal failure (Rowe 2000a).

7.3.1 Materials that are needed for construction and operations but not needed long-term

Many materials used in geoenvironmental applications fall within this category. This includes, for example, geotextiles used as a separator between clay and a granular layer which serves to provide separation during construction and loading but is not essential to the long-term performance. It may include the sacrificial geotextile used in a leachate collection system and the separator geotextile between the waste and the upper gravel layer (Fig. 7.2). An important consideration in these cases is that the geosynthetic has sufficient survivability to withstand conditions during constructions and operations. Specialized testing may be required to verify the suitability of a particular material.

These materials may include the geotextiles on geosynthetic clay liner if those geotextiles are only needed for internal shear strength and/or to contain the bentonite during construction and operations but not long-term.

7.3.2 Materials that must survive for a prescribed period

These materials may include a geomembrane used as a temporary cover during the filling of a landfill and will ultimately be superseded by the final cover. They may include leachate collection pipes in a landfill that are required when infiltration is relatively high during the placement of the waste but are not essential in the long term because of ample capacity in the granular drainage layers. It may include reinforcement and slip layers needed to minimize strain in a geomembrane with consolidation of waste. In these instances, the service life of the component much shorter than the design life of the barrier system may be sufficient.

7.3.3 Materials essential to meeting fitness or purpose for the entire design life

This is the most challenging of the three categories. It typically applies to geomembranes, the bentonite in the GCL, the protection layer above the geomembrane, and the drainage layers essential to maintaining stability and minimizing contaminant transport. Specialized testing is required to verify that materials have the required service life under expected field conditions and given the significance of these facilities a level of redundancy is also advised.

Careful consideration must go to materials selection and/or mitigation of factors that can affect long-term performance. Particular attention MUST be given to what can happen to the materials between when installed and when they become operational (which, in some cases, can be years). Examples, not a complete list, are given in the following sub-paragraphs.

For LCS care is needed to avoid runoff washing fines and sand into the LCS thereby substantially reducing its service life.

Generally, geotextiles should not be left without soil cover for a prolonged period. If they must be left exposed, they must be ballasted against wind, have adequate abrasion resistance, and be stabilized against UV. They should be tested for degradation before ultimately going into service. This exposure period may control the selection of the geotextile.

For GCLs with little or no soil cover consideration must be given to:

> the effect of cation exchange combined with wet/dry and/or freeze/thaw cycles on hydraulic conductivity (e.g. Rowe et al 2017).
> proper hydration of the bentonite and avoidance of issues such as over swelling and fibre pull-out, damage due to traffick by humans or equipment, shrinkage and the formation of gaps between panels, etc.

These issues can be avoided by encapsulation of the bentonite in the GCL in a composite liner with a geomembrane over a multicomponent GCL with the coating facing down. The the trade-off being the effect of hydration of the GCL by the fluid to be contained on GCL hydraulic conductivity.

For GCLs in composite liners that are left uncovered for a prolonged period, consideration must be given to the two items considered in the bullet points above as well as downslope erosion.(Rowe et al 2016, Rowe 2020a). Again the selection of the GCL become critical and again a multicomponent GCL, in this case the coating facing up or down depending on the application and the issues to be addressed is a potential mitigation measure.

For GCLs on steep slopes careful consideration must be given to (a) GCL dimensional instability leading to panel separation, (b) potential gravity driven bentonite migration, and (c) downslope erosion discussed above.

Different considerations may apply to the selection of the geomembrane depending on whether it is intended for long-term exposure (i.e. an exposed geomembrane) or short-term explosion in long-term containment either in the bottom of a containment facility or in a cover. For geomembranes intended in a buried application, long-term exposure prior to burial will require additional consideration due to the effects of UV and elevated temperature on the service life during that exposure period.

Obtaining an adequate service life of the geomembrane is often the critical quantity affecting the viability of the containment system. Geomembranes have come a long way over the last 30 years. Typically, and there have been some exceptions, geomembranes from before about 1990 were relatively poorly stabilized (standard oxidative induction time "Std-OIT" < 100 min.) and had relatively low (< 150 hours) stress crack resistance. Notwithstanding these observations, they have performed extremely well for the past 40 to 50 years and when they have been exhumed they have centuries of remaining service life despite the period of use.

Today most geomembranes contain a medium-density polyethylene resin and premium products have substantially enhanced antioxidant and stabilizer packages. Thus, a high-end (premium) geomembrane will typically have an initial stress crack resistance SCR_o >1000 hours, an initial $Std\text{-}OIT_o$ >160 minutes at 200°C, and an high pressure oxidative induction time (HP-OIT) > 400 minutes. However, while these are "good" numbers they are not especially relevant to the long-term performance of the geomembrane because they don't consider interaction with the chemistry of the fluid to be contained at the operating temperature. The strategy for assessing the performance of several candidate geomembranes for a given application and contact fluid is described by Rowe et al (2020b).

7.4 Landfills

There are now more than 40 years of history of municipal and hazardous waste landfills being constructed with the use of geosynthetics in both drainage and liner systems. The majority of municipal landfills are constructed with single liners and most commonly a single composite liner with a GMB/CCL or GMB/GCL. There has also been an evolution in the use of drainage layers. This section will be broken into two parts. The first deals with bottom liners and the second deals with covers. Leachate lagoons are dealt with in the section on lagoons.

For both bottom liners and covers, the primary consideration is minimizing the leakage of liquid and/or gas. In the case of the cover, the desire is to reduce infiltration into the landfill and hence leachate generation and to collect gas. The liner provides resistance to the advective transport as well as diffusive transport while the drainage layer above minimizes the driving force for leachate infiltration and the gas collection layer below minimizes the driving force for the escape of gas.

7.4.1 Primary leachate collection system (LCS) for putrescible waste

As noted in sections 7.1.2 and 7.1.3, much has been learned about leachate collection systems for putrescible waste over the last 35 years. If leachate was simply water then the medium to coarse sand layer or a geocomposite drain would have adequate capacity to collect the leachate in most municipal solid waste (MSW) applications. However, as previously explained leachate is more than water. The bacteria in the leachate attached themselves to the surface area of granular material and geotextile fibers and feeding on nutrients in the leachate they multiply reducing pore space and hence the hydraulic conductivity. However, the bacteria do more than create organic mass to fill the pores. They also create an environment that encourages the precipitation of calcium carbonate and along with it numerous metals such as iron, zinc, and lead. This can rapidly lead to clogging of the drainage media where clogging is said to have occurred when it leads to a reduction in hydraulic conductivity to below the design value and hence results in a leachate level greater than the design level. It is often said by operators that they have "no clogging" because they are collecting leachate. This is a spurious argument. The primary purpose of the leachate collection system is to control the leachate head at or below the design value and one can still be collecting all the leachate that is not escaping through the liner system in a clogged system. The reduction in hydraulic conductivity, typically to the order of 10^{-8} m/s, does not prevent flow and provides negligible resistance to leakage through the liner (Rowe 2005). However, to maintain continuity of flow with a significant decrease in the hydraulic conductivity of the drainage media, the head must increase to maintain the flow towards the leachate collection sumps. In so doing, the clogging increased the leakage through the underlying liner. As noted in section 7.2.3, even with an electrical leak location survey there will still be holes either initially or subsequently in the geomembrane and there is clear evidence of leakage through primary composite liners. For a given set of holes, the leakage is going to be related to the head and an increased head means more leakage.

Controlling the head in the leachate collection system requires a drainage layer together with a network of pipes. To minimize clogging two strategies can be adopted and they are best adopted together.

1. Minimize surface area for bacterial attachment and maximize pore throats between particles.
2. Provide a sacrificial layer to direct clogging above the primary drainage layer and encourage the bacteria to "clean up" the leachate before it enters the primary drainage layer.

The first strategy is best achieved by using coarse uniform gravel. Ideally, the gravel would be rounded but in reality, it is more often angular. Thus, the best drainage solution must be kept away from the geomembrane liner or the gravel will induce holes in the geomembrane. This then necessitates the provision of a protection layer as discussed below.

The service life of a leachate collection system can be related to the chemistry of the leachate, the particle size, and drainage layer thickness, and the fluid flowing in the system (Rowe and Yu 2013). Two important factors need to be considered in the evaluation of service life for a leachate collection system. First, waste streams differ from region to region. In particular, the amount of organic waste generated and finding its way to landfills in Western Europe is very small compared to North America, and the amount in North America is very small compared to many Middle Eastern and Asian countries. The greater the mass per unit area of organic matter in the landfill, the shorter the service life of a given leachate collection system. Hence a system suitable in Europe may not be suitable in North America and the system suitable in North America may not be suitable in the Middle East and Asia. Second, the volume of leachate generated and flowing through the system is an important factor affecting potential leakage.

Factors to be considered in the design of leachate collection systems include:

a. the waste stream and in particular the chemical oxygen demand and calcium concentrations expected in the leachate.
b. the volume of leachate flowing through the drainage layer (the clogging is related to the product of the volume and the concentration)
c. the grain size and grain size distribution of the proposed drainage material (the more uniform and larger the particle size, the longer the service life other things being equal).
d. thickness of the drainage layer (increasing thickness increases service life if one is prepared to accept ahead more than 0.3 m)
e. the slope of the drainage layer (2-3% preferred).

f. Pipe diameter and the location of access points should allow video inspection and cleaning of the pipe.
g. Pipe wall thickness and the number and location of holes must not compromise the structural integrity of the pipe. The hole should be as large an area as possible but should not exceed the nominal particle size of the drainage layer.
h. Trucks transporting gravel from a quarry to the site generate fines that can accelerate the clogging of the system if they are not separated before placing the gravel in the collection system.
i. The pipe should be video-inspected and cleaned regularly.
j. Clog material is easy to remove from the pipe and perforations where it is relatively new and soft. It can be difficult to remove once it hardens into the solid, predominantly calcium carbonate, state.
k. Without clear evidence to the contrary, geosynthetic drains should not be used as the primary leachate drainage layer for putrescible waste.
l. Leachate that is collected at the sump is not generally representative of the leachate that entered the leachate collection system. This is because during its passage through the collection system has been treated by the bacteria substantially reducing chemical oxygen demand and calcium concentrations, especially when it passes out of the acetogenic and into the methanogenic phase.
m. Studies of the potential for clogging the geocomposite drains that were conducted with leachate from the sump of a landfill well into its methanogenic stage do not correctly represent the likely performance in the landfill's leachate collection system directly below the waste or the pipe. This is especially true if the tests are not conducted at typical landfill temperatures around 30-40° C.

7.4.2 Primary LCS for non-putrescible waste, leak detection and secondary LCS

The primary leachate collection system for non-putrescible waste landfills (e.g. construction and demolition waste, hazardous waste, and low-level radioactive waste landfills) and for the leak detection and secondary leachate collection system is much less prone to biological clogging than municipal solid waste landfills and hence the drainage layer may be finer gravel or in some cases a coarse sand or geocomposite drain. The choice of drainage material will depend on the operating temperature and the required design life. These designs may not need a sacrificial filter layer.

7.4.3 Geosynthetic drains in drainage systems at the bottom of landfills

No form of geosynthetic drainage should be used as the primary leachate drainage system in putrescible landfills. Also, at present there is insufficient evidence to justify the use of a geocomposite drain or geonet in

> the primary leachate systems for nonputrescible waste, or
> the leak detection and secondary leachate collection system

when the design life exceeds 100 years or if the operating temperature will exceed about 20 to 25°C. This situation could change if manufacturers improve the stabilization of drainage geosynthetics in a similar fashion to what that done for geomembranes.

7.4.4 Primary liner systems for landfills with a single composite liner

The basic configuration of a single composite liner system involves a:

i.   protection layer (PROC),
ii. geomembrane (GMB),
iii. clay liner: (a) a GCL, (b) a CCL (c) a GCL+ CCL (Fig. 7.4)
iv. attenuation layer (AL).

7.4.4.1 Protection layer (PROC)

The protection layer separates the leachate collection system drainage layer from the geomembrane. Its primary purpose is to eliminate any measurable strain in the geomembrane due to indentations from the gravel (Rowe and Yu 2019) that will otherwise lead to ultimate failure of the geomebrane by stress cracking (Abdelaal

et al 2014; Ewais et al 2014). This is typically either a needlepunched nonwoven geotextile or a medium sand layer. Each option has advantages and disadvantages.

The needlepunched nonwoven geotextile is the most convenient. Rolls are easily laid out and they use very little airspace compared to a sand protection layer. The suitable geotextile is the ideal solution for situations where it can be shown that the nature of the leachate, operating temperature, gravel, the applied stress, the protection geotextile, and the geomembrane and material below it are such that (a) there is no measurable strain induced in the geomembrane based on a performance test (e.g. Brachman and Gudina 2008a.b; Brachman and Sabir 2010,2013; Hornsey and Wishaw 2012); and (b) the geotextile has a service life that exceeds the design-life of the barrier system. The latter requirement represents the greatest challenge, especially at temperatures greater than 20° C if the design life exceeds 100 years.

To be sufficiently effective, the sand layer only needs to be about 50 to 75 mm thick if it can be placed in a consistent manner. However, for construction reasons it's often necessary to increase the thickness of 150 to 300 mm thereby increasing both cost and loss of airspace. When placed, the sand is in a very effective protection layer that also extends the life of the geomembrane by slowing antioxidant depletion (Rowe et al 2013a). The sand can be a challenge on side slopes since is prone to erosion. It may be necessary either to very quickly cover the sand layer with the geotextile and then with the drainage gravel to avoid notable erosion. Alternatively, a geocell system may be placed and filled with sand. The sand protection layer may be regarded to have an indefinite (millennia) service life.

Geomembrane (GMB)

For landfills, a high density polyethylene geomembrane with either a unimodal or bimodal medium density resin may be used provided it can be demonstrated that has an adequate service life given the design life of the facility. Factors that need to be considered in the selection of the geomembrane include:

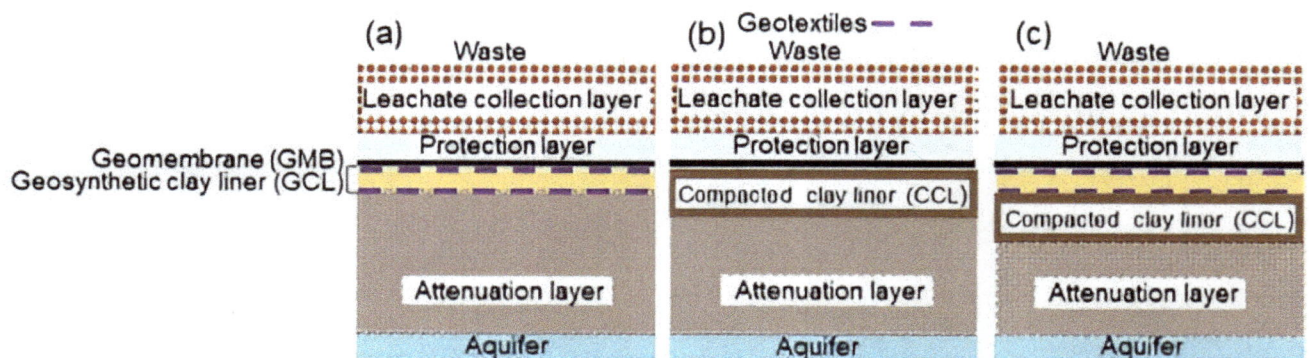

Figure 7.4 Three single composite liners:(a) LCS/PROC/GMB/GCL/AL, (b) LCS/PROC/GMB/CCL/AL, (c) LCS/PROC/GMB/GCL/CCL/AL. Not to scale.

a. does it have an adequate service life? Specialized immersion testing may be required to verify the suitability of candidate geomembranes if the design life exceeds 100 years. This testing should be carried out of well before construction and the geomembranes meeting service life criteria prequalified (Rowe et al 2020b).

b. Smooth, textured one side, textured both sides. Texturing may be required for stability reasons however the texturing can reduce the service life of the geomembrane and unless its use is adequately mitigated, and may increase the risk of damage the geomembrane due to consolidation induced down drag if the slope is steeper than 3 horizontal to 1 vertical. Generally speaking, a textured geomembrane should only be used where the texturing is essential the stability or worker health and safety (Zafari et al 2023a).

c. White or black? White has the advantage of reducing the temperature of the geomembrane and hence wrinkles the impact of temperature on the underlying clay liner. It also creates considerable glare and it needs to establish that the quality of the geomembrane is not compromised by the addition of this thin layer (Zafari et al 2023a).

d. Conductive one nonconductive? In a single liner system over either a GCL or compacted clay, electrical leak location survey can usually be conducted without a conductive geomembrane. It may be argued however that the conductive geomembrane makes it easier to find holes particularly holes in wrinkles.

e. How long will it be exposed to the sun before it is covered with waste?

A white conductive layer geomembrane has a number of advantages for construction provided it can be established that the addition of the surface layers has not compromised long-term performance (e.g. see Zafari 2023a,b,c).

7.4.4.3 Clay liners

In landfill applications of the last century, a geomembrane was most commonly used with a compacted clay liner (CCL) whereas this century it is more often used with a geosynthetic clay liner (GCL). In both cases, the GMB and clay liner were used together to form a single composite liner with the two components complementing each other to provide much better performance than either alone. In both cases, the clay liner is intended to substantially reduce the leakage that will occur through any hole in the GMB just as a finger pressed over a hole in a plastic bottle will substantially reduce the leakage

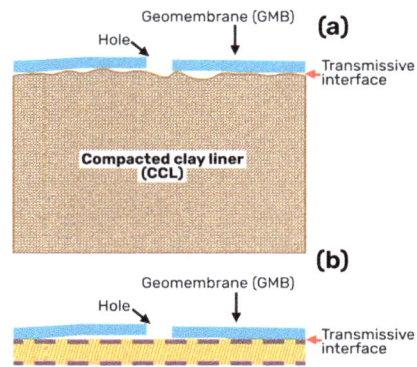

Figure 7.5 Schematic showing the transmissive interface at the boundary between the geomembrane and (a) CCL, (b) GCL. Not to scale.

through the hole in the bottle. In the bottle analogy, one would not be surprised if a little water did leak between the plastic on the thumb and drip despite the thumb substantially reducing the leakage. Similarly, with a composite liner, there is an interface between the geomembrane and the clay liner (Fig. 7.5) that is more permeable than the clay liner, and consequently, water will spread a distance laterally at this interface until there has been sufficient reduction in head that the is flow stopped. The reader is referred to Rowe (2012, 2020a) for more information about interface transmissivity.)

i. Compacted clay liners (CCL)
   Depending on the nature of the clay, the effectiveness of its compaction, the chemical interaction between the clay and leachate, leachate temperature, and the applied stress, the hydraulic conductivity, k, of the clay liner can range from 10^{-9} m/s to 5×10^{-11} m/s. While values within this range can be obtained with good construction (e.g. see Rowe et al 1993, 2004; Benson et al 1999; Rowe 2005; Daniel and Koerner 2007). There are several difficulties with compacted clay:
   a. The availability of suitable clay is often problematic and the CO_2 emissions and financial costs associated with bringing clay from an off-site borrow pit are often high.
   b. Most contractors do not know how to build a compacted clay liner so a great deal of construction quality assurance is needed.
   c. The interface transmissivity between the clay and the GMB is typically 2 to 4 orders of magnitude smaller for the GCL than for an intact CCL in this can increase the leakage that occurs for a given number of holes by 1 to 2 orders of magnitude. Lastly, when covered by a black GMB and to a lesser GMB in the sun, the solar energy heating the geomembrane causes evaporation of moisture in the clay and

desiccation cracking at the surface. While this may not impact the overall hydraulic resistance of the clay itself if the period of exposure is short, this surface desiccation increases the interface transmissivity between the geomembrane and clay by one or more orders of magnitude.

ii. Geosynthetic clay liners (GCLs)

GCLs can have very low hydraulic conductivity, k, if properly selected and installed and can be used effectively both alone and in composite liners. As noted previously, there are distinct advantages to using a GCL in contact with the GMB reducing interface transmissivity, θ, in composite liners and hence leakage. There are many GCLs on the market; one manufacturer alone produces 50 different variants. Thus, while the cheapest GCL meeting specifications may be suitable for some relatively unimportant projects, consideration needs to be given to the questions raised in §7.2. Most GCL's are advertised in the specification sheet as having $k \leq 5 \times 10^{-11}$ m/s and then reference is given to a very specific test to which this applies (e.g. ASTM D5887). This number is used for comparing different products under the same test conditions; it is only useful for that purpose. A more useful number for this purpose is the index flux which represents the flow under these conditions and implicitly incorporates the effect of GCL thickness and mass per unit area of the bentonite. Unfortunately, this can be misinterpreted as a material characteristic. Unless a project happens to be containing distilled water at 20 °C, in an application with an effective stress of 35 kPa and a differential head of 1.5 m, the value is not relevant to design.

The hydraulic conductivity of a GCL in a landfill will depend, inter alia, on the level of interaction with the leachate and subgrade material, temperature, and applied stress. There is a considerable body of literature dealing with GCL hydraulic conductivity but a good overview of the many factors affecting GCL's hydraulic performance in the field can be found in Rowe (2020a). It is, however, important to point out here that the same GCL used at the bottom of the same landfill may have more than one relevant hydraulic conductivity. For example, Figure 7.6 shows a wrinkle in a geomembrane with a hole that permits leachate to enter between the geomembrane and GCL. Some of that leachate will migrate laterally in the interface between the geomembrane and GCL and then migrate vertically through the GCL. At these locations away from the wrinkle there are significant vertical stresses applied to the GCL through the geomembrane from the weight of the overlying material and so there is a relatively low hydraulic conductivity k_a controlling the leakage in this region. However, while the wrinkle gets smaller due to the applied stress it does not go away as much of the vertical stress is transferred around the wrinkle by arching. Thus, the GCL directly beneath the wrinkle has applied an effective stress of essentially zero and relies on the stresses induced by needle-punched bundles restraining the swelling of the bentonite to provide stress and constrain the expansion of the bentonite. Nevertheless, the void ratio and consequently the hydraulic conductivity k_b of the GCL below the wrinkle will be typically be at least one order of magnitude higher than the hydraulic conductivity below the geomembrane away from the wrinkle, k_a.

It is holes in wrinkles that will predominantly control the leakage through a single composite liner and much of that leakage will occur directly beneath the wrinkle and will be directly related to k_b by Darcy's law. Equations for calculating leakage when there are holes and wrinkles can be found in Rowe (2012).

iii. GCL+ CCL

As indicated in the earlier discussion of CCLs there are benefits to combining a GCL with a

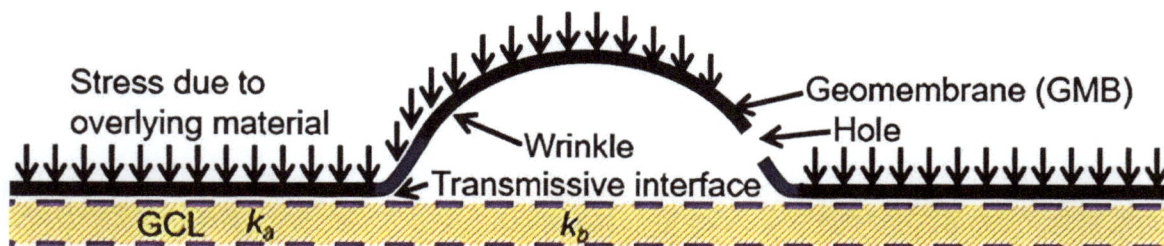

Figure 7.6 Schematic showing a wrinkle in a geomembrane with a hole together with the transmissive interface at the boundary between the geomembrane and GCL. Not to scale.

CCL in terms of mitigating the negative effects of surface desiccation of the CCL on interface transmissivity. For the situation shown in Figure 7.4c, the GCL will partially hydrate by uptake of moisture from the CCL with a degree of saturation the GCL being highly dependent on the mineralogy of water content of the clay liner. Moisture uptake from the CCL will necessarily involve a chemical interaction with the bentonite in the GCL by a process known as cation exchange. This cation exchange will tend to increase the hydraulic conductivity of the GCL although if there is sufficient applied stress the effect can be largely mitigated. However, even with as much is an order of magnitude increase in hydraulic conductivity k_a there is still a significant benefit in terms of reduced leakage due to the much lower interface transmissivity plus the additional hydraulic resistance still provided by the GCL.

7.4.4.4 Attenuation layer

An attenuation layer is typically a natural material (although it could be an engineered fill meeting the same requirements) that is neither an aquifer nor aquitard. It typically lies in the grey area with a hydraulic conductivity between $1\times10^{-7} \leq k \leq 1\times10^{-9}$ m/s. Soil with a hydraulic conductivity of 1×10^{-7} m/s is often considered suitable as a core in an earth dam where the limits on the allowable leakage are very substantially higher than they are for the containment of contaminated water. For example, the observed seepage (leakage) through the core of the Guadalcacín dam near Cadiz, Spainia is about 90 l/min or about 162,000 lphd (liter per hectare per day). While this magnitude of leakage may be quite acceptable for portable water it will be quite unacceptable for contaminated water. Thus, in the barrier system the attenuation layer serves little purpose in controlling advection (leakage) but it does serve an important role with respect to controlling diffusive migration of contaminants to an underlying aquifer and hence its inclusion. A few metres of attenuation layer above a saturated zone has a capacity to retain a significant degree of saturation reducing the potential for both liquid and gaseous migration through the unit.

7.4.5 Leakage through primary liner systems for landfills with a single composite liner

Calculations can be performed to calculate the expected leakage through different composite liner systems using equations and approaches as described in the literature (e.g. Rowe et al 2004, Rowe 2012, 2020b). The leakage through a composite liner is going to be very dependent on the number and size of holes, particularly those intersecting wrinkles (e.g. see Fig. 7.6). The number of holes

can be substantially reduced by an electrical leak location survey (ELLS) both before and after soil is placed over the geomembrane (e.g. Beck 2015, Gilson-Beck 2019). Table 7.1 summarizes the probability of exceeding a given leakage through composite liner based on field able data from a number of cases as discussed in more detail by Rowe and Barakat (2024). The leakage is reduced at the higher levels and low levels of leakage likely because leak location survey detection limit was higher after the soil was placed of the geomembrane in small holes in holes and wrinkles when there was no conductive geomembrane was missed in that survey reflected in the observed leakage through the primary liner.

Table 7.1 Probability of exceeding a given leakage through a composite liner with and without an electrical leak location survey (modified from Barakat and Rowe 2024, based on data from Beck 2015).

Leakage rate, Q (Lphd)	The probability of exceeding Q	
	With an ELLS	Without an ELLS
10	85%	91%
25	61%	80%
50	40%	64%
75	26%	51%
100	21%	41%
150	13%	26%
200	5%	17%

Historically, in the last 50 years of monitoring the leakage less than about 200 L per hectare per day has provided minimal impact that is been observed for the contaminants for which we have normally been monitoring. However, the emergence of significant levels of PFAS in leachate is changing this situation in much lower levels of leakage are going to be required to contain PFAS to current and proposed regulatory limits (see Rowe and Barakat, 2021,2024, Rowe and Zhao 2023). The most appropriate way of reducing leakage when a single composite liner is not adequate is a double composite liner system with a leak detection and secondary leachate collection layer between the primary and secondary liners, as discussed in the following section.

7.4.5 Double composite liners for landfills

While not as commonly used in landfills as single composite liners, double composite liners have been mandated in some jurisdictions mandated, or as an Ontario, mandated for all but small landfills, for more than three

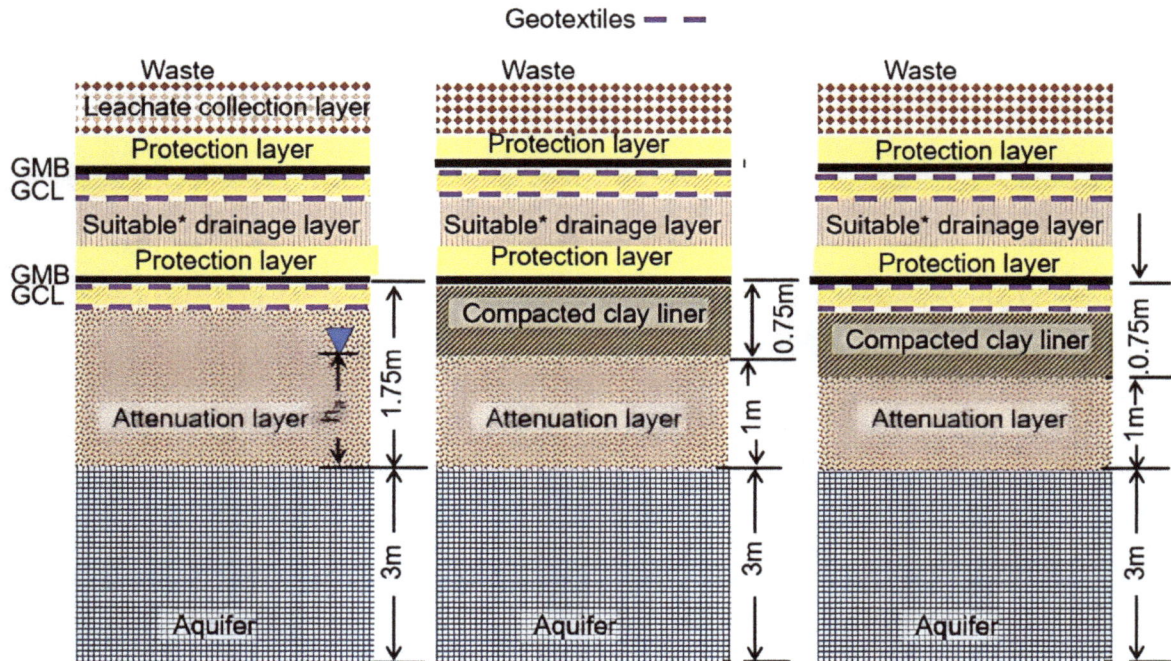

Figure 7.7 Leachate collection system operating but outward diffusion in both cases(a). Hydraulic containment (b) Lined landfill in unsaturated sand. Not to scale.

decades. Most of what is known about leakage through single composite liners from field monitoring is known from monitoring the primary liner in double-lined systems. Three typical double liner systems are shown in Figure 7.7. The attenuation layer shown in Figure 7.7 are based Ontario Reg 232/98 (MoECC 2012) and are not arbitrary. They are based on contaminant transport calculations and dimensions have been selected for controlling the diffusive migration of volatile organic compounds such as benzene, dichloromethane, and vinyl chloride at concentrations typically found in Ontario landfills last century.

The components of the double liner systems have all been described with respect to the single composite liner in the previous section. The selection of appropriate materials for protection layers and drainage layers will depend on the waste being contained. It is essential of these components have a sufficiently long service life to contain the contaminants in landfills, and particularly the concentrations of PFAS found in landfills.

7.4.6 Landfill covers

The final cover in landfills serves to restrict the diffusive migration of gas from the landfill together with the ingress of water into the landfill. In addition, components of the final cover are to protect the critical liner components from accidental, casual, or intentional damage from both humans and animals as well as the environment itself. Specifically, in general, a cover needs to

a. protect against physical damage by humans or animals,
b. protect against windblown damage and major precipitation events,
c. protect clay liner against freeze-thaw and wet-dry cycle as needed,
d. provide adequate drainage of rainwater and snowmelt,
e. provide adequate resistance to infiltration of water and exfiltration of gas, and
f. provide adequate drainage for gas collection.
g. Provide adequate drainage for gas collection

and have a service life, subject to appropriate maintenance, that exceeds the contaminating lifespan of the facility (see §7.2.4). Thus, the final cover has multiple components (Figure 7.8) as described below starting above the waste and finishing at the vegetation.

The intermediate cover is placed over the waste before placement of the final cover. This intermediate cover can act as a foundation layer for the cover system. Also, during the period between the placement of the intermediate cover and the final cover, there may have been differential settlement of the waste and by addition of more material to this layer, an appropriate grade can be obtained for the construction of the other components of the system.

The gas collection layer needs to be sufficiently permeable to collect the gas and direct it to the collection pipes. Sand is generally not suitable as a gas collection layer since it can have a relatively low gas permeability with any moisture uptake. The layer needs to be thick

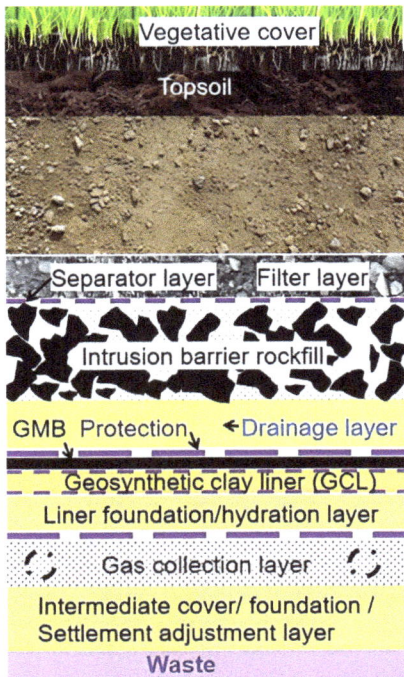

Figure 7.8 Schematic showing the various layers that may be included in a landfill cover. Not to scale.

enough to allow the placement of the pipes for gas collection. The gas collection layer should be sloped such that any water that reaches that layer drains either to a leachate drain or into the waste and hence it can be useful to place windows of granular material through the intermediate cover before placing the gas collection system to both serve as a more permeable conduit for the gas but principally as an escape path for condensation and other moisture that accumulates below the liner system.

The gas collection layer needs to be separated from the geosynthetic clay liner by a suitable geotextile and a liner foundation/hydration layer. This layer should be well-graded-granular material placed at or slightly above optimum water content. It is prevented from spoiling the gas collection layer by the geotextile separator. The primary purpose of this layer is to provide moisture for the hydration of the GCL. GCLs are not good gas barriers at degrees of saturation/hydration less than about 70% (Bouazza et al 2017a,b; Rowe 2020a). The granular foundation layer should, therefore, have a grain size and initial moisture content after placement to allow hydration of the GCL to above 70% degree of saturation. The GCL should be placed with a 300 mm overlap between panels and supplemental bentonite between the panels sufficient to withstand movement of the panels with differential settlement of the waste.

The GCL is overlain by a suitable geomembrane. The geomembrane should be overlain by protection and a drainage layer. Depending on the location, climate, and materials availability the protection and drainage layer could be comprised of a course sand, a geocomposite drain, or a protection geotextile and gravel. While the intrusion barrier rock has a sufficient drainage capacity, due to the risk of damaging the geomembrane, it needs to be separated from the geomembrane by an appropriate protection layer which may also serve as the primary drainage layer.

The intrusion barrier is comprised of relatively coarse rock fill and although it has drainage capacity its primary function is to discourage animals and human beings from disturbing the liner system. Given the nature of the material and the need to be an effective deterrent is usually about 0.4-0.6 m in thickness.

The intrusion barrier needs to be separated from the silty loam soil by a suitable filter-separator which may be either a geotextile or a combination of a geotextile and well-graded granular filter. The silty loam's primary function is to support the vegetation, but it also has an important secondary function in providing separation of the liner system from the local climatic conditions. The silty loam is overlain typically by about 150 mm of topsoil which provides the primary nutrients for the vegetation. The vegetation itself is important for minimizing erosion while the also helps reduce moisture migration through the liner system.

The thickness of many of these layers will depend on the nature of the material that is being used. The choice of material will often depend on the relative service life of the material compared to the design life of the cover system. The thickness of the loam layer will be varied such that the distance from the surface to the GCL is sufficient to minimize the effects of both freeze-thaw cycles and, to a lesser extent, potential wet-dry cycles.

While the cover system shown in Figure 7.8 may be regarded as a gold standard in cover systems, it is not necessarily applicable in areas where temperatures well below freezing are continuously sustained for several months and some of the benefits of the loam layer in providing freeze-thaw protection to the GCL may be lost due to the potential for air entry and atmospheric pumping of the drainage layer and intrusion barrier. Also in regions where this rapid accumulation of snow and or accumulation of freezing of water exiting the pipe can cause a blockage of the drainage outlets and subsequent slow thawing relative to spring thaw higher up the slope because they build up the pore pressures and veneer instability further up the slope. Special attention is needed to design details under the circumstances.

7.4.7 Leachate lagoons

Leachate generated from a landfill is frequently stored in a leachate lagoon prior to treatment and release. Given the potential for groundwater contamination from leachate, especially now it is well recognized that virtually all forms of waste leachate (e.g. municipal solid waste, construct on demolition waste, low-level radioactive waste) will contain per- and poly-fluoroalkyl substances (PFAS) at levels above current drinking water objectives and greatly above proposed new drinking water objectives, there is a need to securely store the leachate in the leachate lagoons. In this context, except for size, leachate lagoons are industrial effluent impoundments with the same basic considerations (§7.6 for more detailed discussion). The primary difference between different contained fluids is the chemistry and both the level of containment required to prevent unacceptable impact and the effect of the interaction between that chemistry and the chemistry of the geomembrane on the liner service life. In the case of leachate lagoons operating a near ambient average temperature, the primary concern is the potential effects of leakage of PFAS contaminated leachate. Where once the single liner system may have been considered adequate for traditional chemicals found in MSW leachate, that changed with the monitoring of leachate for PFAS and the relatively high concentrations typically observed in both MSW and construction and demolition waste leachate (Rowe and Barakat 2024). In this context, a move to double lined ponds for leachate retention requires serious consideration.

7.5 Mining

Mining applications represent a major consumer of geosynthetics. They are widely used in Overarching Design Questions Heap leach pads typically comprise a drainage layer over the geomembrane which is often simply resting on a prepared surface of existing subgrade. In many respects, this is similar to the primary liner in a landfill and a number of similar issues need to be considered with the caveat that except for on-off pads, the loading conditions in heap leach pads can be far more severe than those for a landfill. Survey of 92 heap leach projects 20% had a stress on the liner of 1 - 1.5 MPa, 23% ~ 2 MPa, 14% ~ 3 MPa, 7% with a height above 150 m and stress >3 MPa The 2024 heap leach that failed in Turkey with 257 m high the maximum stress likely the order of 5 MPa. Some factors to consider concerning heap leach pads are discussed below. The factors discussed are not intended to be complete but rather as illustrative of factors to be considered. Every project should be subject to potential failure modes and effects analysis and each failure mode identified and carefully considered. The following represents a starting point for any such analysis and a more detailed discussion liner system design for heap leach pads can be found in Lupo (2010).

7.5.1 Geotechnical/interface stability

These facilities may be built in valleys with steep slopes that require careful consideration from a purely geotechnical perspective even if there is no geomembrane.

The crushed ore is placed and fluid is percolated through the ore. The fluid typically has either a high or low pH (depending on the mineral to be extracted) and when it is collected at the bottom it is often referred to as pregnant liquor. This would is collected and processed for the extraction of the minerals. In addition, rainwater and, in the absence of proper drainage, surface waters can enter the leach pad. In any geotechnical situation where fluid is being added to a granular mass, considerable care is required to ensure stability is maintained given that the fluid serves to both increase the forces tending to cause instability (weight) and decreases the forces resisting instability due to a corresponding buildup in pore pressures.

If, as is very commonly the case, there is a geomembrane on the base and side slopes of the valley then consideration must be given to the potential for the geomembrane intersecting a former drainage path during times of heavy rainfall and adequate drainage provided to prevent a build-up in water pressures beneath the geomembrane with a consequent decrease in stability.

Ensuring adequate drainage is a necessary first step towards maintaining stability. Careful consideration must be given to:

> stability under operating conditions,
> what can go wrong under operating conditions (e.g. excavation at the toe of the slope or placement of excess material near the upper edge of the slope)
> the potential impacts of climate. For example, care needs to be taken with the drainage outlets that can be blocked. Situations have been encountered where these outlets became blocked by debris or, in cold regions, frozen solid at lower elevations where, with the means of escape blocked, melting snow further up the slope leads to a buildup of water that can be a cause of instability.
> the potential impacts of extreme events (e.g. heavy rainfall) that are becoming increasingly common with climate change.
> the potential impact of seismic events or blasting.

The presence of a geomembrane also means the presence of an interface and the potential for failure along that interface both under normal conditions and with an extreme event; both need to be considered.

In situations where the crushed ore is very thick, consolidation under its own weight and the weight of additional fluids (both leaching fluids and infiltration) may lead to down drag forces on the geomembrane which could lead to failure of the geomembrane where covered with ore. Locations just below benches are particularly likely to be problematic under both static and seismic conditions.

In addition, the likely settlement, and in particular differential settlement, of the subgrade under the loaded conditions needs to be considered in the context of how pooling of the pregnant liquor in an area of local depression due to differential settlement will affect leakage through holes in the geomembrane. The effects could be mitigated by appropriate grading of the subgrade prior to placement of the geomembrane.

7.5.2 Geomembrane selection for fitness for purpose

A typical barrier system (Figure 7.9a) involves the ore overlying a drainage layer overlying a geomembrane and underliner. As in the case of liners for landfills, consideration must be given to the potential for holes and excessive strains in the geomembrane as a result of the interaction between the gravel in the drainage layer, the geomembrane, and the underlining. However, there are number of major differences between the geomembrane exposure conditions a landfill and heap leach. Specifically, the geomembranes in a leach pad may be subjected to severe stress (e.g. > 3 MPa in a few cases), chemical exposure (very high or low pH, depending on the mineral being extracted, other chemicals used to aid mineral leaching (e.g. cyanide, kerosene), and in some cases elevated temperatures due to the insulating effect of the overlying waste (see more details see §7.5.2), and/or exothermic chemical reactions, and/or a result of microbial activity used to aid the leaching process using thermophilic iron-oxidizing chemolithotrophs and even higher temperatures may be associated with thermoacidophilic archaeon used for releasing copper (e.g. copper sulfide oxidation) in what is sometimes referred to as biomining. Studies have shown (e.g. Rowe et al 2013, Brachman et al 2014) that that without a protection layer and under 2 MPa pressure, a 1.5 mm-thick LLDPE geomembrane developed 100,000 holes/ha, 1.5 mm thick HDPE geomembrane developed 170,000 holes/ha, and more recent research that a 4.1 mm-thick BGM developed 3,500,000 holes/ha. In each case an adequate protection layer would be appropriate. In some cases a double liner (Figure 7.9b) may be used in this reduces the potential losses but there is significant potential leakage through the primary liner and, with differential settlement, the secondary.

The second issue with respective geomembrane selection is longevity. It is not unusual for heap leach pad to be designed for a specified number of years and then, as that age is approached, the question is raised "can we double that service life?". Once the geomembrane is under a large pile of ore it is not practical to change the geomembrane and so the question can only have a positive answer if the geomembrane that was selected was overdesigned and has far more service life than was originally required. When considering the service life, careful consideration must be given to the compatibility of each specific geomembrane with the relevant chemistry for the application. When considering chemistry, pH is important but so too are other constituents particularly other chemicals used to enhance the leaching of minerals. Temperature is another critical consideration. Elevated temperatures of 60-85°C have been encountered when bacteria have been used to aid mineral extraction. The service life of the geomembrane decreases rapidly with increasing temperature. There is a growing body of data regarding the time to nominal failure (which is related to, but not the same as, service life) for geomembranes in mining solutions (e.g. Rowe and Abdelaal 2016; Abdelaal and Rowe 2017,2023; Zhang et al 2018; Abdelaal et al 2023; Abdelaal and Samea 2023; Samea and Abdelaal 2023a,b). Project specific testing is likely to be required for any application where the geomembrane will be above 40°C or where a service life in excess of 50 years is required.

7.5.3 Geomembrane selection for fitness for sustainability of mining

The substantially increased demand for mining resulting from the movement to electric vehicles has two potentially significant unintended consequences. The first is the CO_2 emissions associated with the increased mining (which already represents 40% of global industrial energy use) will increase substantially and this does not appear to be well considered. The second is the potential environmental impacts of increased mining. Heap leach pads have typically be designed as discussed above (Figure 7.9a,b) for the collection of pregnant liquor. The drainage layer allows the collection of the pregnant liquor. It also offers the potential for both short and long-term damage to the geomembrane. In the short-term the primary concern with the geomembrane selection rests the choice of a geomembrane that will keep the number of holes (i.e. short-term punctures) small enough that it does not significantly impact the recovery operation. However, what is a largely unknown loss from mineral recovery may not be acceptable in terms of long-term environmental impact. The focus on selecting a geomembrane for the operating life needs to be broadened to a focus on providing environmental protection in the long-term. Indentations in the geomembrane due to gravel in either the drainage layer or the subgrade may not cause holes in the short-term can be expected to induce stress cracking and eventually create millions of holes per hectare (Ewais et al 2014). At this point residual chemicals, minerals, and in some cases acid mine drainage will be released. Given the high loads this could be within decades of the closure of the operating heap facility and result major long-term liabilities. Thus, for sustainability, it is critical that the barrier system be designed from both the short-term recovery and the long-term environmental protection perspective.

7.5.4 Clogging and sustainability

The pregnant liquor that enters the drainage system may contain many minerals in addition to those to be recovered. The change in environment in moving from the ore into the drainage layer is a location where bacteria can thrive and where there is a change in the chemical environment that can induce precipitation of some minerals. The combination of bacterial action and chemical precipitation together with some particulates that may be carried in the liquor can result in clogging of the drainage layer. This is analogous to the clogging that can occur in a leachate collection system. Particularly prone to clogging is the area around pipes, especially if a geotextile has been used as a filter around the pipe. The clogging will not prevent the collection of pregnant liquor but will result in the buildup of her head and increase leakage through holes in the geomembrane. Without careful monitoring, this leakage can easily go undetected for decades but ultimately result in environmental problems and liabilities for the mining company. As consideration must be given to the potential for clogging, which will depend on the particular application. Consideration must also be given to minimizing leakage if undetected for unexpected clogging occurs; this highlights the need to design considering not only the punctures but the strains induced in the liner system.

7.5.5 Composite liners

Leakage of pregnant liquor through a liner represents both a lost opportunity and a potential liability for the mining company. If the geomembrane is placed on a low permeability soil and by default the problem is minimized by the combination of the geomembrane and low-probability soil to form a competent liner. However, as the subgrade permeability increases from 10^{-9} m/s the potential leakage through a hole in the geomembrane increases rapidly with no measurable resistance when it reaches 10^{-4} m/s. When there is a composite liner, the presence of wrinkles and the abundance of likely holes combine to substantially increase the hydraulic significance of the wrinkles. The reader is referred to §7.4.5 for a discussion of the significance of wrinkles. The leakage through a composite liner due to wrinkles in a heap leach application is likely to be substantially higher than through a composite liner in a landfill due to the much higher probability of holes to intersect wrinkles. While a leak location survey may still be useful, unless there is an adequate protection layer the vast majority of holes can be expected to occur upon loading, and since they don't exist at the time of the leak location survey they are not detected.

7.5.6 Towards minimal leakage and sustainability

As noted above, excessive leakage is a lost opportunity and increases liability. In general, the leakage from heap leach applications is unknown. The impact from leakage may take decades before it is detected. However, increased need for mining to reduce dependence on carbon fuels in the name of a sustainable environment needs to be accompanied by more emphasis on sustainable mining. There is clear evidence this is occurring. One aspect to be considered within the broader picture of sustainable mining is the need to reduce the known unknown of leakage from heap leach facilities. Considerable progress would be achieved by providing

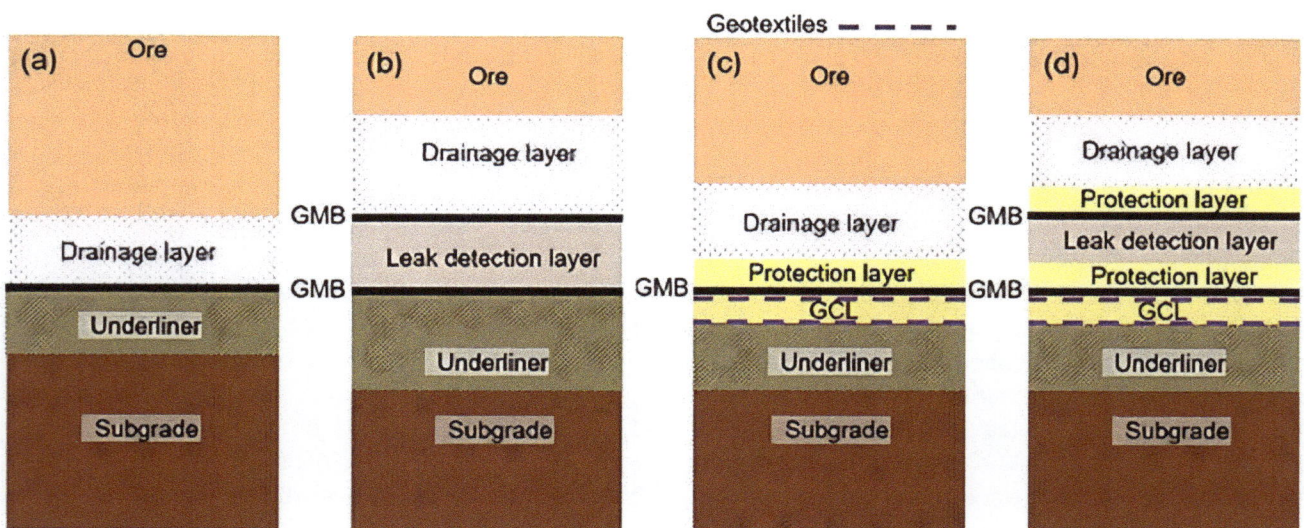

Figure 7.9 Heap leach barrier system (a) Current common single liner, (b) Upgraded single liner, (c) basic double liner upgraded, (d) double liner upgraded. Not to scale.(a) and (c) based on Lupo (2010)

an adequate protection layer suitable the loading conditions thereby addressing the problem by reducing the number of holes and indentations with large strain and hence minimizing the effect of any holes in both the short and longer-term (Figure 7.9c). This combined with the use of the GCL will substantially reduce leakage. The next and most appropriate step provide facility would be a double composite liner (Figure 7.9d). While there will be a notable cost associated with these enhanced barrier systems (Figure 7.9c,d), the cost needs to be weighed against the losses due to lost minerals and the substantial liability, in terms of both cash and social license, associated with having an unacceptable impact on the environment.

7.5.7 Tailings storage facilities

The need for tailings storage facilities is increasing rapidly together with the need for more minerals to support sustainability initiatives such as electric vehicles. In contrast to heap leach application where geomembranes have been used extensively for several decades, their use in tailings storage facilities has been far more limited and mostly focused on applications around the dam itself or lagoons. However, this is also changing with an increased awareness of the potential liabilities associated with leakage and the drive towards more sustainable mining. A number of the issues discussed with respect to heap leach pads in the previous section are directly relevant to tailings and will not be repeated in the reader is advised to read the previous section. The following focuses on issues more specific to tailings.

7.5.7.1 Geotechnical/interface stability

Many of the geotechnical issues related to tailings are similar to those for heap leach and the reader is directed to the discussion of geotechnical issues in the previous section for those issues. Of particular relevance to tailings is the potential for liquefaction under both static and seismic conditions and the potential for internal erosion needs to be evaluated.

7.5.7.2 Geomembrane selection

Major differences between the selection of a geomembrane for heap leach in tailings applications include the much longer service life often required for tailings facilities, the presence of tailings directly over the geomembrane providing a relatively uniform loading from above, different chemical exposure conditions, and less extreme potential thermal conditions. Service lives of millennia are now being requested and are potentially achievable with high-quality modern geomembranes and good design and construction.

Unlike geomembrane selection for heap leach pads, where project-specific testing is only required for the more extreme situations, project-specific testing will generally be required for tailings applications to justify a long service life. The chemistry of the tailings, and in particular any chemicals added to aid in the processing of the tailings need to be considered with the pH being of some importance but also important may be the presence of specialty chemicals used in the processing, especially if they have the characteristics of a surfactant.

The potential for heat generation by the tailings due to exothermic reactions needs to be evaluated in the context particular tailings and site conditions. In many cases there will not be any significant exothermic reaction and the tailings do not directly contribute to heat generation. However, the temperature of the liner needs to be considered in the context of heat flow between the surface of the tailings and either geothermal temperatures below the facility or the temperature at the bottom of earth's crust below the facility. Areas near tectonic plate boundaries and known geothermal energy must be considered on a case-by-case basis. However, the effect of the earth's core may still require consideration well away from plate boundaries and usable geothermal energy. The Earth's temperature increases by about 25 to 30°C per kilometre or about 2.5 to 3° per hundred metres from the surface. The surface of the tailings may be regarded as a boundary at a temperature equal to the annual average temperature at the location (climate change adjusted as needed). As a first approximation the temperature of the liner may be estimated as the annual average temperature plus about 3° per hundred metres of tailings above the liner. Thus, with 300 m tailings, the liner temperature is increased relative to ambient by 7.5 to 9° and that is sufficient to reduce the service life quite substantially. For example, an increase in temperature from 20 to 30° could reduce the time to nominal failure for typical GRI GM 13 geomembrane from about 300 years at 20° to 130 years at 30°. An increase from 30° to 40° would further reduce it to about 50 years (Abdelaal et al 2014). For top-of-the-range premium geomembrane, the same increases could cause a reduction from over 4000 years at 20° to 1500 years at 30° and around 400 years at 40° (Zafari et al 2023). These numbers are dependent on a particular geomembrane and chemistry and should be only used as illustrative of the effect of temperature.

To achieve a long service life for the geomembrane, the subgrade beneath the geomembrane needs to be relatively free of protrusions that can induce tension in the geomembrane (Fan and Rowe 2022c). Ultimately the geomembrane will fail by stress cracking but this can be delayed very substantially by keeping any tensile strains below about 3%. There are some arguments for placing a drainage layer or a series of French drains (periodically spaced piles of gravel) to accelerate the consolidation of the tailings. While it may be a good rationale for providing drainage, it is best kept off the geomembrane. Placing on the geomembrane both increases strains and the potential therefore for holes and provides a conduit of fluid directly to those holes to increase leakage (Fan and Rowe 2024). Any such drains should be located where they do not contribute to increased leakage.

7.5.7.3 Leakage and piping

There's been a considerable body of recent work published regarding the beneficial effects of tailings in contact with the geomembrane for reducing leakage through holes of different sizes and shapes including long thin cracks in the geomembrane both in direct contact with the subgrade (Rowe and Fan 2022,a,b) and in wrinkles (Joshi et al 2017). There is also recent literature on the issue of filter-incompatible foundations and piping (Chou et al 2018; Fan and Rowe 2022a,b).

7.5.2 Mine Waste Covers

Mine waste covers may be required to serve two roles. Like a landfill cover, one objective is to limit infiltration and hence generation of contaminated water. The second objective is to limit gas but, in this case, generally to limit the ingress of oxygen that combined water and certain waste rock or tailings will result in the generation of acid mine drainage. Two paths have been followed in the provision of covers for mine waste. The first can be classified as store and release covers (aka evapo-transpiration covers) which are useful in semi-arid environments but have run into difficulty in humid environments. The second can be classified as barrier-type covers which typically include a geomembrane. This section will focus on barrier-type covers.

In the most basic form, these covers may be an exposed geomembrane over a foundation layer. The biggest challenges and design considerations for exposed geomembranes include:

a. anchorage against wind uplift;
b. relatively short service life due to exposure to the sun and UV, especially for south facing slopes in the northern hemisphere and North facing slopes in the southern hemisphere;
c. potential to damage from animals (bears, caribou, leopards, kangaroos, etc.) and people;
d. potential damage from blown debris;
e. thermal cycles inducing tensions causing stress cracking (this is observed in both hot and cold climates; it is the change in temperature relative to the temperature at which the geomembrane was installed that is critical not the temperature at which it was installed;
f. movement of air and water through holes in the geomembrane; and
g. differential settlement.

"All liners leak". These exposed liners are particularly prone to leak for the reasons itemized above. Provided there is an adequate slope and no differential settlement, the influx of water through holes is relatively small. However, differential settlement even on a 4 horizontal: 1 vertical slope very substantially increases that leakage. Even without differential settlement, a hole will allow air to readily migrate into the waste pile with barometric pumping.

The problems itemized in (a)-(e) can be mitigated by placing cover soil over the geomembrane. While the cover soil substantially improves performance it also introduces holes in the process of covering. The number of holes can be reduced, but not eliminated, with the use of appropriate geotextile or, if readily available, granular protection layer combined with an appropriate leak location survey. To minimize the risk of sliding of the cover soil with heavy rainfall, a suitable drainage layer is required above the protection liner together with a suitable filter between the drainage layer and the overlying soil. Incorporation of all the foregoing improves cover performance, however there is still a significant potential for ingress of oxygen and water with the latter increasing substantially once there is differential settlement (Fan et al 2024).

To reduce infiltration of both water and oxygen through the cover, a composite liner is needed. To be effective the clay component needs to be in as intimate contact as possible with the geomembrane to minimize interface transmissivity between the geomembrane and clay liner. This is extremely difficult to achieve with compacted clay due to irregularities at the surface that invariably occur during construction and are maintained at the low overburden stress, together with desiccation of the surface that is very difficult to avoid. Thus, irrespective of whether compacted clay forms part of the liner system, a GCL is recommended to minimize the ingress of both water and oxygen. To be an effective the gas barrier, the GCL either needs to be well-hydrated from the subgrade to a degree of saturation in excess of 70% (Bouazza et al 2017a,b; Rowe 2020a) or have a robust coating (≥ 200 g/m^2) facing down so that the bentonite is encapsulated between the coating and the geomembrane. In both cases there are challenges. Maintaining the GCL at greater than 70% degree of saturation means (a) the foundation layer below the GCL must have an adequate grading curve and initial water content to adequately hydrate the GCL, and having the GCL below sufficient cover material to protected from wet dry and freeze-thaw cycles. An encapsulated GCL can only be hydrated by leakage of water through a hole in the geomembrane or coating and ideally remains hydrated at least locally as a result. In areas whether the risk of damage to the geomembrane by either burrowing animals or human intervention is significant, an intrusion barrier may be required. Thus, at its highest and rarest level, the cover for a mine waste facility may look very similar to that for a landfill shown in Figure 7.8.

7.6 Industrial Effluent Impoundment-Lagoons

Historically, it has not been uncommon for a leachate lagoons, sewage lagoons in smaller communities, lagoons for runoff that could be contaminated etc. to have an exposed single geomembrane liner or a GCL with 0.3m of cover sand. Numerous problems have been encountered with these lagoons including:

a. excessive leakage through holes in a geomembrane, or
b. excessive leakage due to chemical interaction (cation exchange) with the bentonite in the GCL leading to a high hydraulic conductivity.
c. build-up of biogas beneath the exposed geomembrane causing whales (bubbles in the geomembrane; Fig. 7.10 gas pressures below geomembrane that exceed atmospheric pressure) that displaces leachate (increasing head and leakage) and inducing tensions that can result in stress cracking.
d. potential to damage from animals (bears, caribou, leopards, kangaroos, etc.) and people.

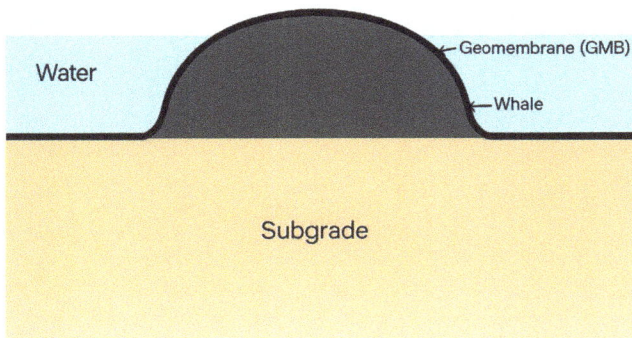

e. potential damage from blown debris.
f. thermal cycles inducing tensions causing stress cracking.
g. damage from pumps and equipment used to clean the pond.
h. damaged by rocks or debris embedded in ice which drags on the geomembrane causing tears or punctures.

The buildup of biogas and differential gas pressures that can cause whales can be mitigated by the installation of a suitable drainage layer and gas vent below the geomembrane (Figure 7.11). Damage from animals, people, windblown debris, thermal cycles, pumps, cleaning, or rocks and debris in the ice can be mitigated by the installation of a suitable protection layer over the geomembrane. The protection layer needs to be placed in such a manner that will not be eroded into the pond or readily picked up and moved by ice.

A primary cause of leaks in ponds and lagoons is due to penetrations through the geomembrane. The maximum extent possible penetration through the geomembrane should be avoided. When penetrations cannot be avoided, the use of shop fabricated components for penetration details such as pipe to plate connections allows better access for welding. It is also possible to ensure better quality control on welding done in the shop than at penetrations in the field (Wall

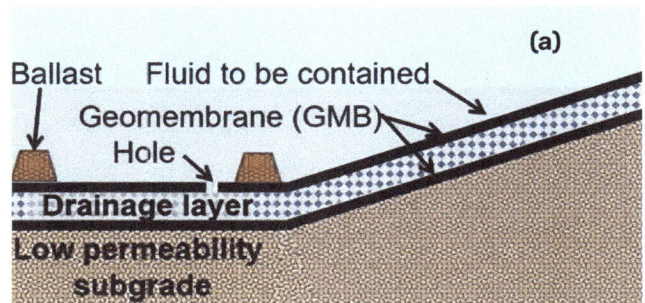

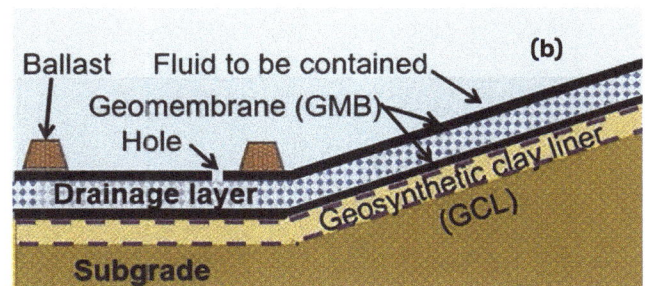

Figure 7.10 Schematic whale in GMB in a pond/lagoon. Not to scale.

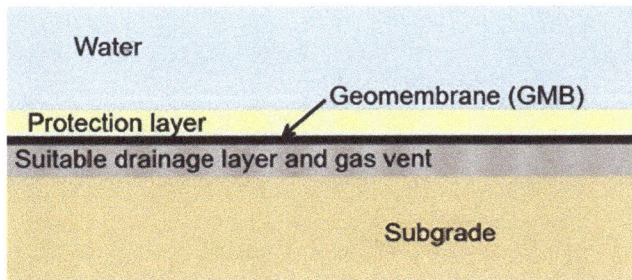

Figure 7.11 Drainage layer and gas vent below GMB and protection layer above in a pond/lagoon. Not to scale.

Figure 7.12 Schematic - double lined lagoon/pond with a single GMB primary liner and secondary liner comprised of a GMB and (a) low permeability subgrade, (b) GCL. Not to scale.

and Rowe 2024). Penetrations are usually associated with pipes running to or from the pond/lagoon. These pipes need good support including compaction below the spring line and testing of the compaction before the geomembranes installed. This generally requires use of smaller equipment such as plate tampers. The phrase "all liners leak" is especially applicable to lagoons with a single geomembrane or GCL liner. Thus, only fluids that would not significantly impact the environment if they were to leak from the lagoons/pond should be contained in single lined lagoons. In general, a double lined system should be adopted. This will commonly involve a primary geomembrane, a drainage layer, and the secondary geomembrane either alone or with the low permeability subgrade or together with the GCL as a composite liner (Fig. 7.12).

With leaks in the primary liner, leaks into the drainage layer can be pumped back into the pond. If pumping becomes excessive then it's a signal that repairs are needed to the primary liner. The primary purpose of the drainage layer is to control the head on the secondary liner to a negligible value. Thus, the drainage layer should not be used as a storage layer of fluids. If the leakage reaches the point that's no longer practical to pump the fluid out into the primary that signals the need to repair the holes in the primary liner. Critical location for monitors include downgradient of the sump and downgradient of any penetrations in the liner.

7.7 Remediation Sites

Geosynthetics play an important role in the remediation of contaminated sites. Some common applications are outlined below.

7.7.1 Vapour barriers between remediated soil and an overlying structure

Contaminated sites (e.g. of railway yards, ports, industrial subdivisions etc.) are often located in areas of potentially high commercial value. The soil that has been contaminated maybe remediated but it is usually too expensive to fully remediate soil and so the concentrations are reduced to a manageable level. When development is proposed over this remediated soil residual concentrations need to be managed. A common means of managing residuals such as volatile organic compounds is to install a suitable vapour barrier between a foundation layer and an overlying concrete floor slab and also providing a pathway to allow vapours to escape outside of the proposed structure (Fig. 7.13a). The large number of vapour barriers on the market which are useful for water vapour but not for volatile organic compounds.

When dealing with soil contaminated by volatile organic compounds, special multilayered vapour barrier (geomembrane) with a layer of ethylene vinyl alcohol (EVOH) represents the best commonly available barrier to toxic vapours. EVOH is a copolymer of ethylene and vinyl alcohol that is long been used in the packaging industry as an oxygen and gas barrier and in the motor industry for minimizing short-chain carbon loss from plastic filtering. The material used as a vapour barrier is typically 0.5 -0.75mm pick and is comprised of an EVOH core, a tie layer on each side and a layer of polyethylene (linear low density or medium density resin). The multilayered system is required because the EVOH layer is hydrophilic and will swell on contact with water. The polyethylene layers provide both physical and hydraulic protection to the thin EVOH core (McWatters et al 2019). The vapour barrier also may need protection by a suitable needle-punched nonwoven geotextile between the foundation layer and the vapour barrier if there is gravel in foundation soil. As with pond liners, penetrations must be very well sealed into the building.

7.7.2 Cutoff walls to control gas migration or fluid migration

Just as vapours may migrate up into a building from remediated soil, they can also migrate laterally in unsaturated gravel or sand layer, especially if overlain by a low permeability (e.g. clay) layer. Similarly, when appropriately designed old dumps may be ignored by planning departments who allow construction up to adjacent to the property boundary of the old dumps (e.g. Ombudsman's Report 2009). In situations like this, there is potential for methane migration from the landfill initially through the unsaturated permeable layer and subsequently through the services network into people's homes where between concentrations of 5% and 15% it represents an explosion risk homes within the subdivision. In these situations, it may be impractical to remediate the old dumps and so there is a need to prevent unacceptable gas migration from the dump into the services network of the subdivision. This can be achieved by constructing a barrier wall (e.g. Figure 7.13b) around the perimeter of the dump (Figure 7.13c) to prevent unacceptable off-site migration of either gas or leachate through the sides of the landfill. The trench and wall should extend below the water or into a low permeability soil. The trench must be adequately vented. For When there is the capacity to collect and either flame or use the gas being generated, a cover needs to be incorporated into the design as shown in Figure 7.13d. Shown here the cover is a basic geocomposite cover overlying the foundation and gas collection layer and covered by soil. Upgrades to the cover would depend on the circumstances but could include a protection layer, drainage layer, and filter between the geomembrane and the cover soil as discussed in the section on covers.

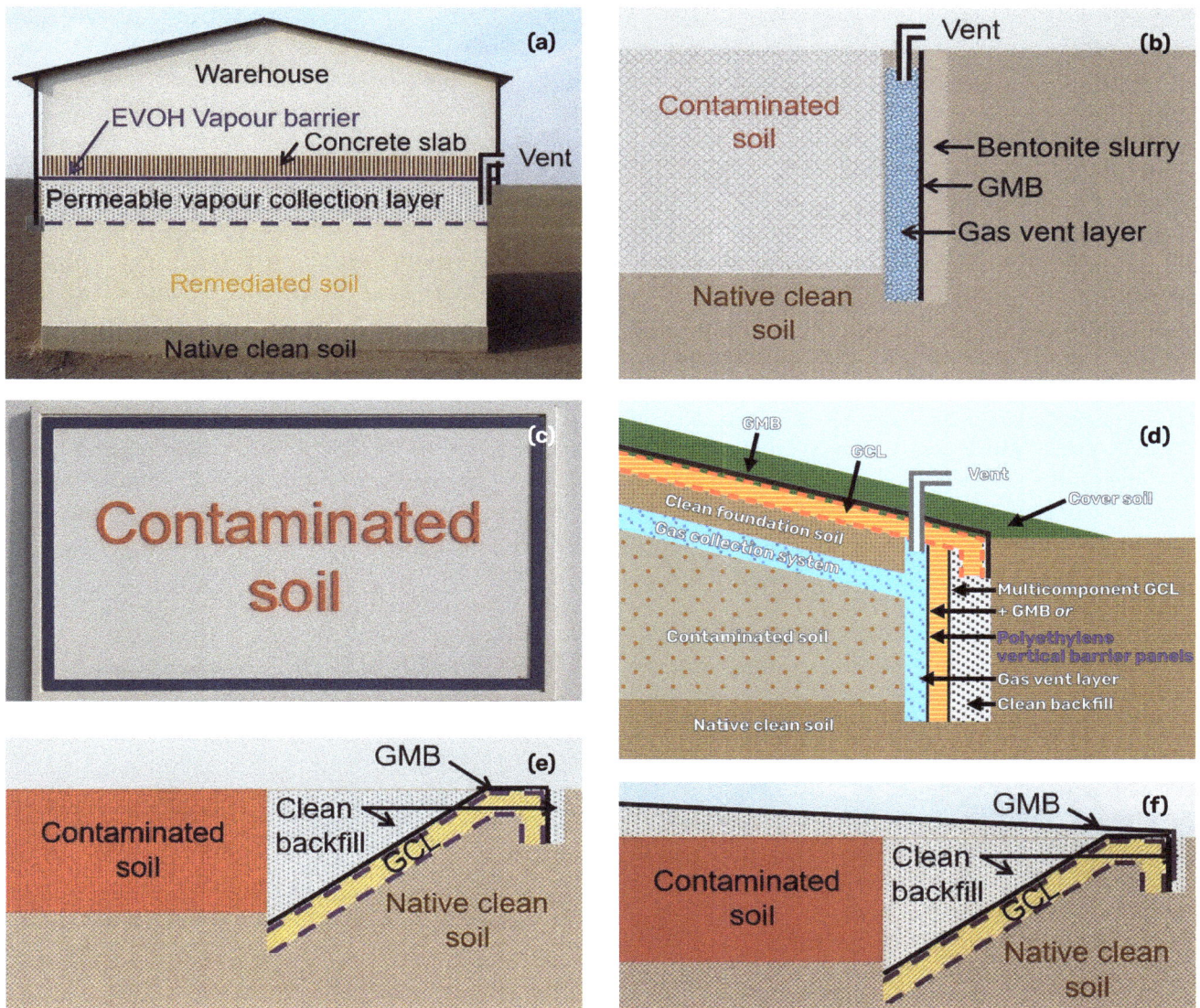

Figure 7.13 Schematic showing potential applications for geosynthetics in the remediation or containment of contaminated soil. (a) Vapour barrier to hydrocarbons. (b) Cutoff venting layer and wall for the containment of contaminated soil. (c) Cut-off drain and wall around perimeter of contaminated zone. (d) Composite cutoff wall and cover over contaminated soil. (e) Composite wall used with clean permeable backfill in funnel and gate system directing hydrocarbon spill to a permeable reactive barrier. (f) Containment of spill for subsequent clean-up. Not to scale.

7.7.3 Containment and remediation of spills

When simply wishing to reduce spreading of the spill and/or desire to direct it to a gate, a lined trench (Figure 7.13e) may serve. The details of the trench which may range from vertical to 3H:1V will depend on local conditions and available equipment. Geosynthetics have a particular advantage that when spills occur at locations well away from large cities they can be shipped by plane relatively quickly and installed with equipment available on the site. Also, emergency conditions were spill is heading towards the critical water resource they can be installed with some effectiveness, though not complete watertightness by unskilled labour. In funnel and gate applications, the wall directs the contaminated water to permeable reactive barrier. The details of the barrier will depend on the nature of this spill. They have for example

been used in Antarctica on a number of occasions to manage fuel spills (Rowe et al 2015; McWatters et al 2016). In the situation shown Figure 7.13e the contaminated soil is left open to precipitation in expectation that it will aid in encouraging migration of this spill towards the gate for treatment. However, there are only certain circumstances in which it is appropriate. In other circumstances it may be desirable to minimize any migration and in that case it may be appropriate to cover the contaminated soil to minimize the ingress of rainwater or snowmelt. For example, when there is going to be a prolonged period between the spill before remediation is desirable to limit the ingress of water, an exposed geomembrane cover is sufficient for the purposes (Figure 7.13f). Bathurst et al (2005) describes one such instance where a system similar to that shown in Figure 7.13f was installed in 2001 to allow cleanup over a 3 year period. Now 23 years later it still has not been cleaned up but the system continues to contain the hydrocarbons.

Jones et al (2023) describe the use of two funnel and gate permeable reactive barriers using exposed composite geomembrane/GCL liners to form the final to direct residual polychlorinated biphenyls (PCBs) moving with the sediment at annual snowmelt towards a sensitive water resource in northern Canada. While the use of an exposed composite liner is not to be encouraged due to the expected degradation in performance of the GCL, when specimens were exhumed after eight years field exposure in the Canadian Arctic there was only a minimal decrease in swell index to 19 mL/2g and the hydraulic conductivity under normal stress was still 5×10^{-11} m/s. Thus even under the most severe of conditions geosynthetics performed remarkably well in protecting the environment.

7.7.4 Liner for remediation of contaminated soil

There are many areas in the world (especially Antarctica and parts of the Arctic) where soil is a very valuable resource and the loss of use of that limited resource due to contamination is a grave loss indeed. The most common source of contamination is fuel spills. It is highly desirable to remediate the soil while ensuring a further contamination. This is best achieved by the construction of geosynthetic lined biopiles. The system is comprised of a GCL, about 1.5 mm-thick geomembrane, a geotextile protection layer, a permeable soil protection and drainage layer, and up to 1.5 m thick layer of contaminated soil and two geotextiles cover this. The bottom barrier system serves to prevent the escape of hydrocarbons into the groundwater that is very close and sometimes intimately in contact with the liner system (McWatters et al 2016). The contaminated soil is mixed with a small quantity of fertilizer when placed in the biopile where it is watered before the biopile is covered.. Periodically, the leachate percolates through the soil to the bottom and is recirculated. A half-life for hydrocarbon degradation is little under two years elapsed time with the bulk of that degradation occurring over two approximate two month-long summer seasons when the daily temperature averages +2 to 4°C , January being the hottest month with an average temperature of 0° C despite the cold ambient conditions , the near 24 hours of sun warms the soil a few degrees during these months. Experience has shown that the GCL's and geomembranes perform well despite the extreme conditions. The most effective geomembrane has been a 1.5 mm thick HDPE with a core of EVOH and two tie layers between EVOH layer and the HDPE. The most challenged component of the system is the geotextile covers provided to prevent contaminants or being blown around the site. Two layers of cover are used so that one sacrificial layer receives abrasion from the soil

while the other receives Inc. significant UV exposure from the sun and both layers are subject to very strong winds . Summer winds are frequently gale force with the highest windspeed recorded at Casey station being 240 km/h. Geotextiles have a service life of about two years.

7.7.5 Liners for secondary containment

While geosynthetics can be effectively used for containment after a spill, the objective of design should be to contain a spill as it happens and prevent the spill from getting into the surface water or groundwater system. This can be achieved by lining a bund around the storage facility with sufficient capacity to retain the complete spill for 24 hours if the storage tank ruptures or, as too frequently occurs, a valve is left open to leak. There are a wide range of possible designs that will meet the need but those involving geosynthetics can be subdivided into three general groups.

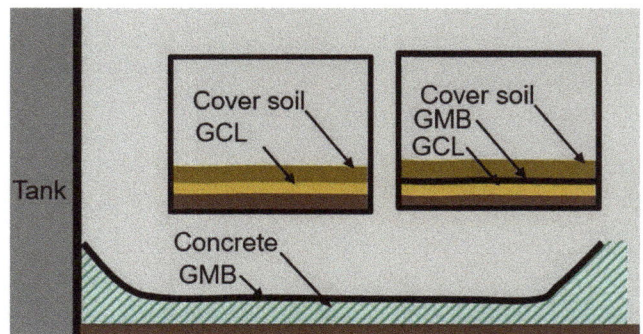

Figure 7.14 Schematic of a bund for secondary containment with liner systems using geosynthetics. Not to scale.

a. A concrete bund that may be lined with a geomembrane to minimize any possible leakage through cracks in the concrete. Special care is needed with the connections to the concrete as important to avoid sharp turns. The geomembrane can be left exposed (with a high risk of damage and effects of the sun) all covered with a soil protection layer.

b. A compacted soil bund with a multicomponent GCL as the liner and a layer of cover soil overlying the GCL. The bund soil should have a grain size distribution and water content sufficient to hydrate the bentonite in the GCL.

c. A compacted soil bund with a composite liner comprised of a GCL and geomembrane and appropriate soil cover.

Within these three groups there are many possible combinations and the details of connections are especially important. Penetrations through the liner should be avoided. Areas where there will be traffic and areas of which pipes pass will need appropriate design. With exposed geomembranes consideration needs to be given to expansion and contraction in the sun.

An appropriate GCL (see Rowe 2020a) hydrated to 70% degree of saturation or above can be an excellent barrier to hydrocarbons for a spill heights of up to about 2 m. A multicomponent GCL or a composite liner are likely to be more cost-effective and with a much smaller carbon footprint than a concrete lined Bund. Bunds can be effective when they kept empty. The bunds need to be regularly drained of precipitation and sediment buildup. This can be challenging environments with strong winds, torrential rain, substantial snowfall or windblown snow, or windblown sediment (e.g. sand) since experience has shown that rupture of tanks or an open valve is far more likely during extreme weather events than at other times.

7.8 References

Abdelaal, F.B and Rowe, R.K. Effect of high pH found in low level radioactive waste leachates on the antioxidant depletion of a HDPE geomembrane", 2017, J. Hazardous, Toxic, and Radioactive Waste, 21(1); 10.1061/(ASCE)HZ.2153-5515.0000262, D4015001

Abdelaal, F.B., Rowe, R.K., Morsy, M.S. and e Silva, R.A. Degradation of high, linear low, and blended polyethylene geomembranes in extremely low and high pH mining solutions at 85oC. 2023. Geotext. Geomembr., https://doi.org/10.1016/j.geotexmem.2023.04.011

Abdelaal, F.B. and Rowe, R.K. Physical and mechanical performance of a HDPE geomembrane in ten mining solutions with different pHs 2023, Can. Geotech. J., 60(7):978-993. http://dx.doi.org/10.1139/cgj-2022-0419

Abdelaal, F.B., Rowe, R.K., Brachman, R.W.I., Brittle rupture of an aged HDPE geomembrane at local gravel indentations under simulated field conditions. 2014. Geosynth. Int. 21 (1), 1–23.

ASTM D5887M-23 Standard Test Method for Measurement of Index Flux Through Saturated Geosynthetic Clay Liner Specimens Using a Flexible Wall Permeameter, ASTM International.

Barakat, F.B, Rowe. R.K., Patch, D. and Weber, K. Transport parameters for PFOA and PFOS migration through GCL's and composite liners used in landfills. Geotext. Geomembr., 2023 (in review)

Bass, J.M. Avoiding Failure of Leachate Collection and Cap Drainage Systems. 1986 U. S. EPA, Land Pollution Control Division, Hazardous Waste Engineering Research Laboratory, Cincinnati, OH. EPA-600/2-86-058.

Bathurst, R.J., Rowe, R.K., Zeeb, B. and Reimer, K. A geocomposite barrier for hydrocarbon containment in the Arctic. 2006 International Journal of Geoengineering Case Histories, 1(1):18-34.

Beck, A. Available technologies to approach zero leaks. In Proc., of Geosynthetics 2015 Conf. Roseville.CA: Industrial Fabrics Association International.

Benson CH, Daniel DE, Boutwell GP. Field performance of compacted clay liners. J. Geotech.Geoenv. Eng. 1999 May;125(5):390-403.

Bouazza, A., Rouf, M.A., Singh, R.M., Rowe, R.K., Gates, W.P., Gas diffusion and permeability for geosynthetic clay liners with powder and granular bentonites. Geosynth. Int. 2017a, 24 (6), 604–614. https://doi.org/10.1680/jgein.17.00027

Bouazza, A., Ali, M.A., Gates, W.P., Rowe, R.K., New insight on geosynthetic clay liner hydration: the key role of subsoils mineralogy. Geosynth. Int. 2017b, 24 (2), 139–150. https://doi.org/10.1680/jgein.16.00022.

Bouchez, T, Munoz, M.L., Vessigaud, S., Bordier, C., Aran, C., and Duquennoi, C. Clogging of MSW Landfill Leachate Collection Systems: Prediction Methods and in Situ Diagnosis. Proceedings Sardinia 2003, 9th International Waste Management and Landfill Symposium, S. Margherita di Pula, Cagliari, Italy, (CD-ROM).

Brachman, R.W.I., Gudina, S., Gravel contacts and geomembrane strains for a GM/CCL composite liner. 2008a. Geotext. Geomembranes 26 (6), 448–459.

Brachman, R.W.I., Gudina, S., Geomembrane strains and wrinkle deformations in a GM/GCL composite liner. 2008b. Geotext. Geomembranes 26 (6), 488–497.

Brachman, R.W.I., Sabir, A., Geomembrane puncture and strains from stones in an underlying clay layer. 2010. Geotext. Geomembranes 28 (4), 335–343.

Brachman, R.W.I., Sabir, A., Long-term assessment of a layered-geotextile protection layer for geomembranes. 2013. J. Geotech. Geoenviron. Eng. 139 (5), 752–764.

Brachman, R.W.I., Rowe, R.K. and Irfan, H. Short-term local tensile strains in HDPE heap leach geomembranes from coarse overliner materials, 2014 ASCE J Geotech. Geoenviron., 140(5):04014011-1 to 04014011-8, https://doi.org/10.1061/(ASCE)GT.1943-5606.0001087

Brune, M., Ramke, H.G., Collins, H.J., and Hanert, H.H. (Incrustation Processes in Drainage Systems of Sanitary Landfills. Proceedings Sardinia 91, 3rd International Landfill Symposium, S. Margherita di Pula, Cagliari, Italy, 1991. pp. 999-1035.

Chou, Y-C., Rowe, R.K. and Brachman, R.W.I. Erosion of Silty Sand Tailings through a Geomembrane Defect under Filter Incompatible Conditions. 2018. Can. Geotech. J. 55(11):1564-1576.

Daniel D.E., Koerner R.M.. *Waste containment facilities: guidance for construction quality assurance and construction quality control of liner and cover systems.* American Society of Civil Engineers; 2007.

E Silva, R.A., Abdelaal, F.B. and Rowe, R.K. "A 9-year study of the degradation of a HDPE geomembrane liner used in different high pH mining application" Geotext. Geomembr, 2025a, 53(1), 230-246

E Silva, R.A., Rowe, R.K. and Abdelaal, F.B. "A 10-year study of HDPE geomembrane longevity in contact with low pH solutions" 2025b, Can. Geotech. J. 62: 1-24, https://doi-org.proxy.queensu.ca/10.1139/cgi-2024-0504

E Silva, R.A., Rowe, R.K. and Abdelaal, F.B. 2025, "Degradation of polyethylene geomembranes exposed to different mine tailings pore waters" Geotext. Geomembr., 2025c, 53(6), 1483-1505.

Ewais, A.M.R., Rowe, R.K., Brachman, R.W.I., Arnepalli, D.N., Service-life of a HDPE GMB under simulated landfill conditions at 85 °C. 2014. J. Geotech. Geoenviron. Eng. 140 (11), 04014060.

Fan, J.-Y. and Rowe, R.K. Seepage through a Circular Geomembrane Hole when covered by Fine-Grained Tailings under Filter Incompatible Conditions" Can. Geotech. J., 2022a. 59(3): 410-423.

Fan, J-Y and Rowe, R.K., Piping of silty sand tailings through a circular geomembrane hole. All. Geotext. Geomembr. 50(1), 183-196.

Fan, J-Y and Rowe, R.K. Effect of Subgrade on Tensile Strains in a Geomembrane for Tailings Storage Applications, 2022c. Can. Geotech. J, 60(1), 18-30.

Fan, J-Y and Rowe, R.K. Effect of a lateral drainage layer on leakage through a defect in a geomembrane overlain by saturated tailings. 2024. Geotext. Geomembr. (in press)

Fan, Y-H, Rowe, R.K., Brachman, R.W.I and Van Gulck, J. Impact of differential settlement on leakage through geomembranes in waste covers. 2024. Geosynth. Int., (in press)

Fleming, I.R. and Rowe, R.K. Laboratory studies of clogging of landfill leachate collection & drainage systems, Can. Geotech. J, 2004, 41(1):134-153.

Fleming, I.R., Rowe, R.K., and Cullimore, D.R. Field Observations of Clogging in a Landfill Leachate Collection System. Can. Geotech. J, 1999, 36(4): 685-707.

Gilson-Beck, A. Controlling leakage through installed geomembranes. Geotext. Geomembr. 2019, 47 (5): 697–710.

Giroud, J. P. 2016. Leakage control using geomembrane liners. Soils Rocks 39 (3): 213–235.

Giroud, J. P. and Bonaparte R. Leakage through liners constructed with geomembranes — Part I: Geomembrane liners. Geotext. Geomembr. 1989a, 8 (1): 27–67. https://doi.org/10.1016/0266-1144(89)90009-5.

Giroud, J. P. and Bonaparte, R. Leakage through liners constructed with geomembranes — Part II: Composite liners. Geotext. Geomembr., 1989b, (2): 71–111. https://doi.org/10.1016/0266-1144(89)90022-8.

Giroud, J. P. Equations for calculating the rate of liquid migration through composite liners due to geomembrane defects. Geosynth. Int. 1997, 4 (3–4): 335–348. https://doi.org/10.1680/gein.4.0097.

Goodall DC, and Quigley RM. Pollutant migration from two sanitary landfill sites near Sarnia, Ontario. Can. Geotech. J. 1977 May 1;14(2):223-36.

GRI-GCL3 Test Methods, Required Properties, and Testing Frequencies of Geosynthetic Clay Liners (GCLs), Revised 2019, Geosynthetic Institute.

GRI-GM13 Test Methods, Test Properties, Testing Frequency for High Density Polyethylene (HDPE) Smooth and Textured Geomembranes , Geosynthetic Institute, Revised 2023.

Hornsey, W.P., Wishaw, D.M., Development of a methodology for the evaluation of geomembrane strain and relative performance of cushion geotextiles. 2012. Geotext. Geomembranes 35, 87–99.

Joshi, P., Rowe, R.K. and Brachman, R.W.I Physical and hydraulic response of geomembrane wrinkles underlying saturated fine tailings. 2017, Geosynth. Int.,24(1):82-94

Koerner, G. R., and Koerner, R. M. Leachate Clogging Assessment of Geotextile (and soil) Landfill Filters, 1995. US EPA Report, CR-819371, March.

Lupo JF. Liner system design for heap leach pads. Geotext. Geomembr. 2010, 28(2):163-73.

McIsaac, R. and Rowe, R.K. Change in leachate chemistry and porosity as leachate permeates through tire shreds and gravel, Can. Geotech. J, 2005., 42(4):1173-1188.

McIsaac, R. and Rowe, R.K. Effect of filter/separators on the clogging of leachate collection Systems, Can. Geotech. J , 2006, 43(7):674-693.

McIsaac, R. and Rowe, R.K.. Clogging of gravel drainage layers permeated with landfill leachate, ASCE J Geotech. Geoenviron., 2007, 133(8):1026-1039.

McIsaac, R. and Rowe, R.K. Clogging of unsaturated gravel permeated with landfill leachate, Can. Geotech. J., 2008, 45(8):1045-1963.

McWatters, R.S., Rowe, R.K., Wilkins, D., Spedding, T., Jones, D., Wise, L., Mets, J., Terry, D., Hince, G., Gates, W.P., Di Battista, V., Shoaib, M., Bouazza, A. and Snape, I. Geosynthetics in Antarctica: Performance of a composite barrier system to contain hydrocarbon-contaminated soil after 3 years in the field. Geotext. Geomembr., 44(5):673-685. http://dx.doi.org/10.1016/j.geotexmem.2016.06.001

MoECC Landfill Standards: A Guideline On The Regulatory And Approval Requirements For New or Expanding Landfilling Sites, 2012 Ontario Ministry of the environment and climate change, PIBS 7792e

Ombudsman's Report 2009. Blooklands Green Estate – Investigation into methane gas leaks, Victorian government printer, Session 2006-09, P.P. No. 237, 14 October, 289p. www.ombudsman.vic.gov.au

Rowe, R.K. Contaminant impact assessment and the contaminant lifespan of landfills. Can. J. Of Civil Engineering, 1991,**18**(2):244-253.

Rowe, R.K. Long-term performance of contaminant

barrier systems, 45th Rankine Lecture, Geotechnique, 2005, 55 (9): 631-678.

Rowe, R.K Short and long-term leakage through composite liners. The 7th Arthur Casagrande Lecture. 2012. *Can. Geotech. J.* **49**(2):141-169.

Rowe, R.K. Geosynthetic clay liners: perceptions and misconceptions, Geotext. Geomembr., 2020a, 48(2):137-156. https://doi.org/10.1016/j.geotexmem.2019.11.012

Rowe, R.K. Protecting the environment with geosynthetics - The 53rd Karl Terzaghi Lecture, 2020b, ASCE J Geotech. Geoenviron., 146(9):04020081. 10.1061/(ASCE)GT.1943-5606.0002239

Rowe, R.K. Design and construction of barrier systems to minimize environmental impacts due to municipal solid waste leachate and gas", 3rd Indian Geotechnical Society: Ferroco Terzaghi Oration, Indian Geotechnical Journal, 2012, 42(4):223-256.

Rowe, R.K. and Barakat, F.B. Modelling the transport of PFOS from single lined municipal solid waste landfill, Computers and Geotechnics, 2021, 137(9):104280-1 to 104280-11. https://doi.org/10.1016/j.compgeo.2021.104280

Rowe, R.K. and Barakat, F.B. Estimating the magnitude of PFAS migration from landfills: approaches and findings. 2024. First Bouazza lecture, GEOANZ #1 Advances in Geosynthetics, Melbourne, July.

Rowe, R.K., Caers, C.J. and Chan, C. Evaluation of a compacted till liner test pad constructed over a granular subliner contingency layer. 1993. Can. Geotech. J. 30(4):667-689.

Rowe, R.K. and Abdelaal, F.B. Antioxidant depletion in HDPE Geomembrane with HALS in low pH heap leach environment, 2016, Can. Geotech. J. 53(10):1612-1627. DOI: 10.1139/cgj-2016-0026.

Rowe, R.K. and Fan, J.-Y. Effect of Geomembrane Hole Geometry on Leakage Overlain by Saturated Tailings, 2021. Geotext. Geomembr., 49(6):1506-1518.

Rowe, R.K. and Fan, J-Y. A general solution for leakage through geomembrane defects overlain by saturated tailings and underlain by highly permeable subgrade. 2022. Geotext. Geomembr., 50(4), pp.694-707.

Rowe, R.K. and Yu, Y. Factors Affecting the Clogging of Leachate Collection Systems in MSW Landfills, Keynote lecture, 6th International Conference on Environmental Geotechnics, New Delhi, November 2010, 3-23

Rowe, R.K. and Yu, Y. Clogging of finger drain systems in MSW landfills, Waste Management, 2012. 23:2342-2352, http://dx.doi.org/10.1016/j.wasman.2012.07.018

Rowe, R.K. and Yu, Y. A practical technique for estimating service life of MSW leachate collection systems, *Can. Geotech. J.* 2013, 50(2):165-178.

Rowe, R.K. and Zhao, L-X. Implications of double composite liner behaviour for PFAS containment, Proceedings of the 9ICEG 9th International Congress on Environmental Geotechnics, June 2023, Chania, Greece, 10p.

Rowe, R.K., Caers, C.J., Reynolds, G. and Chan, C. Design and construction of barrier system for the Halton Landfill", Can. Geotech. J. , 2000, 37(3):662-675.

Rowe, R.K., Quigley, R.M., Brachman, R.W.I., Booker, J.R. Barrier Systems for Waste Disposal Facilities, 2004, E & FN Spon, Taylor & Francis Books Ltd, London, 587p.

Rowe, R.K., Abdelaal, F.B., Brachman, R.W.I., Antioxidant depletion from an HDPE geomembrane with a sand protection layer. 2013a.Geosynth. Int. 20 (2), 73–89.

Rowe, R.K, Brachman, R.R.W., Irfan, H., Smith, M.E., and Thiel, R. Effect of underliner on geomembrane strains in heap leach applications, 2013b, Geotext. Geomembr., 40:37-47.

Rowe, R.K., McWatters, R.M., Jones, D.D., Di Battista, V., and Shoiab, M. Remediation of Emergency Powerhouse Fuel Spill Casey Station Antarctica", 2015. Australian Antarctic Division, Dept. of Environment, Commonwealth of Australia, (236p.).

Rowe, R.K., Brachman, R.W.I, Take, W.A, Rentz, A. and Ashe, L.E., "Field and laboratory observations of down-slope bentonite migration in exposed composite liners", *Geotext. Geomembr.*, 2016),44(5): 686-706. http://doi.org/10.1016/j.geotexmem.2016.05.004

Rowe, R.K., Brachman, R.W.I., Hosney, M.S. Take, W.A. and Arnepalli, D.N. ("Insight into hydraulic conductivity testing of GCLs exhumed after 5 and 7 years in a cover", Can. Geotech. J. 2017, 54(8): 1118-1138, https://doi.org/10.1139/cgj-2016-0473

Rowe, R.K., Priyanto, D., Poonan, R., 2019. Factors affecting the design life of HDPE geomembranes in an LLW disposal facility. In: WM2019 Conference, Phoenix, Arizona, USA, 15p

https://doi.org/10.1016/j.geotexmem.2019.11.012

Rowe, R.K., Abdelaal, F.B., Zafari, M. Morsy, M.S. and Priyanto. An approach to geomembrane selection for challenging design requirements. 2020. Can. Geotech. J., 57(10):1550–1565. doi:10.1139/cgj-2019-0572

Rowe. R.K., Barakat, F.B, Patch, D. and Weber, K. Diffusion and partitioning of different PFAS compounds through thermoplastic polyurethane and three different PVC-EIA liners. 2023*Science of the Total Environment*, 2023, 892:164229.

http://dx.doi.org/10.1016/j.scitotenv.2023.164229

Wall, D. and Rowe, R.K. "Root Cause Failure Assessment of a Water Retention Reservoir at a Mining Industry Site", Proceedings of GeoAmerics 2024, Toronto, May.

Yu, Y. and Rowe, R.K. Co-disposal of MSW and Incinerator Ash in Landfills Affecting the Performance of Leachate Collection Systems", 2021, *Can. Geotech. J.*, 58(1):83-96.https://doi.org/10.1139/cgj-2019-0851

Zafari, M., Abdelaal, F.B. and Rowe, R.K. Degradation Behaviour of two Multi-layered Textured White HDPE Geomembranes and their Smooth Edges. 2023a.

ASCE J Geotech. Geoenviron., 149(5): 04023020_1-15. https://doi.org/10.1061/JGGEFK. GTENG-11101

Zafari, M., Abdelaal, F.B. and Rowe, R.K. Long-term performance of conductive-backed multi-layered HDPE geomembranes. Geotext. Geomembr., 2023b. 51(4) 137-155.

Zafari, M., Rowe, R.K. and Abdelaal, F.B. Longevity of multi-layered textured HDPE geomembranes in low-level waste applications. 2023c. Can. Geotech. J, (in press)

Zhang, L., Bouazza, A., Rowe, R.K. and Scheirs, J. Effects of a very low pH solution on the properties of HDPE geomembrane 2018, Geosynth. Int., 25(2):118-131. https://doi.org/10.1680/jgein.17.00037

Chapter 8

Geosynthetics Support Systems

8.0 Overview

This is a catch all chapter illustrating the support systems utilized by the geosynthetic community. These systems have been created over the past fifty years and have evolved over time. It should be clearly indicated that many of these programs and systems described in this chapter are international and may vary due to regional and regulatory jurisdiction. With that said, they are part of a complex support system that is immensely interrelated and applicable to most projects. Without care and dedication to best practice, the best designs and regulatory requirements will not always translate into a successful project involving the use of geosynthetics.

It is therefore strongly suggested that a quality system approach be embraced when embarking on projects that utilize geosynthetics. This process-approach is outlined below and described in detail by Koerner (2024).

> Appropriate regulations and code giving guidance to best practices
> Good design by professional engineer with sustainability considerations
> Quality materials covered by generic specifications and MQC/MQA
> Product certification
> Accredited testing
> Best available installation by certified installers
> Construction Quality Assurance and Quality Control (CQA/CQC)
> Proof testing prior to commissioning the facility
> Careful operations and maintenance

In most cases, one is selecting geosynthetics based on individual properties and how they interact with other materials on the project. The intrinsic properties define the function of the material, (strength, elongation, transmissivity, etc.) where the "interaction properties" (friction angle, creep, survivability etc.) define the selection of specific materials with relation to other components in the design. It is the geosynthetic manufacturer who makes the material and can define the stand-alone properties, while it is the engineer's role to understand these properties and incorporate them into the design. What the design consultant expects from the manufacturer is consistent quality. Even when the manufacturer has been able to "meet the specification", it is much more difficult to maintain a consistent product over time. Delivering produced material to the site and properly installing it can also be challenging. This overall goal (good materials, on time, properly installed and great benefit/cost) of all successful projects is an achievement possible with the systems described in this chapter.

A geosynthetic specification is a written document describing in detail the scope of work, materials to be used, methods of installation, and quality of workmanship for each item placed under contract. It is usually utilized in conjunction with construction projects both private and public. With the geosynthetic community, *specifications* provide material requirements for various types of geosynthetics used in geotechnical, environmental and transportation applications. Generally, the geosynthetic is not permitted for use on a site until certificates of compliance covering the material have been submitted and approved. This is needed for each type of geosynthetic used on the jobsite. Once verified, the geosynthetic can be unloaded and properly protected against damage until it is installed.

Specifications in construction documents typically include information on the project's necessary materials, the timeline, methods and requirements. There are three primary types of specifications: proprietary, performance and prescriptive. Most often, the geosynthetic community utilizes generic prescriptive specifications available through regulatory agencies, institutes or a standards organization. The Geosynthetic Institute, ASTM International, ISO, CEN and many other state and federal agencies provide specifications for different types of geosynthetics. They are available at links listed at the end of this chapter. The specifications are specific to material type, function and application.

8.1 Stakeholders

Many individuals are involved directly or indirectly with an entire project. The individuals, their affiliation, responsibilities and authority are discussed below. In addition, the principal organizations and individuals involved in designing, permitting, constructing, and inspecting a project or facility are also described below:

> *Permitting Agency*. The permitting agency is usually a government regulatory agency but may include local or regional agencies and/or the national entities. It is the responsibility of the permitting agency to review the owner/operator's permit application, including plans, specifications and the site-specific manufacturing quality assurance (MQA)/construction quality assurance (CQA) document, for compliance with the agency's regulations and to make a decision to issue or deny a permit based on this review.

> *Owner/Operator*. This organization (private or public) will own and operate the facility, site or structure. The owner/operator is responsible for the design, construction, and operation of the project. This responsibility includes complying with the requirements of the permitting agency, submitting MQA/CQA documentation, and assuring the permitting agency that the facility was constructed as specified in the construction plans and specifications and as approved by the permitting agency. The owner/operator has the authority to select and dismiss organizations charged with design, construction, and MQA/CQA. If the owner and operator of a facility are different organizations, the owner is ultimately responsible for these activities.

> *Owner's Representative*. The owner/operator has an official representative who is responsible for coordinating schedules, meetings, and field activities. This responsibility includes coordination between all parties involved, i.e. owner's representative, permitting agency, material suppliers, general contractor, specialty subcontractors or installers, and the MQA/CQA engineer.

> *Design Engineer*. The design engineer's primary responsibility is to design a facility that fulfills the operational requirements of the owner/operator, complies with accepted design practices, and meets or exceeds the minimum requirements of the permitting agency. The design engineer may be an employee of the owner/operator or a design consultant hired by the owner/operator. The design engineer may be requested to change some aspects of the design if unexpected conditions are encountered during construction (e.g. a change in site conditions, unanticipated logistical problems during construction, or lack of availability of certain materials). Because design changes during construction are not uncommon, the design engineer is often involved in the MQA/CQA process. The plans and specifications will generally be the product of the design engineer. The design engineer is a major and essential part of the permit application process and the subsequently constructed facility.

> *Manufacturer*. Many components of a structure, including all geosynthetics, are manufactured materials. The manufacturer is responsible for the manufacture of its materials and for quality control during manufacture (i.e. MQC). The minimum or maximum (when appropriate) acceptable properties of materials should be specified in the permit application. The manufacturer is responsible for supplying materials complying to those specifications included in the contract of sale to the owner/operator or its representative. The quality steps taken by a manufacturer are critical to overall quality management in construction. Such activities often take the form of process quality control, computer-aided quality control, and the like. All efforts at producing better quality materials are highly encouraged. If requested, the manufacturer should provide information to the owner/operator, permitting agency, design engineer, fabricator, installer, or MQA engineer that describes the quality control (MQC) steps that are taken during the manufacturing of the product. The manufacturer may sometimes be requested to offer access to the manufacturing facility to stakeholders of the project to observe the manufacturing process and quality control procedures if they so desire, and often under specific conditions such as signed non-disclosure agreements. Random samples of materials should be available for subsequent analysis and/or archiving. However, the manufacturer should retain the right to insist that any proprietary information concerning the manufacturing of a product be held confidential. Signed agreements of confidentiality are at the discretion of the manufacturer. The owner/operator, permitting agency, design engineer, fabricator, installer, or MQA engineer may request that they be allowed to observe the manufacture and quality control of some or all of the raw materials and final products to be utilized on a particular job. The manufacturer should be willing to accommodate such requests. Note that these same comments apply to sales organizations which represent a manufactured product made by others, as well as the manufacturing organization itself.

> *Fabricator.* Some geosynthetic materials are fabricated from individual manufactured components. For example, certain geomembranes are fabricated by seaming together smaller, manufactured geomembrane panels at the fabricator's facility to create large panels and minimize onsite assembly. Some large geotextiles are stitched together to create "geotubes", large, flexible containers made of permeable geotextile fabric, commonly used for dewatering and erosion control applications. The minimum characteristics of acceptable fabricated materials are specified in the permit application.

> *Project Manager.* For large facilities, a project manager may be hired by the owner/operator to control and monitor the construction activities. One of the main tasks in this regard is the decision as to whether to contract with a general contractor or to hire individual subcontractors (e.g. separate contractors for installation of geosynthetics, earthwork placement, etc.). Even further, the project manager may decide to take on some of the activities typically done by contractors (e.g. procurement of materials). These decisions are made by the owner/operator working with the identified project manager. Important in this regard is that the activities described below for the general, installation, and/or earthwork contractors must be carefully coordinated by the project manager if the site has such an individual or company acting as the project manager.

> *General Contractor.* The general contractor has overall responsibility for construction and for construction quality control (CQC) during construction. The general contractor arranges for purchase of materials that meet the plans and specifications, enters into a contract with one or more fabricators (if fabricated materials are needed) to supply those materials, contracts with one or more installers (if separate from the general contractor's organization), and has overall control over the construction operations, including scheduling and CQC. The general contractor has the primary responsibility for ensuring that a facility is constructed in accordance with the plans and specifications that have been developed by the design engineer and approved by the permitting agency. The general contractor is also responsible for informing the owner/operator and the MQA/CQA engineer of the scheduling and occurrence of all construction activities. For example, a waste containment facility may be constructed without a general contractor. An owner/operator or project manager may arrange all the necessary material, fabrication, and installation contracts. In such cases, the owner/operator's representative or project manager will serve the same function as the general contractor.

> *Installation Contractor.* Manufactured products (such as geosynthetics) are placed and installed in the field by an installation contractor who is a general contractor, a subcontractor to the general contractor, or a specialty contractor hired directly by the owner/operator. The installer must install the geosynthetics as defined in the specification prepared by the design engineer. The installer's personnel may be employees of the owner/operator, manufacturer, fabricator, or they may work for an independent installation company hired by the general contractor, the owner/operator, or the project manager. The installer is responsible for handling, storage, placement, and installation of manufactured and/or fabricated materials. The installer should have a CQC plan to detail the proper manner that materials are handled, stored, placed, and installed. The installer is also responsible for informing the owner/operator and the MQA/CQA engineer of the scheduling and occurrence of all geosynthetic construction activities. The installer and representative employees of the installer may also be credentialed by the International Association of Geosynthetic Installers (IAGI).

> *Earthwork Contractor.* The earthwork contractor is responsible for grading the site to elevations and grades shown on the plans and for constructing earthen components of the facility according to the specifications, including requirements for planarity and gradations. The earthwork contractor may be hired by the general contractor or, if the owner/operator serves as the general contractor, by the owner/operator directly. In some cases, the general contractor's personnel may serve as the earthwork contractor. The earthwork contractor is responsible not only for grading the site to proper elevations but also for obtaining suitable earthen materials, transport and storage of those materials, preprocessing of materials (if necessary), placement and compaction of materials, and protection of materials during and (in some cases) after placement. Earthwork functions must be carried out in accord with plans and specifications approved by the permitting agency. The earthwork contractor should have a CQC plan (or agree to one written by others) and is responsible for CQC operations aimed at controlling materials and placement of those materials to conform with project specifications. The earthwork contractor is also responsible for informing the owner/operator and the CQA engineer of the scheduling and occurrence of all earthwork construction activities.

> *CQC Personnel.* Construction quality control personnel are individuals who work for the general contractor, installation contractor, or earthwork contractor and whose job it is to ensure that construction is taking place in accordance with the plans and specifications approved by the permitting agency. In some cases, CQC personnel, perhaps even a separate company, may also be part of the installation or construction crews. In other cases, supervisory personnel provide CQC or, for large projects, separate CQC personnel, perhaps even a separate company, may be utilized. It is recommended that a certain portion of the CQC staff should be certified. Such a program is available through the International Association of Geosynthetic Installers (IAGI).

> *MQA/CQA Engineer.* The MQA/CQA engineer has an overall responsibility for manufacturing quality assurance and construction quality assurance. The engineer is usually an individual experienced in a variety of activities, although particular specialists in soil placement, polymeric materials, and geosynthetic placement will typically be involved in a project. The MQA/CQA engineer is responsible for reviewing the MQA/CQA plan as well as general plans and specifications for the project so that the MQA/CQA plan can be implemented with no contradictions or unresolved discrepancies. Other responsibilities of the MQA/CQA engineer include education of inspection personnel on MQA/CQA requirements and procedures and special steps that are needed on a particular project, scheduling and coordinating of MQA/CQA inspection activities, ensuring that proper procedures are followed, ensuring that testing laboratories are conforming to MQA/CQA requirements and procedures, ensuring that sample custody procedures are followed, confirming that test data are accurately reported and that test data are maintained for later reporting, and preparation of periodic reports. The most important duty of the MQA/CQA engineer is overall responsibility for confirming that the project was constructed in accord with plans and specifications approved by the permitting agency. In the event of nonconformance with the project specifications or CQA Plan, the MQA/CQA engineer should notify the owner/operator as to the details and, if appropriate, recommend work stoppage and possibly remedial actions. The MQA/CQA engineer is normally hired by the owner/operator and functions in a separate and independent manner. The MQA/CQA engineer must be a registered professional engineer who has shown competency and experience in similar projects and is considered qualified by the permitting agency. It is recommended that the person's resume and record on similar projects be submitted

in writing and accordingly accepted by the permitting agency before activities commence. The permitting agency may request additional information from the prospective MQA/CQA engineer and his/her associated organization including experience record, education, registry, and ownership details. The permitting agency may accept or deny the MQA/CQA engineer's qualifications based on such data and revelations. If the permitting agency requests additional information or denies the MQA/CQA engineer's qualifications, it should be done prior to construction so that alternatives can be made which do not negatively impact the progress of the work. The MQA/CQA engineer is usually required to be at the construction site during all major construction operations to oversee MQA/CQA personnel. The MQA/CQA engineer is usually the MQA/CQA certification engineer who certifies the completed project.

> *MQA/CQA Personnel.* Manufacturing quality assurance and construction quality assurance personnel are responsible for making observations and performing field tests to ensure that a project is constructed in accord with the plans and specifications approved by the permitting agency. MQA/CQA personnel normally are employed by the same firm as the MQA/CQA engineer, or by a firm hired by the firm employing the MQA/CQA engineer. Construction MQA/CQA personnel report to the MQA/CQA engineer. A relatively large proportion (if not the entire group) of the MQA/CQA staff should be trained specifically for MQA/CQA purposes. In this regard there are professional courses available, many of which offer continuing education units (CEU's). Certification of CQA personnel for both geosynthetic materials and compacted clay liners is available from the Geosynthetic Certification Institute's - Inspectors Certification Program (GCI-ICP).

> *Geosynthetic Testing Laboratories.* Many MQC/CQC and MQA/CQA tests are performed by commercial laboratories. The testing laboratories should have their own internal QC plan to ensure that laboratory procedures conform to the appropriate American Society for Testing and Materials International (ASTM International), ISO, CEN standards or other applicable testing standards. The testing laboratories are responsible for ensuring that tests are performed in accordance with applicable methods and standards, for following internal QC procedures, for maintaining sample chain-of-custody records, and for reporting data. The testing laboratory should be accredited by an accrediting authority recognized by the International Laboratory Accreditation Cooperation (ILAC). In addition, for geosynthetic materials specifically, such accreditation is available through

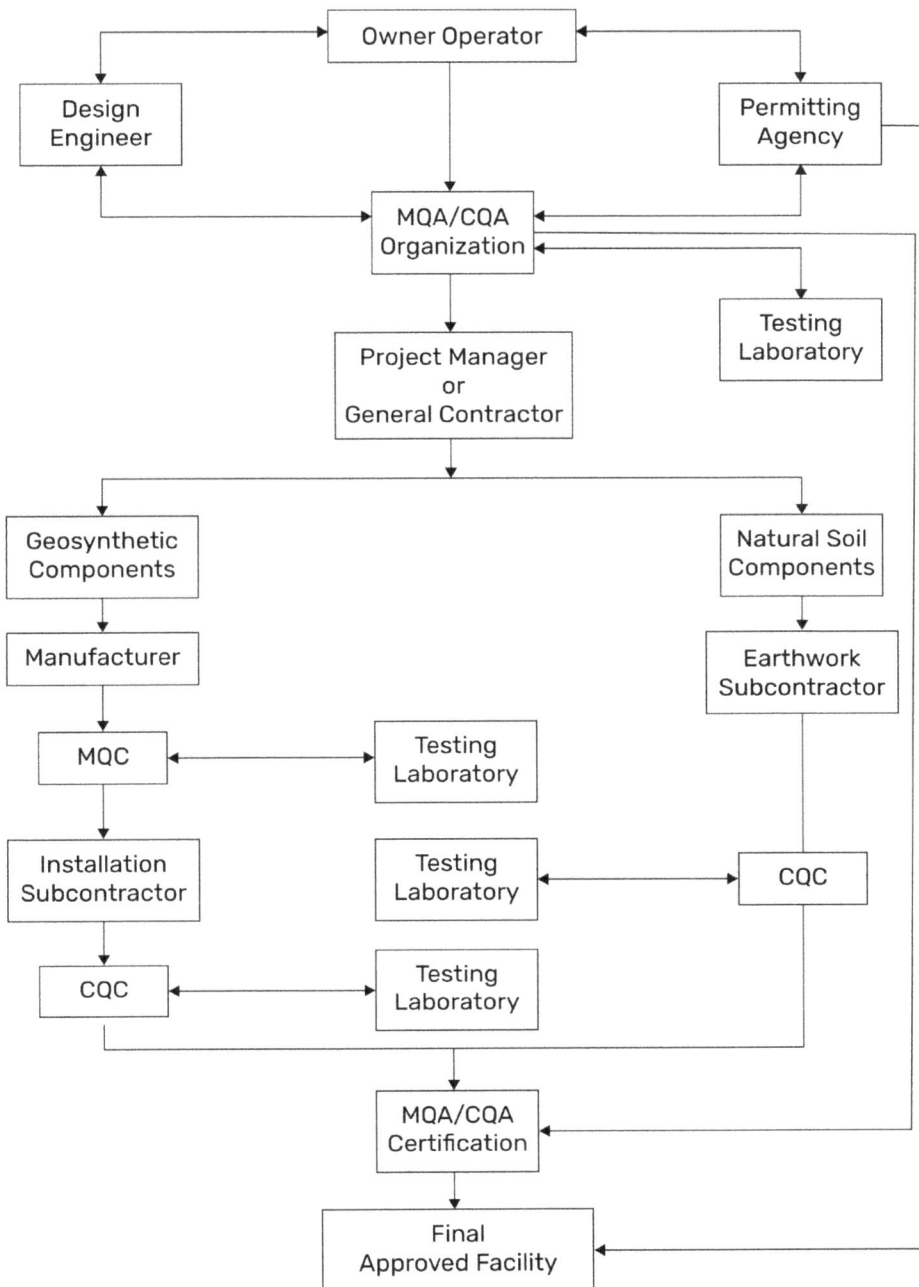

Figure 8.1 Organizational Structure of Quality Control and Quality Assurance Activities

the Geosynthetic Accreditation Institute-Laboratory Accreditation Program (GAI-LAP). The testing laboratory must be willing to allow project stakeholders to observe the sample preparation and testing procedures, or record-keeping procedures, if they so desire. The owner/operator, permitting agency, design engineer, or MQA/CQA engineer may request that they be allowed to observe some or all tests on a particular job at any time, either announced or unannounced. The testing laboratory personnel must be willing to accommodate such a request, but the observer should not interfere with the testing or slow the testing process.

> *MQA/CQA Certifying Engineer.* The MQA/CQA certifying engineer is responsible for certifying to the owner/operator and permitting agency that, in his/her opinion, the facility has been constructed in accordance with plans and specifications. Also, that the MQA/CQA document has been approved by the permitting agency. The certification statement is normally accompanied by a final MQA/CQA report that contains all the appropriate documentation, including daily observation reports, sampling locations, test results, drawings of record or sketches, and other relevant data. The MQA/CQA certifying engineer may be the MQA/CQA engineer or someone else in the MQA/CQA engineer's organization who is

a registered professional engineer with experience and competency in certifying such installations.

Like the above, the following flow chart (Daniel and Koerner 2006) describes the interactions of these different entities as they apply to a particular project so as to produce an appropriate level of quality.

8.2 Manufacturing

Geosynthetic are products of the petrochemical industry. They are generally made of polymers, "resins" (polypropylene (PP), polyethylene (PE), polyester (PET), poly vinyl chloride (PVC), polyamide (PA) etc.) used as raw materials. The base resin is blended with additives and stabilizers to enhance performance and are called the "let down" or formulation. This formulation is subsequently fabricated into the engineered final products. Details about the manufacture of various geosynthetics can be found in the following references (Koerner 2006)

Geosynthetic manufacturing is varied and evolving. It borrows form century old technology such as traditional weaving methods to fully automated manufacturing systems complete with extruders, conveyors and robots.

ISO 9001 is an ISO standard from the ISO 9000 family. This standard is the most popular standard from the ISO 9000 family and also the most popular among different industries. ISO 9001 provides the requirements for a quality management system (QMS) and the latest version of ISO 9001:2015 Quality management systems — Requirements. ISO 9001 discusses the seven Quality Management Principles (QMP). These seven QMPs are the core of the ISO 9001, since it gives a guide for continual improvement of an organization. The seven QMPs are namely, Customer Focus, Leadership, Engagement of People, Process Approach, Improvement, Evidence-based decision making, and Relationship management.

The geosynthetic community is also heavily involved with ISO 14001. ISO 14001:2015 Environmental management systems (EMS) – Requirements for guidance, use EMS, which is from a family of standards which helps an organization create a systematic approach to managing environmental issues while integrating them into the organization's processes. ISO 14001 also provides legal compliance with government regulations for businesses to reduce pollution and increase the sustainability of their products.

Geosynthetic manufactures and distributors and may provide geosynthetics from roll goods to entire prefabricated systems. The geosynthetics industry is constantly evolving, innovating and adapting to specified project or regulatory requirements.

8.3 Standardization

Two consensus-based standardization organizations with global membership are described below.

8.3.1 ISO TC221

ISO (International Organization for Standardization) is an independent, non-governmental international organization with a membership of 170 national standards bodies. Technical Committee 221 is focused on geosynthetic related standards. Through its members, it brings together experts to share knowledge and develop voluntary, consensus based, market relevant international standards that support innovation and provide solutions to global challenges. It is based in Geneva, Switzerland and is involved in the standardization of all geosynthetic products. ISO employs a one country-one vote model of voting.

The following is a list of TC21 Working Groups.

ISO/TC 221/WG2
Terminology, identification and sampling

ISO/TC 221/WG3
Mechanical properties

ISO/TC221/WG4
Hydraulic properties

ISO/TC 221/WG6
Design and sustainability using geosynthetics

8.3.2 ASTM International

ASTM International Committee D35 on Geosynthetics was formed in 1984. The committee meets twice each year, in January and June, with members participating in project task group meetings over three days. ASTM D35 on Geosynthetics, with a 2025 membership of 335 members, currently has jurisdiction of over 150 approved standards that are published in the Annual Book of ASTM Standards, Volume 04.13. ASTM International employs a one member–one vote model of voting with membership balance requirements.

The following is a list of D35 subcommittees;

D35.01 Mechanical Properties
D35.02 Endurance Properties
D35.03 Permeability and Filtration
D35.04 Geosynthetic Clay Liners
D35.05 Geosynthetic Erosion Control
D35.06 Geosynthetic Specifications
D35.10 Geomembranes
D35.40 Sustainability
D35.90 Executive

D35.93 Editorial and Terminology
D35.95 Awards
D35.96 US TAG to ISO/TC22I on Geosynthetics

The portfolio of approved standards and work Items under construction are available on the ASTM website which formulates test methods, specifications, guides, practices, terminology, and the dissemination of knowledge dealing with geosynthetics. This would include but not be limited to applications such as roadway stabilization and repair, erosion control, soil drainage and reinforcement, as well as hydraulic barriers composed primarily of man-made polymer sheets or spray applied systems.

8.4 Regulations

Regulations are authoritative rules or orders issued by a government agency and often have the force of an administrative procedure backing them up. These rules can be given on a national, state or regional level and are generally site or application specific. Regulations are generally presented in the form of building codes, which are a set of regulations written by an official entity with the help of construction pros that govern the design, construction and modification of facilities or structures in their jurisdiction. Violating these regulations can result in fines and/or penalties. There are many regulations involving geosynthetics on a federal, state and regional level.

8.5 International Geosynthetics Society (IGS)

With more than 40 years of operation, the International Geosynthetics Society (IGS) is a learned society dedicated to the scientific and engineering development of geosynthetics, related products, and associated technologies. It is a registered not-for-profit corporation with a global community with thousands of members including corporate, individual and student members, with a shared passion for what geosynthetics can achieve.

The IGS has individual chapters located throughout the world, making it easy to connect with colleagues locally. The IGS provides greater understanding of geosynthetic technology and promotes its appropriate use throughout the world. The society believes that geosynthetics can make a fundamental contribution to meeting societal challenges through sustainable technological and engineering solutions. It is governed by the membership for the purpose of having a greater scientific and technological understanding, and promoting responsible practices, which ultimately benefit the industry and the communities they serve.

The IGS can trace its roots back to 1977 with the first

International Conference on Geotextiles held in Paris. Every four-year international conference, along with the regional and chapter conferences, is a vital forum to discuss the continuous advancement and transfer information regarding geosynthetics. The conference proceedings are a treasure trove of information. Below is a listing of the IGS international conferences known as the ICG to date;

1977, 1. Paris France: Geotextiles
1982, Las Vegas, NV, U.S.A.: Geotextiles & 1984, 2. Denver, CO, U.S.A.: Geomembranes
1986, 3. Vienna, Austria: Geosynthetics
1990, 4. The Hague, Netherlands: Geosynthetics
1994, 5. Singapore: Geosynthetics
1998, 6. Atlanta, GA, U.S.A.: Geosynthetics
2002, 7. Nice, France: Geosynthetics
2006, 8. Yokohama, Japan: Geosynthetics
2010, 9. Guaruja, Brazil: Geosynthetics
2014, 10. Berlín, Germany: Geosynthetics
2018, 11. Seoul, South Korea: Geosynthetics
2022, 12. Rome, Italy: Geosynthetics
2026, 13. Montreal, Canada: Geosynthetics

In 2025, the IGS has four Technical Committees: Barrier Systems, Soil Reinforcement, Hydraulics, and Stabilization. Each committee hosts both regional and international workshops. The IGS also has an online forum providing a space where IGS members can interact with their colleagues worldwide to discuss all topics and applications related to geosynthetics.

The International Geosynthetics Society (IGS) has grown enormously, both in the number of members and in the scope and impact of its activities over the past forty plus years. In 2025, the IGS has over 4,500 members, including over 175 corporate members and 750 student members. During this period, the IGS organized a remarkable number of international conferences and regional (continental) conferences as well as hundreds of national (chapter) conferences. In addition, it has formed chapters in 45 countries or groups of countries, and implemented numerous educational, technical and outreach programs. The IGS website, geosyntheticssociety.org provides an overview of the history of the IGS, its standing within the context of international learned societies, an overview of its chapters, of its conferences and of ongoing initiatives aimed at providing continued advancement of geosynthetics. It also links to the IGS Digital Library where a wealth of technical content is preserved, much of it in different global languages to serve our global membership. It is strongly recommended to review the "IGS History Archives" available on the website or Zornberg (2013) for more details on the IGS and impact on geosynthetics globally. All engineers and other infrastructure related professionals are invited to join the IGS as members.

8.6 Geosynthetic Institute (GSI)

The Geosynthetic Institute (GSI) and its interrelated institutes are positioned to maintain a balance of activities between the agencies, producers and general interest groups involved with geosynthetic materials. GSI is the umbrella organization housing the following institutes;

> Geosynthetic Research Institute (GRI)
> Geosynthetic Information Institute (GII)
> Geosynthetic Education Institute (GEI),
> Geosynthetic Accreditation Institute (GAI)
> Geosynthetic Certification Institute (GCI)

GSI was incorporated in the State of Delaware on December 16, 1991 and holds an independent, not-for-profit 501(c) (3) organization status as of July 21, 1994.

The institute operates under bylaws and is advised by a nine-person board of advisors (BOA) which is elected by the membership and serve a three-year term. Operations are overseen by a six-person board of directors (BOD). GSI is supported by annual membership subscription, income from its interrelated institutes, and research contracts/grants/awards. The Geosynthetic Institute (GSI) is about 3 miles (4 km) from the Philadelphia International Airport, located at 475 Kedron Avenue Folsom, PA 191033-1208 USA.

The GSI mission is to develop and transfer knowledge, assess and critique geosynthetics, and provide service to the member organizations. Geosynthetics consist of the following polymeric materials used in environmental, geotechnical, transportation, hydraulic engineering, and private development applications. The goals and objectives of GSI are to develop, investigate and implement the various facets of geosynthetics, recognizing them as engineering materials.

8.7 Accreditation(s)

There are numerous laboratory accreditation schemes active within the geosynthetics testing laboratory and measurement industry. Key among these is the following.

ISO 17025, is an international standard that specifies the general requirements for the competence, impartiality, and consistent operation of laboratories. It essentially provides a framework for laboratories to demonstrate their ability to generate valid and reliable testing and calibration results, ensuring confidence in their work. A geosynthetics testing laboratory may be accredited for ISO 17025 by an accrediting authority recognized by the International Laboratory Accreditation Cooperation (ILAC).

Geosynthetic Accreditation Institute-Laboratory Accreditation Program (GAI-LAP) is focused on a Laboratory Accreditation Program (LAP) for all geosynthetic test methods. The GAI-LAP was developed for accrediting geosynthetic testing laboratories on a test-by-test basis. GAI-LAP suggests that laboratories use ISO 17025 as their quality system model. In addition, the program uses the GSI lab as the reference test lab and operates as an ISO 17011 enterprise. In short, this means that the GSI lab does not conduct outside commercial testing. The GAI-LAP does not offer ISO certification, nor does it "certify" laboratory results. GAI-LAP provides accreditation to laboratories showing compliance with equipment and documentation for specific standard ASTM International or ISO test methods. In addition, GAI-LAP verifies that an effective quality system exists at accredited laboratories by way of proficiency testing.

8.8 Certifications / Credentials

8.8.1 Inspector Certification

The Geosynthetic Certification Institute-Inspectors Certification Program (GCI-ICP) is administered by the Geosynthetic Certification Institute which is a branch of the Geosynthetic Institute. The GCI-ICP focuses on CQA but can (and should) pertain to CQC as well. Three different programs are offered; one is focused on geosynthetic materials, the second on inspection of walls and slopes and the third on compacted clay and GCL liners. While the overall performance of containment systems and MSE structures using geosynthetic has been considered good, it has certainly been less than perfect. The aim of this program to improve the status quo in regards to overall project quality.

8.8.2 Installer Certification

The international Association of Geosynthetic Installers (IAGI) operates a Certified Welding Technician (CWT) program that recognizes the knowledge, experience and skill of those technicians who hold the certification. Engineers benefit from IAGI's CWT program because certification verifies that those performing thermal welding in during geomembrane installation have experience in welding and meet industry standards of skill for those geomembranes that they are certified in. IAGI encourages all engineers to require that any welding done on their jobsite be done with CWTs.

IAGI developed the Polyethylene, PVC, Reinforced and EPDM geomembrane welder's certification programs so installers could define standards of proficiency, recognize the knowledge, experience, skills of installers, and reward those who qualify with industry recognition. Member companies who have invested resources in training and testing for their welding technicians take pride in the skill of their welding technicians.

IAGI also has an Approved Installation Contractor (AIC) certification which recognizes geosynthetic installation companies that meet a minimum level of professionalism, ethics and business practices. Approved Installation Contractors must meet requirements in the following areas: corporate history and business practices, insurance verification, safety training, and professional competence and experience.

8.8.3 Product Marking (CE marking)

The CE Mark is the 'passport' which enables a product to be sold in all European Union Countries. The Manufacturer is responsible for affixing the CE Mark. In order to be CE Marked, a product must have the 'Manufacturers Declaration of Performance' and a 'Certificate of Conformity of the Factory Production Control'.

To support CE marking of Geotextiles and geotextile-related products, there are ten harmonized Technical Specifications that currently exist. These are also commonly known as the required characteristic documents or the application standards. The ten application standards are as follows.

EN 13249: The construction of roads and other trafficked areas (excluding railways and asphalt inclusion);

EN 13250: The construction of railways;

EN 13251: Earthworks, foundations and retaining structures;

EN 13252: Drainage systems;

EN 13253: Erosion control works (coastal protection, bank revetments);

EN 13254: The construction of reservoir and dams;

EN 13255: The construction of canals;

EN 13256: The construction of tunnels and underground structures;

EN 13257: Solid waste disposals;

EN 13265: Liquid waste containment projects".

Each standard not only specifies the application of the geosynthetic in the construction but also gives the possible function or functions that it may fulfil. The functions are identified by the following letters in each standard.

F = filtration – This is the restraining of soil or other particles subjected to hydrodynamic forces, while allowing the passage of fluids into or across a geosynthetic.

S = separation – is the prevention of intermixing of adjacent dissimilar soils and/or fill materials by the use of a geosynthetic.

R = reinforcement – is the use of the stress-strain behavior of a geosynthetic to improve the mechanical properties of soil or other construction materials.

D = drainage – is collecting and transporting of precipitation, ground water and/ or other fluids in the plane of the geosynthetic.

P = protection – is preventing or limiting of local damage to a given element or material by the use of a geosynthetic.

Each Application Standard tells us the main function of the geosynthetic used in the application, the characteristics, their relevancy to the conditions of use.

Depending on the standard(s) and the function(s) to be attested to, there are numerous mechanical and hydraulic that have to be performed on each material to be CE marked, these tests are:

EN ISO 10319: "Geosynthetics. Wide-width tensile test"

EN ISO 10321: "Geosynthetics. Tensile tests for joints/seams by wide-width strip method"

EN ISO 12956: "Geotextiles and geotextile-related products. Determination of the characteristic opening size"

EN ISO 12236: "Geosynthetics. Static puncture test (CBR test)"

EN ISO 11058: "Geotextiles and geotextile-related products. Determination of water permeability characteristics normal to the plane, without load"

EN ISO 12958: "Geotextiles and geotextile-related products. Determination of water flow capacity in their plane"

EN ISO 13433: "Geosynthetics. Dynamic perforation test (cone drop test)"

EN 13719: "Geotextiles and geotextile-related products. Determination of the long-term protection efficiency of geotextiles in contact with geosynthetic barriers"

EN 14574: "Geosynthetics. Determination of the pyramid puncture resistance of supported geosynthetics"

EN ISO 12957-1: "Geotextiles. Determination of friction characteristics. Direct shear test"

EN ISO 12957-2: "Geotextiles. Determination of friction characteristics. Inclined plane test"

EN ISO 10722: "Geosynthetics. Index test procedure for the evaluation of mechanical damage under repeated loading. Damage caused by granular materials"

EN ISO 13427: "Geosynthetics. Abrasion damage simulation (sliding block test)"

EN ISO 13431: "Geotextiles and geotextile-related products. Determination of tensile creep and creep rupture behavior"

EN ISO 25619-1: "Geosynthetics. Determination of compression behavior. Compressive creep properties"

It is essential that a geotextile performs effectively for the required duration of the design, therefore the manufacturer must also make a durability assessment. This will require tests to look at characteristics influencing durability:

Weathering (EN 12224);

Hydrolysis (of Polyester materials) (EN 12447);

Oxidation (for polypropylene or polyethylene materials) (EN ISO 13438);

Liquids (both acid and alkali) (EN 14030);

Microbiological degradation (EN 12225).

These test results will form the "Manufacturers Declaration of Conformity."

For the sake of completeness, reference needs to be made that many countries have their own certification systems for geosynthetics. For example, the government of the United Kingdom has laid legislation to continue recognition of current EU requirements, including the CE marking (Conformité Européane, or European Conformity marking). The legislation will apply indefinitely for a range of product regulations. This means businesses will have the flexibility to use either the UKCA (UK Conformity Assessed) or CE marking to sell products in Great Britain (GB).

8.9 Geosynthetics Links

The following is a list of links that may be helpful resources in your geosynthetics journey, enjoy!

https://www.geosyntheticssociety.org/
https://geosynthetic-institute.org/
https://www.geosynthetica.com/
https://www.geosyntheticnews.com.au/
https://geosynthetics.textiles.org/
https://geosyntheticsmagazine.com/
https://gma.now.org/
https://www.astm.org/
https://www.iso.org/
https://www.cencenelec.eu/

8.10 Definitions and References

MQC: Manufacturing Quality Control. Manufacturer testing to monitor and control factory-made product and to ensure compliance with specified values

MQA: Manufacturing Quality Assurance. Independent inspections, verifications, audits, and evaluations of raw or produced materials

CQC: Construction Quality Control. Installer or contractor activities to control the construction process and to comply with specified requirements for materials and workmanship

CQA—Construction Quality Assurance. Independent activities that provide the owner and permitting agency assurance that the facility was constructed as specified.

AASHTO M288-06-UL (2006), Geotextile Specification for Highway Applications, *American Association of State Highway and Transportation Officials*, Washington, D.C., USA, pp 45.

Association of the Nonwoven fabric industry (INDA) (1978), Guide to Nonwoven Farics, New York NY USA pp.32

ASTM International, Annual Book of ASTM Standards (2024), Section 4 Construction, Volume 04.13 Geosynthetics, Stock number S041321, West Conshohocken, PA USA

CEN European Committee for Standardization, Central Secretariat: Rue de Stassart 36, B-150 Brussels Belgium

Daniel, D. E. and Koerner, R. M. (2006), 2nd Edition, Waste Containment Facilities: Guidance for Construction, Quality Assurance and Quality Control of Liner and Cover Systems, ASCE Press, New York, New York, 354 pgs. ISBN 0-7844-0003-2

Daniel, D. E. and Koerner, R. M. (1993), US EPA Technical Guidance Document: "Quality Assurance and

Quality Control for Waste Containment Facilities", EPA/600/R-93/182, Washington, DC, Office of R & D, Cooperative Agreement No. CR-815546-01-0, Project Officer: D. Carson, 305 pgs.

Koerner, G. R. (2024), "A Quality System Approach to Waste Containment Facilities," ASCE's GeoEnviro-Meet, Portland Oregon USA pp. 9

Koerner, R. M. (2012), Designing with Geosynthetics, 6th Edition, Xlibris Corporation, Thorofare, NJ U.S.A., pp.914.

Koerner, R. M. and J. P. Welsh (1980) Construction and Geotechnical Engineering using Synthetic Fabrics, John Wiley & Sons, New York, pp. 267.

ISO, case postale 56, CH-1211, Geneva Switzerland 20

International Geosynthetics Society. www.geosynthet-icssociety.org

GS History Archives

Richardson, G. N. (1992), US EPA Technical Guidance Document: "Construction Quality Management for Remedial Action and Remedial Design of Waste Containment Systems", EPA/650/R-92/073, Washington, DC, Office of R & D, Contract No. 68-CO-0068, Project Officer: R. Landreth, 113 pgs.

U. S. Environmental Protection Agency, Technical Guidance Document, "Inspection Techniques for the Fabrication of Geomembrane Field Seams," EPA/530/SW-91/C51, May, 1991, 174 pages.

Zornberg, J.G. (2013). "The International Geosynthetics Society (IGS): No Borders for the Good Use of Geosynthetics." 25-Year Retrospectives on the Geosynthetic Industry and Glimpses Into the Future, Twenty-fifth Geosynthetic Research Institute Conference (GRI-25), April 01-02, Long Beach, California, pp. 342-357

9 798999 976918